石油高等院校特色规划教材

Recent Advances in Oil and Gas Production Engineering

油气开采新技术

（英汉对照·富媒体）

周德胜　马先林　刘娅菲　等编著

石油工业出版社

内 容 提 要

本书以英汉对照结合富媒体展示的方式系统介绍了当前采油气工程中的新理论、新技术、新工艺及应用，内容包括排液采气技术、水力压裂技术新进展、水平井压裂和体积压裂技术、连续油管技术、智能完井技术、非常规油气开采技术等。

本书可作为石油工程专业高年级本科学生和留学生的教材，也可供油气田开发工程研究生和从事油气田开发的科研人员参考。

图书在版编目(CIP)数据

油气开采新技术：富媒体：英汉对照/周德胜等编著. —北京：石油工业出版社，2020.6

石油高等院校特色规划教材

ISBN 978-7-5183-4042-2

Ⅰ.①油… Ⅱ.①周… Ⅲ.①油气开采—高等学校—教材—英、汉 Ⅳ.①TE 3

中国版本图书馆 CIP 数据核字(2020)第 085802 号

出版发行：石油工业出版社

(北京市朝阳区安华里 2 区 1 号楼　100011)

网　址：www. petropub. com

编辑部：(010)64523579　图书营销中心：(010)64523633

经　销：全国新华书店

排　版：北京密东文创科技有限公司

印　刷：北京中石油彩色印刷有限责任公司

2020 年 6 月第 1 版　2020 年 6 月第 1 次印刷

787 毫米×1092 毫米　开本：1/16　印张：20.75

字数：520 千字

定价：49.90 元

(如发现印装质量问题，我社图书营销中心负责调换)

前　　言

Petroleum production engineering is a discipline which deals with how to cost-effectively recover hydrocarbons from subsurface reservoirs. The subject is closely related to other branches of petroleum industry such as reservoir engineering, drilling engineering, and oil & gas storage and transportation engineering, and has a highly comprehensive and scientific nature. Upon the birth of the petroleum industry, petroleum production engineering has undergone many transformations. At present, the production challenges faced by today's petroleum production engineers are extremely complicated including either high liquid loading in gas wells or low to extra-low permeability reservoirs, as well as different types of unconventional oil and gas extraction. All of them have posed further challenges and requirements for the petroleum production engineers. The fundaments of petroleum production engineering are introduced in a compulsory course *petroleum production engineering*. As a technology-driven subject, the techniques in the petroleum production engineering are constantly evolving. Therefore, it is necessary to write a textbook to reflect the technical advances in the field during past few years. This textbook is intended to fill a significant gap to cover the advanced topics.

油气开采工程是石油与天然气工程的重要组成部分。其目标是将油气经济有效地从地下采至地面,与石油工业中的油藏工程、钻井工程和油气储运工程紧密衔接,具有较高的综合性与科学性。从石油工业诞生之日起,油气开采科学与技术便也不断前进发展,至今已经历了多次变革。如今,现场工程师所面对的油气开采条件极其复杂,不论是高含液气井,或是致密低渗透储层,乃至各类非常规油气的开采,都对油气开采提出了新的挑战与要求。油气开采科学与技术的基础内容,本科生已在必修课"采油工程"中习得。但如前所述,作为一门技术驱动学科,油气开采不断革新发展,以应对油气开采工程中出现的新问题。因此,有必要编写一本介绍近年来油气开采科学与技术进展的教材,本书从而应运而生。

This book was revised based on internal textbook *Recent Advances in Oil and Gas Production Engineering* coauthored by Desheng Zhou and Haiyan Jiang and years of teaching practices. In addition to updating materials of each chapter in the book, the revised edition of book has included much more advanced topics in each area of petroleum production engineering bilingually in English

本书是在周德胜、蒋海岩主编的《采油气科学与技术新进展》讲义及编者多年教学科研实践的基础上编写的。本书以双语形式介绍当前油气开采工程系统各个分支的新理论、新技术和新工艺,可作为国内石油工程专业学生及国际留学生油气

and Chinese. The book was written as a college textbook for senior undergraduates and international students, but also it can be used as a reference book for the relevant professionals.

The main characteristics of the book include: discussing various advanced topics by closely related to conventional ones; introducing each of new technology in order of principles, characteristics, applications and future development; learning and teaching bilingually; describing the topics with lots of colorful illustrations.

This book was written by Dr. Desheng Zhou (Chapter 1), Dr. Xianlin Ma (Chapter 2), Dr. Yafei Liu (Chapter 5 and Chapter 6), and Dr. Wenbin Cai (Chapter 3 and Chapter 4). English audios in the book are recorded by Luxiaohe Zhang, Wen Wen and Fengqi Ma. In particular, postgraduate students Zexuan He, Jingwen Yang, Haiyang Wang, Meng Li, and Peiyao Xiao participated in the text compilation, collection and arrangement of some materials. We appreciate them for their hard work and dedication.

In addition, in the process of editing this book, Dr. Haiyan Jiang, Dr. Guanzheng Qu have made their contributions. This book is supported by the Excellent Academic Publication Fund of Xi'an Shiyou University, the National Natural Science Foundation of China (51874242, 51934005, 51904244) and The Ministry of Education's Brand Course (Oil and Gas Production Science and Technology) in English for international students, the Shaanxi Science and Technology Coordination Innovation Project Plan (2012KTZB03 - 03 - 03 - 02), the National Science and Technology Major

开采工程的课程教材，同时还可以供相关专业技术人员学习参考。

本书的特点是：紧密衔接采油工程的基础理论与技术，介绍各类新理论与技术；在介绍每类新技术时，秉持技术原理、技术特点、技术应用和未来发展的主线；采用中英文本对照，适合双语教学；插入大量的图片，在避免文字介绍单一的同时有利于读者加深理解。

本书由周德胜、马先林、刘娅菲和蔡文斌共同编写完成。其中周德胜负责第 1 章的编写，马先林负责第 2 章的编写，蔡文斌负责第 3 章和第 4 章的编写，刘娅菲负责第 5 章和第 6 章的编写。本书音频由张露小荷、温雯和马凤岐录制。特别需要指出的是，研究生何泽轩、杨静雯、王海洋、李萌、肖沛瑶参与了资料的收集整理与文字整理等工作，对他们的辛勤付出在此表示真诚的感谢。

在本书编写过程中得到了西安石油大学蒋海岩、曲冠政等专家的支持和帮助，得到了西安石油大学优秀学术著作出版基金、国家自然科学基金项目(51934005,51874242,51974253,51904244)、教育部来华留学英语授课品牌课程(采油气科学与技术)、陕西省科技统筹创新工程计划项目(2012KTZB03 - 03 - 03 - 02)、国家科技重大专项资助项目(2016ZX05050 - 009)的资助，

Project of China (2016ZX05050 - 009). Here, we would like to gratefully acknowledge the colleagues and the organizations who contributed to our efforts.

Finally, it would be impossible to bring out this textbook without reference to related books, technical papers and manuals, etc. We are very thankful to many authors and publishers for the use of their materials, draws, photographs, videos, etc. in this textbook which may not directly be cited in - text but a list of comprehensive literature citations provided at the end of each chapter, and special thanks go to the creators of colorful illustrations from the Internet. It is impractical to list all available literature, we apologize sincerely for any omissions.

Due to lack of time and limited ability of the author, there will certainly be errors and omission that you, the reader, will find and we certainly want to hear from you about those.

在此表示感谢。

本书引用了大量前人的研究成果(包括图、照片和视频等),在此一并表示感谢。有些引用没有直接在文字中给出,而是列在每章结束的参考文献中。部分材料来自互联网,对此表示特别感谢。由于未能给出所有的参考文献,对于那些遗漏的我们深表歉意。

由于编者水平有限,书中难免存在不足之处,欢迎各位读者批评指正。

编著者

2020 年 2 月

Contents/目录

Introduction
绪论

As a strategic resource, petroleum plays an important role in economic development and national energy security. The top priority of oil industry is how to maximize oil and gas production in a safe and cost-effective manner. Therefore, the importance of petroleum production engineering is self-evident. The textbook *Petroleum Production Engineering* has already presented in details about theoretical knowledge of well production design and analysis, oil well production system design and operating mode analysis, water injection and profile control technology, acidizing and hydraulic fracturing technology and other basic knowledge of production engineering.

石油作为一种战略资源,对于经济发展和国家能源安全起着十分重要的作用。对于整个石油工业来说,油气开采是其中的核心部分,如何以经济有效的方式实现石油和天然气产量的最大化广为整个石油工业所关注。因此,油气开采科学与技术的重要性不言而喻。本科生必修课"采油工程"中已经详细介绍了采油工程涉及的基础理论与技术,包括:油井生产设计与分析、油井生产系统设计与工况分析、注水调剖技术、酸化与压裂技术等。

With the increased global demand in oil and gas, traditional production methods can no longer keep up with the pace. Therefore more and more new production technologies have emerged. In addition, the proved reserves of unconventional oil and gas resources are huge. The proved reserves of unconventional oil including tight oil, heavy oil and shale oil are about 329.7 billion tons and the unconventional gas such as tight gas, coalbed methane and shale gas are about 343 trillion cubic meter. However, due to difficult reservoir characteristics for example, high viscosity, complex oil and gas migration mechanism, and

随着全球油气需求量的增长及非常规油气资源的开发,越来越多的油气开采新技术成为研究热点。此外,当前全球非常规油气资源的已探明储量巨大,其中,致密油、重油、页岩油等非常规油的资源量约为 3297×10^8t,致密气、煤层气与页岩气等非常规气的资源量为 343×10^{12} m³。但这些非常规油气资源具有储层条件差 、黏

low permeability, which have made the traditional production technologies no longer suitable. The new petroleum production technologies such as massive fracturing have become the main means to extract unconventional oil and gas resources.

度高、油气运移机理复杂、渗透率低等特性,也使得传统的采油工程技术不再适用,大规模压裂及其配套技术成为开发非常规油气资源的有效技术手段。

At present and for a long time in the future, China will face low permeability, high water cut, high wax content and various types of oil and gas reservoirs that have reached the middle and late stage of development. At the same time, unconventional oil and gas reservoirs will gradually become the main resources in China because all easy oil and gas has been found and is gone.

我国现阶段和未来很长一段时间都将面临低渗、高含水、高含蜡等问题,大多数油气藏已经步入开发中后期,开发难度低的、技术要求低的油气藏早已被开发殆尽,非常规油气藏逐步成为我国油气田开发的主力军。

The oil and gas have become more difficult to extract, which implies that the advanced technologies are required. In particular, the emergence and application of the new technologies and procedures, which have been widely accepted and widely applied in recent years, will undoubtedly have an important impact on oil and gas production and provide a forward-looking inspiration.

随着开采难度的增大,油气开采对技术的要求越来越高。近些年来被广泛接受和大规模应用的采油气新技术、新工艺,对油气开采有着重要的影响和前瞻式的启发。

Therefore, in terms of the challenges of China's oil and gas production discussed above, the book presents the recent advances in petroleum production in the following chapters.

针对现阶段我国油气田开发的上述特点,本书通过以下章节对油气开采新技术新进展进行介绍:

The first chapter covers gas well liquid unloading methods to remove the liquid accumulated at a wellbore. The chapter firstly discusses the influence of liquid loading on gas well production and the identification of liquid loading, mainly deals with different types of unloading methods and their applicability. Finally, the selection of the unloading methods is described.

第 1 章针对含水气井的生产问题,系统介绍了排水采气技术。首先论述了井底积液对气井生产的影响及井底积液的识别,之后阐述了不同类型的排水采气工艺及其适用性,最后讲述了排水采气方式的选择。

The second chapter mainly covers the hydraulic fracturing technology that is widely used in low permeability and tight reservoirs. After reviewing the traditional hydraulic fracturing, the chapter presents the new development of hydraulic fracturing technology in three different areas.

第 2 章主要讲述了广泛应用于低渗致密储层的水力压裂技术。在综述了传统的水力压裂技术后,从三个方面介绍水力压裂技术的新进展。

In the third chapter, the advanced horizontal well fracturing and volume fracturing are introduced. After reviewing the concepts of horizontal well fracturing and volume fracturing, the chapter describes the principles, operation steps and tools for open hole, cased hole, hydrajet, just-in-time fracturing and volume fracturing.

第3章介绍了更为先进的水平井压裂和体积压裂技术。从二者的定义出发,先后介绍了套管完井水力压裂技术、裸眼完井水力压裂技术、水力喷射射孔技术、即时压裂技术和体积压裂技术等技术的原理、操作步骤和使用的工具。

The fourth chapter presents coiled tubing technology. The manufacturing process and development history of the coiled tubing is introduced, and then the application of coiled tubing in oil production and its mechanical performance are discussed. Finally, the future development of the coiled tubing is analyzed.

第4章着眼于在现场应用越来越多的连续油管技术,首先介绍了连续油管的制造过程和发展历史,之后对连续油管在采油工程的应用和连续油管的机械性能进行了详细论述,最后分析了连续油管的市场发展情况。

The fifth chapter, starting with the definition, characteristics and applications of intelligent well completion, summarizes the current development of the global intelligent well market, and introduces the products of the service companies.

第5章从智能完井技术的定义、特征、用途出发,介绍了智能完井技术的发展历史,概述了当前全球智能完井市场发展情况,并对智能完井工具进行了介绍。

The sixth chapter mainly describes the production technologies of unconventional oil and gas. The definitions, properties and distributions of unconventional oil and gas including oil sand, oil shale, tight oil, coalbed methane and shale gas are discussed. The methods and technologies used in the production are particularly introduced. Finally, several new technologies for unconventional oil and gas production are briefly summarized.

第6章讲述了非常规油气资源的开采技术,主要论述了油砂、油页岩、致密油、煤层气和页岩气这几类非常规油气资源的定义、特征、分布等,介绍了开采过程中用到的技术方法,最后对非常规油气资源开采的几种新技术进行了介绍。

Gas Well Unloading Technologies 1

排液采气技术

Audio 1.1

Gas well liquid loading is one of the most serious problems in gas-well production. It is estimated that at least 90% of the producing gas wells in the U. S. are operating in liquid loading regime. During gas production, some liquids may not be carried out by the gas stream, but accumulate at the well bottom. The process of liquid accumulation in a gas well is called gas well liquid loading. As liquids accumulate at the well bottom, the flowing bottom. hole pressure will increase, and increased water saturation around a wellbore will reduce the effective gas permeability near the wellbore, therefore reducing gas-production rate. Decreased gas-production rate deteriorates the loading problem, and eventually the loaded liquids will kill the gas well. Better predictions of liquid loading will help operators in reducing costs and improving revenue.

气井积液是气井生产中最严重的问题之一。据估计,美国至少有90%的气井在积液状态下生产。生产期间,一些液体可能不会被气流携带出,而是积聚在井底。气井中液体在井底积聚的过程称为气井积液。随着液体在井底积聚,井底压力会不断增加,并且井眼周围含水饱和度的增加会降低井筒附近的气体渗透率,造成产气量降低。产气量的降低会加重气井积液问题,最终造成气井停产。因此预测并预防气井积液现象有助于降低成本和提高收入。

This chapter discusses gas well unloading techniques. First of all, discusses the influence of gas well liquid loading on production and the identification of gas well liquid loading. Secondly, introduces different types of unloading methods and their applicability. Finally, describes the selection of gas well unloading methods.

本章主要介绍了排液采气技术。首先论述了气井积液对气井生产的影响及气井积液的识别,之后介绍了不同类型的排液采气方法及其适用性,最后讲述了排液采气方法的选择。

1.1 Gas Well Liquid Loading

Many gas wells cease producing economically long before their reservoirs have depleted due to liquid loading problem. The methods of removing liquids from gas well bottom hole are becoming more and more important around the world. This section presents the flow patterns in a gas well, the occurrence of liquid loading and the problems caused by liquid loading. It focuses on the recognition of the occurrence of liquid loading by calculating the critical velocity, and briefly describes several models that are currently widely used and a new model.

1.1.1 Flow Patterns in a Gas Well

The flow pattern in a vertical production conduit of a gas well is usually illustrated by four basic flow patterns or flow regimes as shown in Fig. 1.1. The flow regimes are largely classified with bubble flow, slug flow, slug-annular transition flow and annular mist flow, which are determined by the velocity of the gas and liquid phases and the relative amounts of gas and liquid at any given point in the flow stream.

If the flow pattern is an annular mist type, the well still may have a relatively low gravity pressure drop. However, as the gas velocity begins to drop, the well flow can become a slug type and then bubble flow. In these cases, a much larger fraction of the tubing volume is filled with liquid.

A gas well may go through any or all of these flow regimes during its lifetime. The general progression of a typical gas well from initial production to its end of life is shown in Fig. 1.2.

Initially, the well may show the annular mist flow regime that brings a high gas rate and then transit into slug-annular transition, slug, and bubble flow with time. Liquid production may also increase as the gas production declines.

1.1 气井积液

许多气井在气藏废弃之前很早由于积液就已经达不到经济产能,气井排液采气工艺在世界范围内变得越来越重要。本节介绍气井流态、气井积液现象的出现及气井积液引起的问题,重点介绍通过计算临界流速来识别积液现象的发生,并简单介绍目前普遍应用的几种计算模型和一个新模型。

1.1.1 气井流态

气井垂直管中的流动通常用四种基本流态来说明,如图1.1所示。流态主要分为泡流、段塞流、过渡流、环雾流,它们由气相、液相的流速及任何给定点的气相、液相的相对量来决定。

如果流态是环雾流,则井仍具有相对低的压降。然而,随着气体速度下降,流态将变成段塞流,然后是泡流。在这些情况下,油管的大部分容积充满液体。

气井在其整个生产周期会经历一种或几种流态。典型气井从生产开始到废弃的一般过程如图1.2所示。

最初,气井处于环雾流状态,气体流速较高,然后逐渐变为过渡流、段塞流和泡流。随着天然气产量下降,液体产量也可能增加。

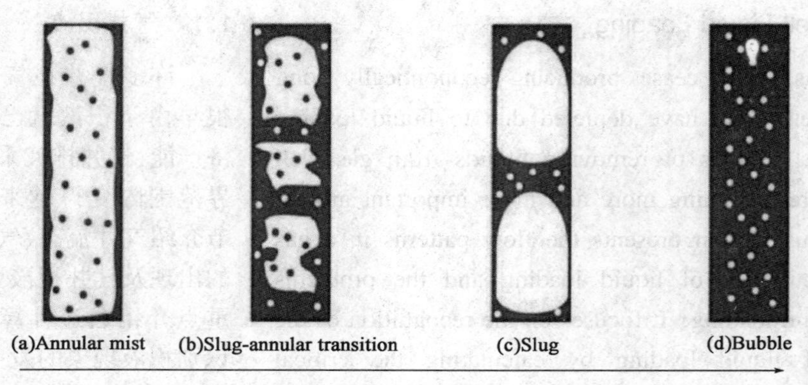

Fig. 1. 1 Flow Regimes in Vertical Multiphase Flow
(Source: http://www. energyresources. com)

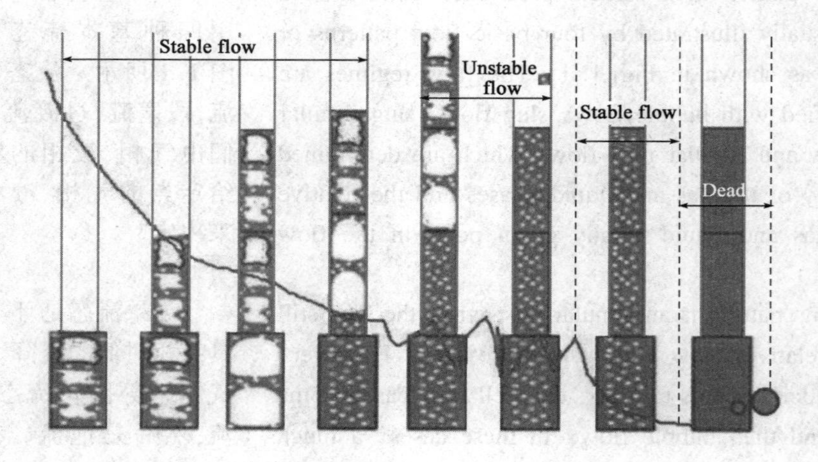

Fig. 1. 2 Life History of a Gas Well
(Source: http://www. petrosol. com)

1.1.2 Occurrence of Liquid Loading

Gas and liquid are both produced to surface if the gas velocity is high enough to lift or carry liquid. The problem happens because the velocity of the gas in the tubing drops with time, and the velocity of the liquids decline even faster as the production goes on. As a result, the liquid begins to accumulate in the bottom of the well and liquid slugs are formed in the conduit, which increase the percentage of liquids in the conduits while the well is flowing. The bottom hole pressure increases and gas production decreases until gas flow stops. In other words, the liquid loading process occurs when the gas velocity within the well drops

1.1.2 积液现象的出现

如果气体速度足够高,可以提升或携带液体,则气体和液体都可以被举升到地面。之所以出现气井积液,是因为管柱中的气体速度会随着时间的推移而下降,而且随着生产的进行,液体的速度会下降得越来越快。所以液体开始积聚在井底,液体段塞在管内形成,当井内流体流动时,管柱中的液体百分比增加,井底压力增加,气体产量减

below a certain critical gas velocity. The gas is then unable to lift the water coproduced with the gas (either condensed or formation water) to surface. The water will fall back and accumulate downhole. A hydrostatic column is formed that imposes a back pressure on the reservoir and hence reduces gas production. The process eventually results in intermittent gas production and well die-out. Fig. 1. 3 shows the liquid loading phenomena in gas wells.

少,直到停止生产。换句话说,当井内的气体速度下降到某一临界速度以下时,就会发生气井积液现象。此时,气体不能将与其共同产生的水(凝析水或地层水)提升到地面。水将回落并积聚在井下,形成静压柱,对储层施加回压并因此减少气体产量。该过程最终导致气井间歇性生产和死井。图 1. 3 显示了气井中的积液现象。

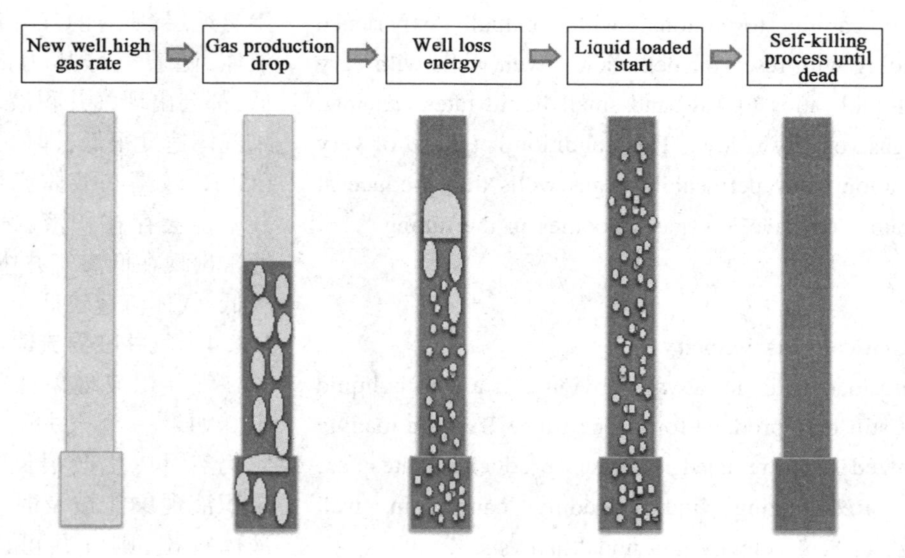

Fig. 1. 3 Liquid Loading Phenomena in Gas Wells
(Source: http://www. ukessays. com)

1.1.3 Problems Caused by Liquid Loading

Liquid loading can lead to erratic, slug flow and decreased production. The well may eventually die if the liquids are not continuously removed. Often, as liquids accumulate in a well, the well simply produces at a lower rate than expected.

If the gas rate is high enough to remove most or all of the liquids, the flowing tubing pressure at the formation surface and production rate will reach a stable equilibrium.

1.1.3 气井积液引起的问题

气井积液可能导致不稳定流或段塞流并减少产量。如果液体不能被连续排出,则井最终可能会停止生产。通常,当井底存在积液时,气井仅表现出比预期的产量低的现象。

如果气体速率足够高,可以携带出井底大部分或全部液体,则井底流动压力和产量都将处于较稳定的平衡状态。

If the gas rate is too low, the pressure gradient in the tubing becomes large due to the liquid accumulation, resulting in increased pressure on the formation. As the back-pressure on the formation increases, the rate of gas production from the reservoir decreases and may drop below the critical rate required to remove the liquid. More liquids will accumulate in the wellbore and the increased bottom hole pressure will further reduce gas production and may even kill the well.

All gas wells that produce some liquids, whether in high or low permeability formations, will eventually experience liquid loading with reservoir depletion. Even wells with very high gas-liquid ratios (GLR) and small liquid rates can load up if the gas velocity is low. This condition is typical of very tight formation (low permeability) gas wells that produce at low gas rates and have low gas velocities in the tubing.

如果气体速率太低,由于存在积液,管柱中的压力梯度变大,导致对地层的压力增加。随着地层回压的增加,气藏气体流动速率降低并且可能降至低于携带液体所需的临界速度。这样更多的液体将积聚在井筒底部,造成井底压力增加。井底压力的增加将进一步减少气体产量甚至导致死井。

无论是高渗透地层还是低渗透地层,所有的气井都会产出液体,随着气藏压力的逐渐降低,都会出现气井积液现象。如果气体速度很低,即使在气液比(GLR)较高和液体流量很小的井中也会存在积液。这种情况通常出现在低渗气井中,其产量较低,气体流速较低。

1.1.4 Critical Gas Velocity

Liquid loading is not always obvious. If a well is liquid loaded, it still may produce for a long time. If liquid loading is recognized and reduced, higher producing rates are achieved. Recognizing liquid loading can from well symptoms, critical velocity and nodal analysis.

1.1.4 气井临界流速

气井积液现象往往不易识别。如果一个气井存在积液,它仍然可以生产很长时间。如果积液现象能被及时发现并将液体排出,则气井可以获得更高的产量。可以通过一些气井生产动态,计算临界流速,并用节点分析方法来识别气井积液。

1.1.4.1 Well Symptoms

Symptoms indicating liquid loading include the following.

(1) Tubing and casing pressure differential. In wells with open ended completions when the wells begin to liquid load we will see a decrease in tubing pressure and an increase in casing pressure. The increase in casing pressure is in indication of the increased FBHP due to the accumulated fluids in the tubing.

1.1.4.1 气井积液迹象

表明气井存在积液的迹象包括以下几点。

(1)油管和套管压差。当井底产生积液,在油套管连通的情况下,可以发现油管压力降低,套管压力增加。套管压力的增加表明由于积液的产生导致井底流动压力增加。

(2) Sharp drops in a decline curve (Fig. 1.4).

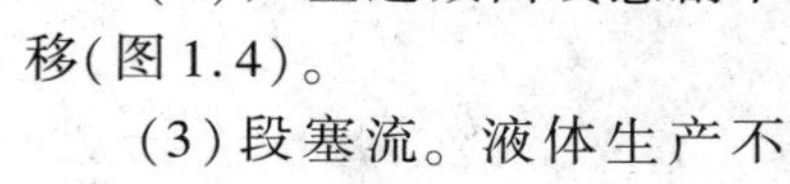
(2)产量递减曲线急剧下移(图1.4)。

(3) Liquid slugging. Liquid production does not arrive to surface in a steady continuous flow but instead in slugs of fluid or heading.

(3)段塞流。液体生产不会以稳定的连续流动状态而是以段塞流到达地面。

(4) Fluctuating gas production. Daily gas production is dramatically different with no changes to the flowline pressure.

(4)日产气量存在波动。日产气量波动较大,而管内压力没有发生变化。

(5) Variance from decline curve. Typical gas production wells will follow an exponential type curve, as the liquid loading occurs there will generally be deviation from the curve with a lower than predicted production rate.

(5)产能递减曲线的变化。典型的气井产能将遵循指数型递减,当产生积液时,通常会出现偏离曲线且低于预测的产量。

(6) Pressure survey reveals a heavier gradient in the tubing pressure.

(6)压力测量显示油管压力梯度较大。

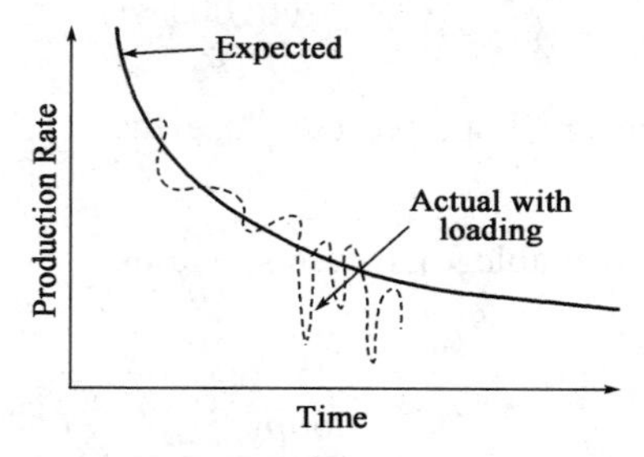

Fig. 1.4 Decline Curve Showing Onset of Liquid Loading
(Source: https://core.ac.uk)

1.1.4.2 Critical Velocity

The controlling factor to inhibit liquid buildup in a gas well is the gas velocity (gas-flow rate) in the well. It is believed that there is a critical velocity (or critical gas rate) below which a static liquid column will develop in the well. To keep a well producing without liquid-loading problems, gas-flow velocity must be maintained above the critical velocity. This section begins with a brief review of the existing liquid loading prediction models for gas wells, and then introduces a new model for calculating critical velocity.

1.1.4.2 临界流速

气井积液现象的控制因素是井中的气体速度。人们认为存在一个临界流速,若气体速度低于该临界流速,会在井中形成积液并造成井底新增液压。为了使气井在没有积液的情况下进行生产,气体速度必须保持在临界流速以上。本部分首先简单描述了几种目前存在的用于气井积液预测的模型,然后介绍了一种计算临界流速的新模型。

1. Review of the Existing Liquid Loading Prediction Models for Gas Wells

The most popular model for the prediction of critical velocity for liquid loading in gas wells is the entrained-droplet model presented by Turner et al. Turner's model is

1. 现有气井积液预测模型回顾

用于预测气井积液临界速度的最常用的模型是 Turner 等人提出的液滴模型。Turner 模型是

$$v_{crit-T} = 1.593\frac{\sigma^{\frac{1}{4}}(\rho_L-\rho_g)^{\frac{1}{4}}}{\rho_g^{\frac{1}{2}}} \tag{1.1}$$

where v_{crit-T} is the critical velocity from Turner's model, ft/s; σ is surface tension, dyn/cm; ρ_L is liquid density, lb/ft^3; and ρ_g is gas density, lb/ft^3.

其中，v_{crit-T}是 Turner 模型计算的临界流速，ft/s；σ 是表面张力，dyn/cm；ρ_L 是液体密度，lb/ft^3；ρ_g 是气体密度，lb/ft^3。

The equation can be expressed as a function of pressure, where p is in units of psi.

该等式可以表示为压力的函数，其中 p 以 psi 为单位。

$$u_{water} = 5.34\frac{(67-0.0031p)^{\frac{1}{4}}}{(0.0031p)^{\frac{1}{2}}} \tag{1.2}$$

$$u_{condensate} = 4.02\frac{(45-0.0031p)^{\frac{1}{4}}}{(0.0031p)^{\frac{1}{2}}} \tag{1.3}$$

When water and condensate are both present, use the water equation.

当水和冷凝物都存在时，使用水的公式[式(1.2)]。

Once a critical velocity is available, its corresponding critical rate can be calculated as

只要确定了临界流速，相应的临界流量为

$$q_{crit-T} = \frac{3060pv_{crit-T}A}{Tz} \tag{1.4}$$

where q_{crit-T} is the critical rate from Turner's model, 10^6ft^3/d, p is in-situ pressure, psi; A is the flow area of a conduit, ft^2; and T is in-situ temperature, °R; z is the compressibility factor.

其中，q_{crit-T}是 Turner 模型计算的临界流量，10^6ft^3/d；p 是现场压力，psi；A 是管道的流动面积，ft^2；T 是现场温度，°R，z 是压缩因子。

Turner et al. (1969) found that the entrained-droplet model still underestimated the critical rates of their wells. They adjusted 20% upward from the droplet model and found that with the 20% adjustment, the model covered most of their well situations. With the 20% adjustment, the critical velocity and critical rate from the entrained-droplet model are, respectively

Turner 等人(1969)发现液滴模型仍低估了携液临界流速。他们把液滴模型向上调整了20%，发现通过20%的调整，该模型涵盖了大多数井的积液情况。调整后液滴模型的临界流速和临界流量分别为

$$v_{crit-T20\%} = 1.2v_{crit-T} \tag{1.5}$$

$$q_{crit-T20\%} = 1.2q_{crit-T} \tag{1.6}$$

注：1ft = 0.3048m；1dyn/cm = 1mN/m；1lb = 0.454kg；1lb/ft^3 = 16.02kg/m^3；145psi = 1MPa。

where $v_{crit-T20\%}$ and $q_{crit-T20\%}$ are the adjusted critical velocity and critical rate in ft/s and $10^6 ft^3/d$, respectively.

其中，$v_{crit-T20\%}$ 和 $q_{crit-T20\%}$ 分别代表调整后的临界流速和临界速率，单位为 ft/s 和 $10^6 ft^3/d$。

However, since Turner's data sets were mostly consisting of high wellhead pressure ($p_{wh} > 1000psi$), they could not determine that 20% adjustment gives incorrect results for low wellhead pressure wells. Another drawback of Turner's model was an assumption that the shape of droplet is spherical and does not change while flowing in the well.

然而，由于 Turner 的数据主要由高井口压力（$p_{wh} > 1000psi$）组成，他们无法确定 20% 的调整是否适用于低井口压力井。Turner 模型的另一个缺点是假设液滴的形状是球形的，并且在井中流动时形状不会改变。

Coleman et al. (1991) firstly reported that Turner's 20% adjustment does not work with low-rate and low wellhead pressure wells and should be applied without adjustment. Coleman's non-adjusted droplet model is widely used for wells with wellhead pressures less than 500 psi.

Coleman 等人首次认为 Turner 的 20% 调整不适用于低井口压力井，低井口压力井不需要调整。Coleman 的非调节液滴模型广泛用于井口压力小于 500psi 的井。

Li et al. unlike Turner's spherical-droplet model introduced new flat-shaped droplet model. They explained that in high velocity gas flow, the fore and apt portions of droplet have a pressure difference. This pressure difference changes droplet's shape from spherical to like a convex bean form, which has unequal sides. Fig. 1.5 shows the droplet's shape changes from spherical to flat in a high velocity. Compared with spherical droplets, flat ones need low gas velocity and flow rate due to having more efficient area.

与 Turner 的球形液滴模型不同，Li 等人引入了新的扁平形状液滴模型。他们解释说，在高速气流中，液滴的前后具有压差。这种压差将液滴的形状从球形变为凸起的豆形，其具有不相等的侧面。图 1.5 显示了液滴的形状在高速下从球形变为扁平形。与球形液滴相比，由于扁平形液滴具有更大纵向投影面积，气体携带所需要的气体速度和流量较低。

$$v_{crit-T} = 0.7241 \sqrt[4]{\frac{(\rho_L - \rho_g)^{\sigma}}{\rho_g^{0.5}}} \tag{1.7}$$

$$q_{crit-T} = 3060 \frac{Apv_{crit-T}}{zT} \tag{1.8}$$

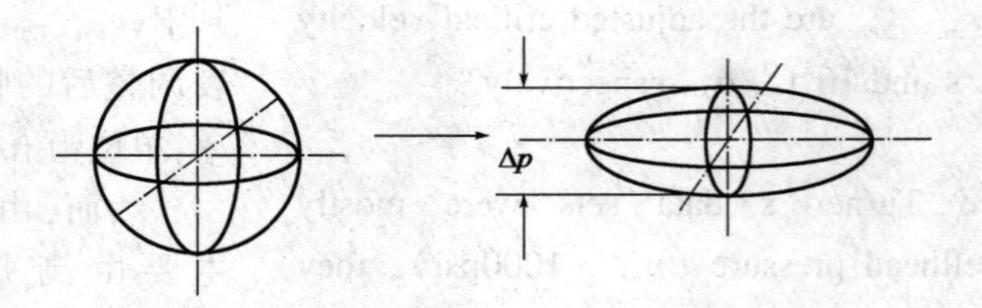

Fig. 1.5 Entrained Droplet's Shape in a High Velocity Gas Stream

Although, Li introduced the new shaped droplet model and showed Turner's spherical droplet model is not completely accurate, they made the correlation based on the assumption that the shape of droplet is constant and does not convert. They did not consider if droplets coalesce, how it can change their shape. In reality, experiments show that droplets coalesce and separate many times in the wellbore. Consequently, they will need more gas velocity to move upward.

虽然 Li 等人的新形状的液滴模型表明 Turner 的球形液滴模型并不完全适用,但是这些模型都是基于液滴形状恒定不变的假设得出的,没有考虑气流中大量液滴是否聚并,且如何改变它们的形状的问题。实际上,实验表明,液滴在井筒中聚结、分离很多次。因此,有大量液滴时将需要更大的气体速度才能向上移动。

2. New Model

Previous models for liquid loading are independent of the liquid amount in a gas stream. When gas velocity is higher than critical velocity in a gas well, no liquid loading exists following previous models. A new model was presented by Zhou et al., which is the first model to include the amount of liquids in a gas stream in the calculation of its critical velocity. According to the new model, critical gas velocity is not a single value; it varies with the liquid holdup in the gas stream of a gas well once the holdup exceeds a threshold value.

2. 新模型

以前用于预测积液现象的模型均与气流中的液体量无关,由此得出当气体速度高于计算的临界流速时,气井液体全部被携带出井口,井底不存在积液。Zhou 等人认为气体携液速度与气流中液体的多少有关,提出了一个在计算临界流速时考虑液体量的新模型。根据该模型,临界流速不是一个单一值, 一旦气井中的持液率超过一定值,临界流速就会随着气井中的持液量而发生变化。

Two mechanisms have been proposed for predicting liquid loading in gas wells: entrained liquid-droplet model and liquid-film model. As concluded by Turner et al., the liquid-droplet model represents the liquid-loading problem well, but the liquid-film model fails.

以前预测气井积液有两种模型:液滴模型和液膜模型。正如 Turner 等人所总结的那样,液滴模型可以表示积液问题,但液膜模型却不能。

As shown in Fig. 1. 6 (a), Turner ' s liquid-droplet model is based on the force balance on a single liquid droplet. Three forces act on the single liquid droplet: upward drag force, F_D; upward buoyant force, F_B; and downward gravity force, F_G. The droplet will speed up in the upward direction if the upward forces ($F_D + F_B$) are greater than the downward force, F_G, and it will accelerate downward if the upward forces are smaller than the downward force. The balance of the forces ($F_D + F_B = F_G$) yields the droplet model (Eq. 1. 1), at which the droplet will keep its velocity, and gives the critical gas velocity to sustain the droplet. According to the model, a liquid droplet will be carried out by a gas stream if the stream flows faster than the critical gas velocity.

如图 1. 6(a)所示,Turner 的液滴模型基于气流中作用在单个液滴上的力平衡。共有三个力作用在单个液滴上:向上曳力 F_D,向上浮力 F_B 和向下重力 F_G。如果向上的力($F_D + F_B$)大于向下的力 F_G,则液滴将向上加速;相反,如果向上的力小于向下的力,则液滴将向下加速。通过力的平衡($F_D + F_B = F_G$)得出液滴模型[式(1. 1)],液滴将保持其速度,还给出了维持液滴的临界流速。根据该模型,如果流体流动的速度高于临界流速,则液滴将随着气流流出。

Turner's model is based on the force balance on a single liquid droplet, but what if there is more than one droplet in the gas stream? As shown in Fig. 1. 6 (b), there are two droplets (A and B) in a gas stream (assuming all the gas streams in Fig. 1. 6 and Fig. 1. 7 satisfy its critical gas velocity). For a laminar gas stream, the two droplets may move in the same direction in the well and will flow out of the well. For gas wells, gas velocity is usually very high and the flow is turbulent. In a turbulent gas stream, liquid droplets move not only upward with the gas stream, but also in all directions irregularly. The nearby liquid droplets may encounter each other and coalesce into a bigger droplet. As illustrated in Fig. 1. 6 (b), liquid droplet A and droplet B move in different ways and may encounter and coalesce into a bigger droplet AB at the top of the well.

Turner 模型是基于单个液滴上的力平衡,但是如果气流中有多个液滴会怎么样?如图 1.6(b)所示,在气流中有两个液滴(A 和 B)(假设图 1.6 和图 1.7 中的所有气流满足其临界流速)。对于层流,两个液滴可以在井中沿相同方向移动并且流出。然而,对于通常的气井,气体速度通常较高,其流动是湍流。在湍流中,液滴不仅随气流向上移动,而且在所有方向上不规则地移动。附近的液滴可能彼此相遇聚并成更大的液滴。如图 1. 6(b)所示,液滴 A 和液滴 B 以不同的方式向上运动,可能在井中某处相遇并聚结成更大的液滴 AB。

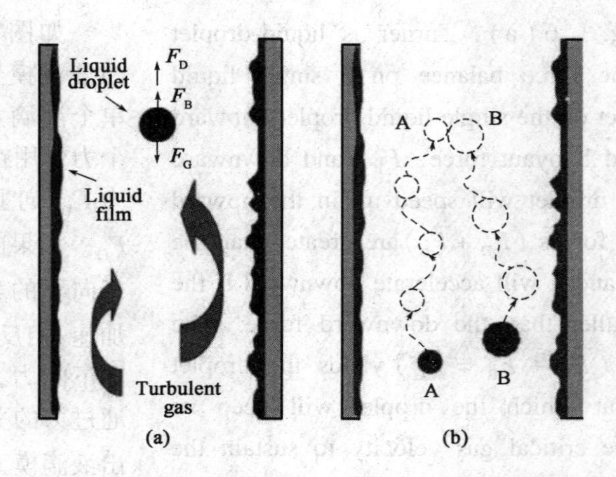

Fig. 1. 6　Encountering Two Liquid Droplets in Turbulent Gas Stream

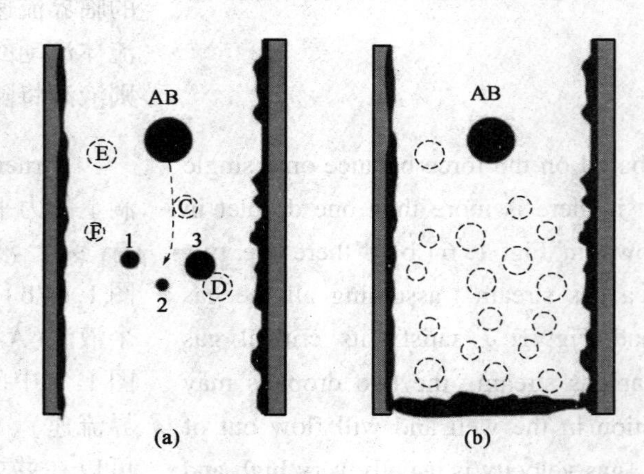

Fig. 1. 7　Liquid Loading When Liquid – Droplet Number Reaches a Threshold Value

The newly formed droplet [droplet AB in Fig. 1. 7(a)] may fall down in the gas stream because its bigger size needs higher gas velocity to suspend it.

During falling, the newly formed bigger droplet may shatter into small droplets [droplet AB shatters into droplets 1, 2, and 3 in Fig. 1. 7(a)] by velocity pressure, and the small droplets may be picked up again by the drag forces from the gas stream. If there are other droplets in the upstream [droplets C, D, E, and F in Fig. 1. 7(a)], the bigger droplet or its shattered droplets may encounter other droplets during their falling and the encountered droplets may

新形成的液滴——图1.7(a)中的液滴AB,因为尺寸较大需要较高的气体速度才能使其悬浮,因此可能会在气流中向下运动。

在新形成的较大液滴下落时,受速度压力作用被破碎成小液滴[液滴AB破碎成图1.7(a)中的液滴1、2和3],新形成的多个小液滴因受到气体的向上曳力和浮力将再次向上运动。如果在气流中存在其他液滴[图1.7(a)中的液滴C、D、E和F], coalesce and keep falling. As shown in Fig. 1.7(a), droplet AB may encounter droplet C and coalesce into a new droplet ABC before droplet AB shatters, or its shattered droplet 3 may encounter upstream droplet D. Because only a few droplets are in the well, the droplets will be carried out finally.

则较大的液滴或较大液滴破碎后形成的多个较小液滴可能与其他液滴相遇，相遇的液滴会聚并成较大液滴并继续下落。如图 1.7(a)所示，液滴 AB 可能遇到液滴 C 并在破碎之前合并成新的液滴 ABC，或者其破碎的液滴 3 可能遇到上游液滴 D。因为井中只有少量液滴，液滴最终将会被携带出来。

As shown in Fig. 1.7(b), if there are more liquid droplets in the gas stream, the chance of the process of liquid-droplet encountering, coalescing, falling, and shattering increases. As the number of liquid droplets in a gas stream, called liquid-droplet concentration here, increases to a threshold value , the process of droplets encountering, coalescing, falling, and shattering will continue and bring those liquid droplets down to the well bottom.

如图 1.7(b)所示，如果气流中存在更多的液滴，则液滴相遇、聚并、下落和破碎的可能性都会随之增加。当气流中的液滴数量(称为液滴浓度)增加到一定数量时，液滴相遇、聚并、下落和破碎的过程将大面积产生，这些大液滴掉落到井底，形成井底积液。

Liquid holdup can be used to represent the liquid droplet concentration in a gas stream. Liquid holdup is defined as

气流中的持液率用来表示气井中的液滴浓度。持液率定义为

$$H_L = \frac{v_{sL}}{v_{sg} + v_{sL}} \tag{1.9}$$

where H_L is liquid holdup and v_{sL} and v_{sg} are superficial liquid and gas velocities, respectively.

其中，H_L 是持液率，v_{sL} 和 v_{sg} 分别代表液体和气体的表观速度。

Liquid-droplet concentration is the control factor in droplet encounters. The higher the concentration of liquid droplets in a turbulent gas stream, the greater the chance that the droplets will combine and fall.

液滴浓度是液滴碰撞的控制因素。湍流中液滴浓度越高，液滴相互聚并和下降的可能性就越大。

The concentration of liquid droplets in a gas stream may be the third mechanism contributing to liquid loading, in addition to the liquid-film mechanism and Turner's liquid-droplet mechanism.

除了液膜机理和 Turner 的液滴机理之外，气流中的液滴浓度是导致气井积液的第三机理。

Turner's entrained liquid-droplet model is based on the force balance on a single droplet and does not include the encounter effect. For low liquid-droplet concentration, the chance of encounters is low and Turner's model works well. However, when the liquid concentration reaches a certain value, the encounter coalescing falling process of liquid droplets in a gas stream will dominate the entrained-liquid-droplet movement, and hence Turner's single liquid-droplet model losses its function, even with gas-stream flows faster than critical velocity.

Turner 的液滴模型基于作用在单个液滴上的力平衡,不包括液滴相遇效应。低液滴浓度的情况下,液滴相遇的概率较低,Turner 的模型效果很好。然而,当液滴浓度增加到一定值时,气流中液滴的聚并下降过程将主导液滴的运动,因此单液滴模型失去其适应范围,即使气体速度大于 Turner 临界流速,大量液滴的聚集及下落依然会造成气井积液。

As discussed previously, there is a threshold value of liquid-droplet concentration. Below it, the entrained liquid-droplets do not encounter, or they do encounter, coalesce, fall, and shatter, but still are brought out of the well by the gas stream before they accumulate at bottom hole. Turner's model can be used in this situation.

如前所述,存在一个液滴浓度的临界值。当液滴浓度小于该临界值时,液滴不会相遇,或者它们确实相遇、聚结成大液滴,下落,再分散成小液滴,但仍在井底积聚之前被气流带出井口。Turner 模型在这种情况下可以使用。

Above the threshold concentration value, higher gas velocity is needed because higher gas velocity provides higher drag force and can bring bigger droplets upward. Also, higher gas velocity has higher velocity pressure that prevents bigger liquid-droplet formation and shatters bigger droplets faster. Therefore, the critical velocity for liquid loading is not a single value. It varies with the liquid-droplet concentration in a gas stream once the concentration exceeds the threshold value.

当液滴浓度高于临界值时,需要更大的气体速度才能将液滴携带出井口。因为更大的气体速度能提供更高的拖曳力,使更大的液滴向上移动。此外,较高的气体速度可产生较高的速度压力,防止较大的液滴形成,并更快地破碎较大的液滴形成小液滴。因此,气体携液的临界流速不是单一值。一旦液滴浓度超过其极限值,临界流速就会随着气流中的液滴浓度而发生变化。

According to the liquid droplet-concentration mechanism, Zhou proposed an empirical correlation to estimate the critical velocities for gas-well liquid loading as

根据液滴浓度机理,Zhou 等人提出了一个经验关系式来估算积液的临界速度:

$$v_{\text{crit-N}} = v_{\text{crit-T}} = 1.593\frac{\sigma^{\frac{1}{4}}(\rho_L - \rho_g)^{\frac{1}{4}}}{\rho_g^{\frac{1}{2}}} \quad (H_L \leqslant \beta) \tag{1.10}$$

$$v_{\text{crit-N}} = v_{\text{crit-T}} + \ln\frac{H_L}{\beta} + \alpha \quad (H_L > \beta) \tag{1.11}$$

where v_{crit-N} is the critical velocity from the new model in ft/sec, H_L is liquid holdup that reflects the liquid-droplet concentration, β is the threshold value of liquid-droplet concentration for petroleum production wells, and α is a fitting constant. The maximum liquid holdup is 0.24. When liquid holdup becomes higher than 0.24, the two-phase flow changes to slug or churn flow pattern.

其中，v_{crit-N}是新模型的临界速度，单位为 ft/s；H_L 是反映液滴浓度的持液率；β 是气井液滴浓度的极限值；α 是一个拟合常数。最大持液率为0.24，当液体滞留量高于 0.24 时，两相流变为段塞流动模式。

The new model is composed of two parts. When liquid holdup is less than or equal to the threshold value, β, the critical-velocity model is the same as Turner's model. When liquid holdup is greater than the threshold value, β, the critical velocity varies with the liquid holdup and can be calculated from the new model.

新模型由两部分组成。当持液率小于或等于极限值时，临界流速模型与 Turner 模型相同。当持液率大于极限值时，临界速度随持液率变化，并且可以通过新模型计算。

The critical-rate correlation for the new model is the same as that by Turner et al., as shown in Eq. 1.4

新模型的临界流速等式与 Turner 等人的等式［式(1.4)］相同：

$$q_{crit-N} = \frac{3060 p v_{crit-N} A}{Tz} \tag{1.12}$$

3. Nodal Analysis

Liquid loading can be easily observed through Nodal analysis by plotting IPR & OPR curves. In Fig. 1.8, the point A shows that the well is already loaded and prevention methods must be considered to remove liquid from the wellbore and make the well healthy (point B). Nodal analysis clearly shows that liquid loading problem is very important for gas wells and immediate measures should be taken as early as possible.

3. 节点分析

绘制 IPR 和 OPR 曲线，应用节点分析很容易预测积液的产生。在图 1.8 中，A 点表明井已经产生积液并且必须考虑采取措施从井筒中移除液体并使井恢复正常(B 点)。节点分析清楚地表明，积液问题对于气井非常重要，应尽早采取相应的措施。

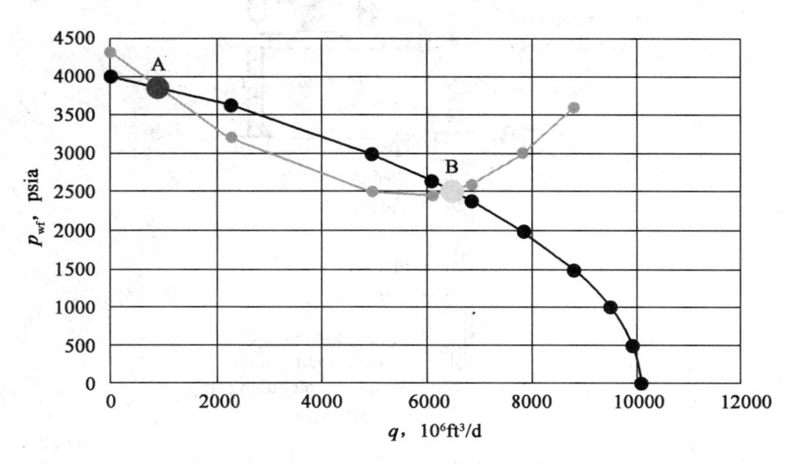

Fig. 1.8 Detected Liquid Loading through Nodal Analysis

1.2 Gas Well Unloading Methods

Audio 1.2

Liquid loading has been a problem in natural gas wells for several decades. With gas fields becoming mature and gas production rates dropping below the critical rate, deliquification becomes more and more critical for continuous productivity and profitability of gas wells. The liquid unloading methods include plunger lift, velocity string, surfactant, foam, well cycling, pumps, compression and gas lift. All of these methods are applied to lift liquids up to surface, which increases the operating cost both onshore and offshore. This section will focus on the plunger lift, foaming, velocity string and artificial lift technology to remove the bottom hole liquids from gas wells.

1.2 气井排液采气方式

几十年来,气井底积液一直是天然气生产的一个问题。随着天然气田的开发,气井内气体速度降至临界流速以下,排液采气对于气井连续生产和盈利变得越来越重要。目前排液采气的方法包括:柱塞法,速度管柱法,表面活性剂法,泡沫法,井循环法,泵、井口增压法和气举法。这些方法都是为了将液体举升到地面,这增加了生产成本。本节重点介绍柱塞法、泡沫法、速度管柱法和人工举升法。

1.2.1 Plungers

1.2.1.1 Introduction

Plunger systems utilize a physical plunger that travels from the bottom of the well to the surface. At the bottom of the well, the pressure must build enough to lift the plunger and liquid that has accumulated above the plunger to the surface (Fig. 1.9).

1.2.1 柱塞法

1.2.1.1 引言

柱塞系统的主要部件是从井底上升到地面的物理柱塞。在井底,压力必须足够大以将柱塞和积聚在柱塞上方的液体提升到地面(图1.9)。

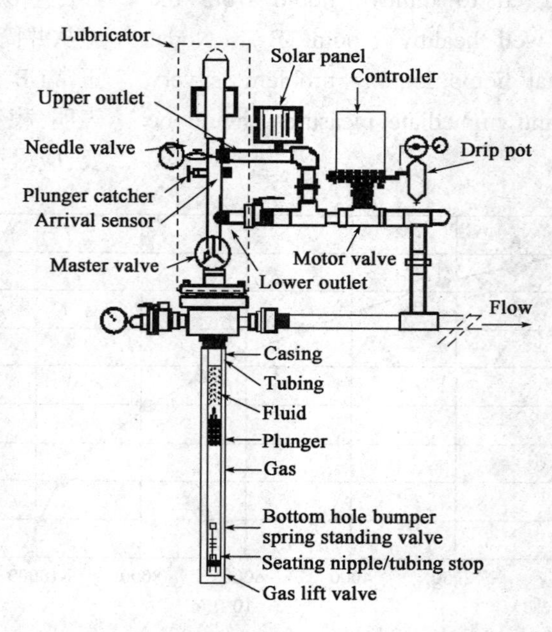

Fig. 1.9 A Typical Conventional Plunger Lift Installation
(Source: http://www.epa.gov)

The plunger placed in the tubing contains a valve, which controls the fluid flow. A cushion seat, containing an opening, at the bottom and a rubber or spring bumper at the upper end of the tubing helps the plunger valve to open and close. Gravitational force pulls the plunger down, and rise of bottom hole pressure with production from formation lifts it up. As decreasing depth and increasing productivity the efficiency of plunger lift decreases. Plungers can be used to aid intermittent gas lift or to lift liquids accumulating in gas wells, Fig. 1. 10. In suitable wells plunger lift is an efficient method which is trouble free and cheap.

放置在油管中的柱塞包含一个控制流体流动的阀门。底部的开口缓冲座和管上端处的橡胶或弹簧缓冲器,有助于柱塞阀打开和关闭。重力使柱塞向下运动,井底压力不断增加,将柱塞段液体举升到地面。柱塞举升的效率随着深度的降低和生产率的提高而降低。如图1.10所示,柱塞除可直接用于举升积聚在气井中的液体外,还可用于辅助间歇性气举。对适宜的气井,柱塞举升是一种无故障且经济有效的方式。

Fig. 1. 10 Real-Life Example of Plunger Lift Well
(Source: http://www.petroskills.com)

Traditionally, plunger lift was used on oil wells. After that, plunger lift has become more common on gas wells for de-watering purposes. A plunger lift system uses gas pressure buildup in a well to lift a column of accumulated liquid out of the well. Basically, the plunger lift system utilizes a plunger traveling up and down inside the tubing to lift the liquid. Thus, the operation of the plunger system relies on the natural buildup of pressure in a gas well while the well is shut-in.

传统上,柱塞举升用于油井。但随后,柱塞举升逐渐用于气井排液采气中,并得到越来越广泛的应用。柱塞举升系统利用井中气体压力的积聚来将积液提升到地面。基本来说,柱塞举升系统利用油管内柱塞的上下移动来举升液体。因此,柱塞系统的操作依靠关井时气井中压力的自然累积。

1.2.1.2 Type of Plungers

Fig. 1.11 shows some typical plungers that were tested to provide data for developing plunger lift system models. These shown are typical but do not include all types of plungers available to the industry.

In this figure, the plungers are identified from left to right as:

(1) Capillary plunger, which has a hole and orifice through it to allow gas to "lighten the liquid slug above the plunger".

(2) Turbulent seal plunger with grooves to promote the "turbulent seal".

(3) Brush plunger used especially when some solids or sand is present.

(4) Another type of brush plunger.

(5) Combination grooved plunger with a section of "wobble washers" to promote sealing.

(6) Plunger with a section of turbulent seal grooves and a section of spring-loaded expandable blades. Also a rod can be seen that will open/close a flow-through path through the plunger depending on whether it is traveling down or up.

(7) Plunger with two sections of expandable blades with a rod to open flow-through plunger on down stroke.

(8) Mini-plunger with expandable blades.

(9) Another with two sections of expandable blades and a rod to open flow-through passage during plunger fall.

(10) Another with expandable blades and a rod to open a flow through passage during the plunger fall and close it during the plunger rise.

(11) Wobble washer plunger and a rod to open flow passage during the plunger fall.

1.2.1.2 柱塞类型

图1.11展示了一些典型的柱塞,这些柱塞经过测试可为建立柱塞举升系统模型提供数据。这些是典型的柱塞,但不包括业内可用的所有类型的柱塞。

在该图中,柱塞从左到右依次为:

(1)毛细管柱塞,其上有孔口允许气体通过,可以减少柱塞上方的液塞。

(2)带有凹槽的湍流密封柱塞,以促进"湍流密封"。

(3)当存在一些固体或砂子时,需特别使用刷子柱塞。

(4)另一种刷子柱塞。

(5)带凹槽的柱塞和带有一段"摆动垫圈"的组合来促进密封。

(6)带有一段湍流密封槽和一段弹簧加载的可膨胀叶片柱塞,同时还有杆用于打开或关闭通过柱塞的流通路径(取决于它是向下还是向上运动)。

(7)有两段可膨胀叶片的柱塞,带有一个杆,用于在向下行程中打开流通柱塞。

(8)带有可膨胀叶片的微型柱塞。

(9)另一种具有两个可膨胀叶片的柱塞,用于在柱塞下降期间打开流通通道。

(10)另一种具有可膨胀叶片的柱塞,它的杆用于在柱塞下降期间打开流过通道并且在柱塞上升期间关闭它。

(11)摆动垫圈型柱塞,它的杆在柱塞下落期间打开流动通道。

(12) Expandable blades with a rod to open a flow-through passage on the plunger fall that could fall against the flow and operate as continuous flow.

(12)带有杆的可膨胀叶片柱塞,杆用于打开柱塞上的流通通道,并逆流下落以连续流动运行。

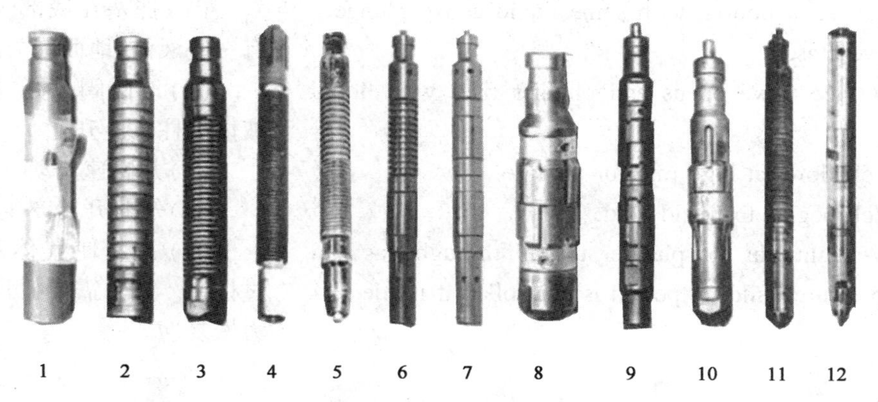

Fig. 1. 11　Various Types of Plungers

Several of these plungers have a push rod to open a flow passage through the plunger to allow flow through the plunger when falling to increase the fall velocity. When the plunger arrives at the surface, the push rod forces the flow passage open for the next fall cycle. When the plunger hits on bottom, the push rod is pushed upward to close the flow passage for the next upward cycle.

其中一些柱塞中具有推杆,可以打开穿过柱塞的流动通道,以允许在下降时流体流过柱塞来增加下降速度。当柱塞到达表面时,推杆迫使流动通道打开以进行下一个下降循环。当柱塞撞击底部时,推杆被向上推动以关闭流动通道并进行到下一个向上循环。

The brush plunger was found in testing to show the best seal for gas and liquids, but it typically wears sooner than other plungers. The brush plunger is the only plunger that will run in wells, making a trace of sand or solids. Plungers with the spring-loaded expandable blades showed the second best sealing mechanism and they do not wear nearly as fast as the brush plunger.

在测试中发现刷子柱塞对气体和液体的密封效果最佳,但它通常比其他柱塞磨损更快。刷子柱塞是唯一可以在井中运行并留下砂粒运动痕迹的柱塞。带有弹簧式可膨胀叶片的柱塞具有第二好的密封效果,其磨损速度几乎与刷子柱塞一样快。

1.2.1.3　Plunger Cycle

1.2.1.3　柱塞循环

Plunger lift is a preferred method by many Gas Production Operators. Plunger operation may be classified as conventional plunger lift (maybe 90% + of applications) and free cycle plunger lift where the well flows all the time except for a brief shut-in time at the surface to allow the plunger or part of the plunger system to fall against the flow.

柱塞举升是许多生产商生产天然气的首选方法。柱塞操作可以分为传统的柱塞举升(可能是 90% 以上的应用)和自由循环柱塞举升,除了在地面处的短暂关井以允许柱塞或柱塞系

Fig. 1. 12 shows a conventional plunger cycle:

(1) Plunger at bottom with some liquid above plunger. Surface valve closed.

(2) Surface valve opens and plunger rises with liquid above it.

(3) Well flows at high rate for a while.

(4) Well begins to liquid load.

(5) Well shut in for plunger to fall through gas and liquid. A pressure build up period is controlled if needed.

统的一部分反流外,井一直处于生产状态。

图 1. 12 展示了传统的柱塞循环:

(1)柱塞在底部,柱塞上方有一些液体,地面阀关闭。

(2)地面阀打开,柱塞上的液体随柱塞上升。

(3)井高速流动一段时间。

(4)气井开始产生积液。

(5)关井使气体和液体通过柱塞。如果需要,可以控制压力恢复期。

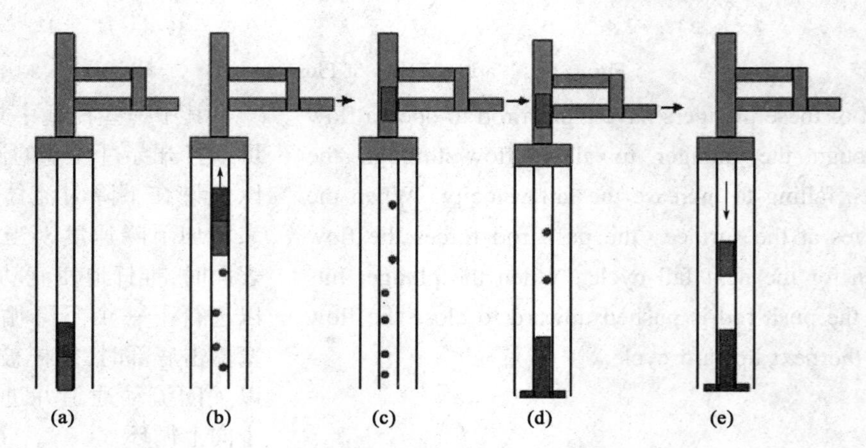

Fig. 1. 12 Conventional Plunger Cycle
(Source: http://www. nanopdf. com)

Plungers used for this cycle include brush plungers, grooved plungers, wobble washer plungers, padded plungers, and other special types.

Fig. 1. 13 shows a freecycle plunger cycle or continuous plunger cycle. It can be achieved by use of a two piece plunger as below or other contained ball and seat or contained valve type plungers.

可用于该循环的柱塞包括刷子柱塞、带槽的柱塞、摆动垫圈柱塞、带衬垫的柱塞和其他特殊类型的柱塞。

图 1. 13 显示了自主往复柱塞循环或连续柱塞循环。它可以通过使用如下的两件式柱塞或其他包含球座、阀式柱塞来实现。

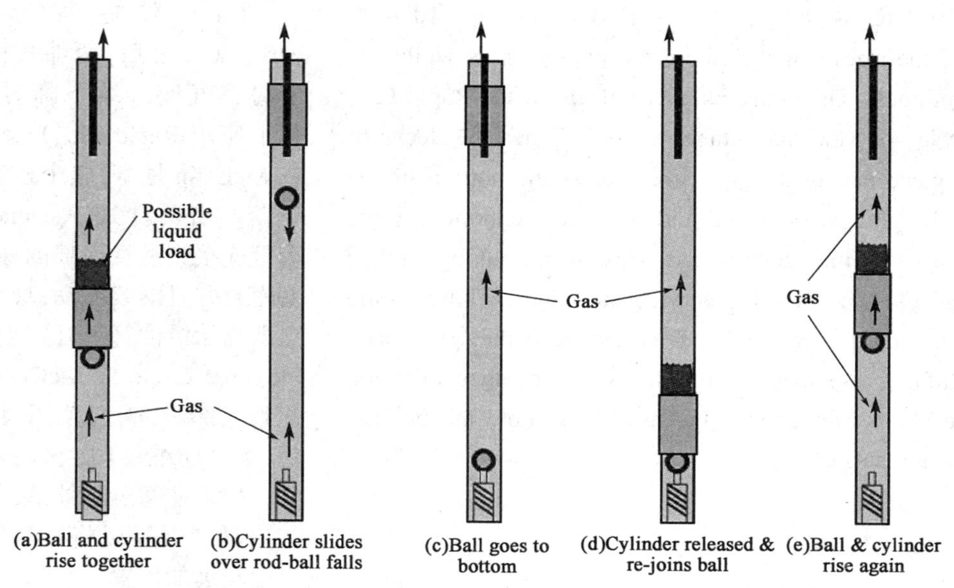

Fig. 1.13 A Free Cycle Plunger Cycle or Continuous Plunger Cycle
(Source: http://www.nanopdf.com)

Fig. 1.14 shows a twopiece plunger (sleeve and ball) but this cycle could also work with a valved plunger such as the Weatherford Rapid Flo™ Plunger, the FB Free Cycle™ Plunger, the McClain™ Plunger, or as shown above, the Pacemaker™ (two-piece) plunger. Other plungers that are not mentioned here may also work on this cycle.

图1.14显示了一个两件式柱塞(套筒和球)循环,但这个循环也适用于带阀门的柱塞,如Weatherford Rapid Flo™柱塞、FB Free Cycle™柱塞、McClain™柱塞,或Pacemaker™(两件式)柱塞。此处未提及的其他柱塞也可用于此循环。

(1) The plunger is sealed by the ball on the bottom of a sleeve or a valve is closed in the plunger. It is carrying up a slug of liquid with the surface valve open.

(1)柱塞由套筒底部的球密封,或阀门在柱塞中关闭。它在地面阀打开的情况下携带液体。

(2) The plunger arrives at the surface. If the two-piece plunger is used, the sleeve is held on a receiving rod and the ball falls against the flow. If a valve-type plunger is used, the plunger is held on a receiving rod with the flow.

(2)柱塞到达地面。如果使用两件式柱塞,则套筒固定在接收杆上,并且球反流下落。如果使用阀式柱塞,则柱塞顺着流动固定在接收杆上。

(3) The gas continues to flow with the sleeve of the two-piece plunger. If the two-piece plunger is used, the ball can fall with flow continuing.

(3)气体继续与表面两件式柱塞的套管一起流动。如果使用两件式柱塞,球会随着流动继续下降。

(4) After a short shut-in period of perhaps 10 ~ 20 seconds, the sleeve of the two-piece plunger falls while gas flow continues. The entire plunger will fall if the Rapid Flo, Free Cycle, or McClain plunger is used. When the sleeve of the two-piece Pacemaker joins the ball at the bottom of the tubing, the plunger is sealed and with flow continuing, the ball and sleeve immediately start back up the tubing. When the Rapid Flo, Free Cycle, or McClain plungers hit bottom, a valve is closed, the plunger is sealed, and they start back up the tubing as flow continues. If the plunger falls too soon, an Auto catcher may be installed to hold the plunger longer at the surface.

（4）在短暂的关井（10～20s）之后，两件式柱塞的套筒在气流继续下降时下降。如果使用 Rapid Flo、Free Cycle 或 McClain 柱塞，整个柱塞将会掉落。当两件式 Pacemaker 的套管连接到油管底部的球时，柱塞密封并且继续流动，球和套筒立即开始沿油管上升。当 Rapid Flo、Free Cycle 或 McClain 柱塞撞到底部时，阀门关闭，柱塞密封，当流动继续时，它们开始上升。如果柱塞过早掉落，可以安装自动捕捉器以将柱塞保持在表面较远的位置。

(5) The sealed continuous flow plunger is rising with the flow with a slug of liquid being carried.

（5）密封的连续流动柱塞随着流动而上升，同时携带出液体。

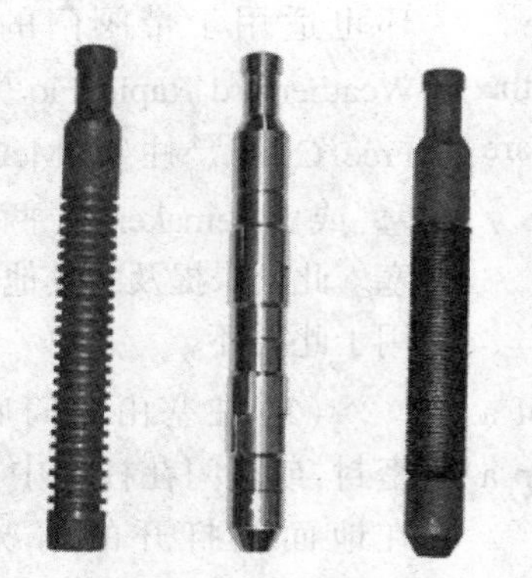

(a)Conventional Grooved,Padded and Brush Plungers

(b)Two-Piece Plunger

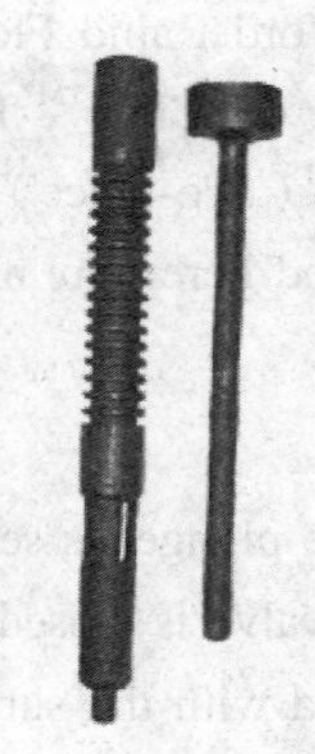

(c)Continuous Rapid Flow Plunger

Fig. 1. 14　Typical Plungers

(Source: http://www.nanopdf.com)

1.2.2　Foaming

1.2.2.1　Introduction

Foams have several applications in oil field operations. They are used as a circulation medium for drilling wells, well cleanouts, and as fracturing fluids. These applications differ slightly from the application of foam as a means of removing liquid from producing gas wells. The former applications involve generating the foam at the surface with

1.2.2　泡沫法

1.2.2.1　引言

泡沫在油田作业中具有多种应用。它们可用作钻井液、洗井液和压裂液。这些应用与泡沫用于气井排液采气略有不同，前者通常用水在地面通过泡沫发生器产生泡沫。在用于气井排

controlled mixing and using only water. In gas well liquid removal applications, the liquid-gas-surfactant mixing must be accomplished downhole and often in the presence of both water and liquid hydrocarbons. The use of foam produced by surfactants can be effective for gas wells that accumulate liquid at low rates (Fig. 1. 15).

液采气时,液—气—表面活性剂必须在井下混合,并且井下通常同时存在水和液态烃。利用表面活性剂产生的泡沫对于积液速度较低的气井是有效的(图 1.15)。

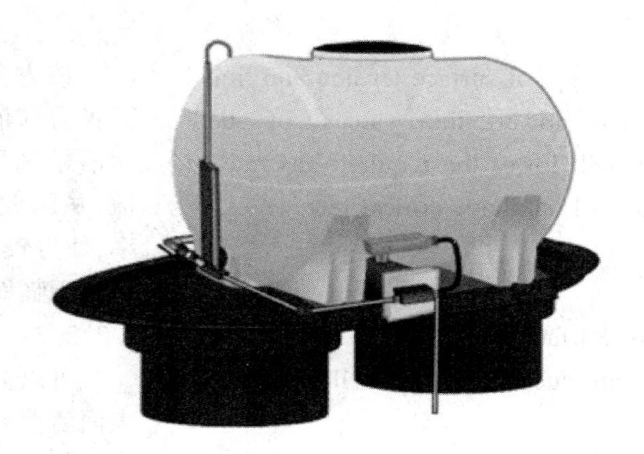

Fig. 1. 15 Liquid Foaming Agent

The principal benefit of foam as a gas well dewatering method is that liquid is held in the bubble film and exposed to more surface area resulting in less gas slippage and a low-density mixture. The foam is effective in transporting the liquid to the surface in wells with very low gas rates when liquid holdup would otherwise result in sizable liquid accumulation and/or high multiphase flow pressure loses.

泡沫法排液采气的主要优点是液体保持在气泡膜中并且表面积大,所以很少发生气体滑脱,并产生较低密度的混合物。当气井产量很小时,液体滞留量大,会导致相当多的液体积聚和(或)高的多相流压力损失,泡沫可以有效地将液体举升到地面。

Foam is a particular type of gas and liquid emulsion. Gas bubbles are separated from each other in foam by a liquid film. Surface active agents (surfactants) generally are employed to reduce the surface tension of the liquid to enable more gas-liquid dispersion. The liquid film between bubbles has two surfactant layers back to back with liquid contained between them. This method of tying the liquid and gas together can be effective in removing liquid from low volume gas wells.

泡沫是一种特殊类型的气液混合物。气泡通过液膜在泡沫中彼此分离。表面活性剂通常用于降低液体的表面张力,以达到更好的气液分散效果。气泡之间的液膜具有两个背对背的表面活性剂层,它们之间存在液体。这种将液体和气体结合在一起的方法可以有效地从低产量气井中排出液体。

Foam is used with best results in gas wells making only water. It can handle deep wells and wells with moderate liquid rates.

泡沫在仅产水的气井中使用效果最佳。它可以处理深井和中等产液量的井。

From the discussion of critical rate, an intermediate equation for critical velocity was shown as:

从临界流速的角度来看,临界流速的方程为:

$$u = 1.92\frac{\sigma^{\frac{1}{4}}(\rho_L - \rho_g)^{\frac{1}{4}}}{\rho_g^{\frac{1}{2}}} \tag{1.13}$$

This equation is a function of surface tension and liquid density. Surfactants will lower the values of both parameters, and as such will lower the required gas rate so that a lower gas rate can still be above critical rate.

该方程表明临界流速是表面张力和液体密度的函数。表面活性剂可以降低这两个参数值,并因此降低所需的气体速率,使得较低的气体速率仍然可以高于临界流速。

1.2.2.2 Application of Surfactants

1.2.2.2 表面活性剂的应用

Surfactants can be introduced into the well using three methods:

可以使用以下三种方法将表面活性剂注入井中:

(1) Soap sticks dropped down the tubing.

(1)将肥皂棒沿着油管投入。

(2) Surfactants dropped or pumped down the casing/tubing annulus (no packer).

(2)将表面活性剂注入或泵入油套环空(无封隔)。

(3) Lubricate a cap string through the wellhead down the tubing to the perforations.

(3)通过井口沿油管向下润滑管柱直至穿孔处。

Soap sticks can be launched down the tubing in a variety of ways, Fig. 1.16 shows one type of an automated soap stick launcher.

肥皂棒可以通过多种方式沿油管投入,图1.16展示了一种自动肥皂棒投入器。

If no packer is present, batching down the annulus is a very acceptable way of surfactant injection (see Fig. 1.17).

如果没有封隔器,那么通过环空中注入是一种可接受的表面活性剂加入方式(图1.17)。

There is evidence that sometimes soap sticks do not find their way to the bottom of the well. New weighted and shaped sticks may help, but use of the capillary string to inject chemicals to the bottom of the tubing is a sure way of getting the chemicals to the pay zones.

有现象表明有时肥皂棒无法顺利到达井底,加重和改变其形状之后可能会有所改善,但使用毛细管柱将化学品注入管的底部是将化学品送到井底的可靠方法。

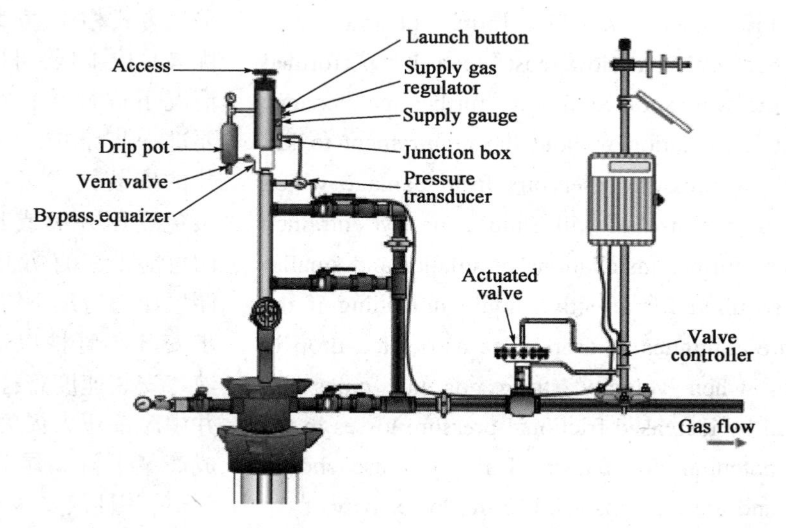

Fig. 1. 16 Soap Stick Launcher with Automatic Controller
(Source: http://www. arab-oil-naturalgas. com)

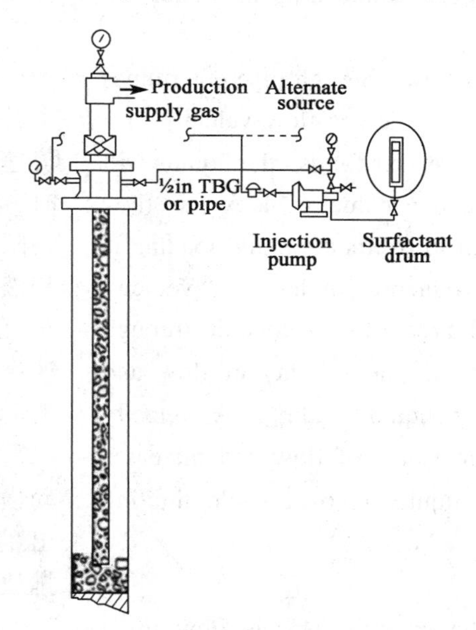

Fig. 1. 17 Batch Treating Down the Annulus

1.2.3 Velocity Strings

Liquid loading is detrimental to well performance because it causes a drop in production rates, liquid slugging at the surface and can eventually lead to a well dying if left untreated. The technologies available to remediate liquid loading have also been well documented, and include strategies such as wellhead compression, plunger lift, surfactant injection, gas lift, ESPs and progressive cavity pumps. The installation of a velocity string can be advantageous

1.2.3 速度管柱法

井底积液对井的性能会产生不利影响,因为它会导致产能下降、液体堵塞,如果不加以处理,最终会导致死井。排液采气技术广泛应用于此类问题的井中,包括井口增压法、柱塞举升、表面活性剂注入、气举、电潜泵和螺杆泵等技术。与其他排液

compared to the other available forms of gas well deliquification because it is low cost, can be performed without killing the well and requires no further maintenance after installation. Installation without the requirement to kill a well protects low pressure reservoirs from damage which can occur due to the use of kill fluids, or by entrained solids. However, further installations of smaller and smaller tubing will be required for effective liquid unloading if the reservoir pressure continues to drop. The associated drop in production rates when velocity strings or tail pipes are installed (caused by increased frictional pressure losses in the well) are also potential drawbacks. Tail pipes use shorter lengths of pipe and are only installed in the lower part of the completion where the fluid velocity is below the critical velocity, and so the frictional losses and drop in production rates are lower.

采气方式相比,安装速度管柱的优势是成本低,可以在不关井的情况下进行并且在安装后不需要进一步维护。这种不需要关井的安装技术可以保护低压储层免受压井液或其携带的固体而可能带来的伤害。然而,如果储层压力持续下降,则需要进一步安装小管柱以进行有效的排液。安装速度管柱或尾管时,由井中摩擦压力损失增加,可能造成产量下降。尾管使用较短的管道,并且仅安装在流体速度低于临界流速的完井管柱下部,这样摩擦损失和生产率下降较低。

The velocity at which gas flows through pipe determines the capacity to lift liquids. When the gas flow velocity in a well is not sufficient to move reservoir fluids, the liquids will build up in the well tubing and eventually block gas flow from the reservoir. One option to overcome liquid loading is to install smaller diameter production tubing or "velocity tubing". The cross-sectional area of the conduit through which gas is produced determines the velocity of flow and can be critical for controlling liquid loading. A velocity string reduces the cross-sectional area of flow and increases the flow velocity, achieving liquid removal while limiting blowdown to the atmosphere.

气体流过管道的速度决定了其举升液体的能力。当气井流速不足以携带出储层流体时,液体将在井筒中积聚并最终阻止气体从气藏流出。解决积液问题的一种方式是安装较小直径的生产油管或速度管。速度管的横截面积决定了气体流动速度并对控制井底积液至关重要。速度管柱可以减小流动横截面,增加流速,将液体排出的同时限制了对大气的排放。

Fig. 1. 18 shows that the conduit for gas flow up a wellbore can be either production tubing, the casing-tubing annulus or simultaneous flow through both the tubing and the annulus. As a rule of thumb, gas flow velocity of 1,000 feet per minute is needed to remove liquid. These figures assume used pipe in good condition with low relative roughness of the pipe wall.

图 1. 18 说明气体向井筒流动时可以通过生产油管、油套环空或同时流过油管和油套环空。根据经验,气流速度需要达到 1000ft/min 来携带液体。这个数字是假设使用的管道状况良好且管壁的粗糙度相对较低。

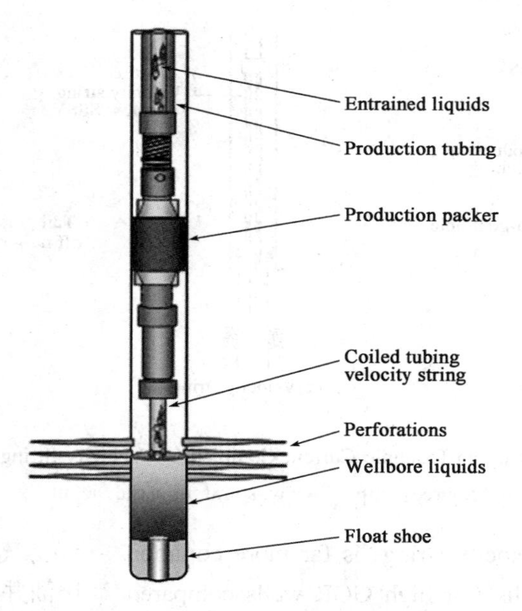

Fig. 1. 18 Velocity String
(Source: http://www. arab-oil-naturalgas. com)

The installation of a velocity string is relatively simple and requires calculation of the proper tubing diameter to achieve the required velocity at the inlet and outlet pressures of the tubing. Velocity tubing to facilitate liquid removal can be successfully deployed in low volume gas wells upon initial completion or near the end of their productive lives. Candidate wells include marginal gas wells producing less than $60 \times 10^6 ft^3/d$. Installation of velocity tubing requires a well workover rig to remove the existing production tubing and place the smaller diameter tubing string in the well.

速度管柱的安装相对简单，需要计算入口和出口处压力达到所需速度的管道直径。速度管柱适用于完井初期或低产气井接近停产时。可供选择的井包括天然气产量低于 $60 \times 10^6 ft^3/d$ 的井。安装速度管柱时需要用修井机来移动现有的生产油管且在井中放置较小直径的油管管柱。

Velocity strings and tail pipes tend to come in standardized sizes. The requirement for a fully operational SSSV means that velocity strings must be set below the SSSV. This can be achieved using new packers. Tail pipes however, can be hung off existing no-go nipples present in the current completion. No-go nipples are placed deep in the production tubing and provide a reduced diameter, preventing tools of a certain size falling within the tubing. Fig. 1. 19 shows the difference between a current completion, velocity string and tail pipe installation.

速度管柱和尾管往往采用标准尺寸。速度管柱必须安装在井下安全阀以下，这可以通过安装新的封隔器来实现。尾管可以悬挂在无齿状接头上。若无接头，可放置在生产管的深处，并选用较小的直径，防止工具落入管内。图 1.19 显示了完井、速度管柱和尾管安装之间的差异。

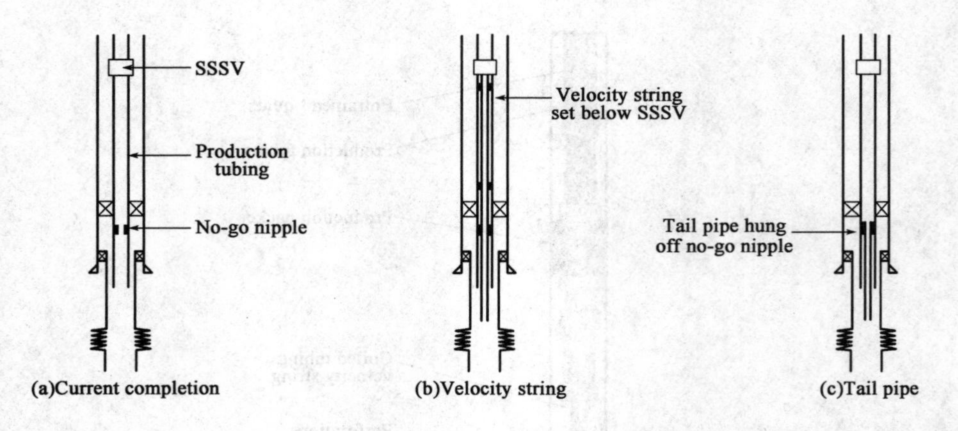

Fig. 1. 19　Diagram Showing an Example Current Completion, Velocity String and Tail Pipe Installation
(Source: http://www.spiral.imperial.ac.uk)

The application of velocity strings is far more common for gas wells than oils wells (or high GOR wells compared to low GOR wells). This is because ESPs have proved to be a very effective method for adding pressure to oil wells. Their application to gassy wells is less popular due to their inability to handle large volumes of gas.

气井中速度管柱的应用远比油井(或高气油比井比低气油比井)中更常见。这是因为电潜泵可以增加油井压力,但无法处理大量气体,它们在气井中不适用。

SSSVs must be fully operational at all times, providing a technical challenge in the use of technologies such as velocity strings or plunger lift, though new systems allowing the use of plungers below the SSSV have been developed. Chemical injection using surfactant via capillary tubes can also be used, but surfactants need rigorous testing before they can be approved. Velocity strings therefore provide a simple, quick and relatively low cost solution to problems regarding liquid loading.

井下安全阀必须始终正常运行,这对速度管柱或柱塞举升等技术提出了技术挑战,尽管已开发出允许在井下安全阀以下使用的新柱塞系统,也可以通过毛细管注入表面活性剂,但表面活性剂在应用之前需要经过严格的测试。因此,速度管柱为井底积液问题提供了简单、快速和相对低成本的解决方案。

1.2.4　Artificial Lift

Forty percent of the wells in the world are on some kind of artificial lift. Fig. 1. 20 shows the six primary methods of artificial lift. The commonly used artificial lift methods for gas well unloading include the following: Sucker rod pumping, Gas lift, Hydraulic piston pumping, Hydraulic jet pumping, Progressing cavity pumping, Electrical submersible pumping.

1.2.4　人工举升

世界上有40%的井都在使用人工举升。图1.20显示了人工举升的六种主要方法。气井常用的人工举升方法包括:有杆泵、气举、水力活塞泵、水力射流泵、螺杆泵、电潜泵。

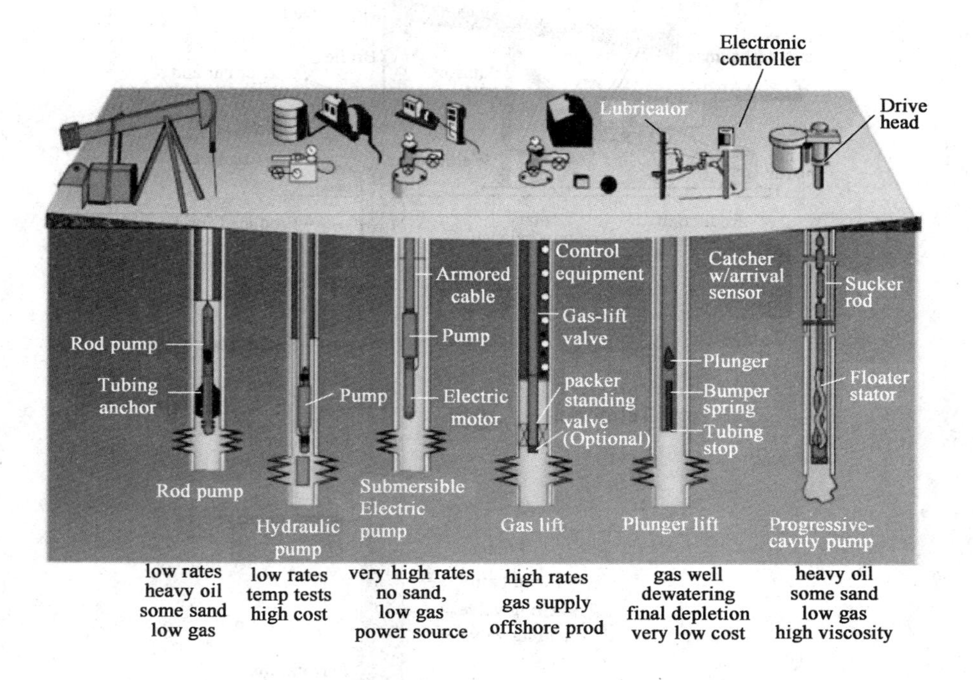

Fig. 1.20 Examples of the Primary Methods of Artificial Lift
(Source: http://www.petroskills.com)

This section first describes the sucker rod pumping, gas lift and hydraulic pumps briefly, and then focuses on the description of progressing cavity pumping and electrical submersible pumping.

本部分首先简要介绍有杆泵、气举和水力泵(水力活塞泵、水力射流泵)方法,然后重点介绍螺杆泵和电潜泵。

1.2.4.1 Sucker Rod Pumping

1.2.4.1 有杆泵

Sucker Rod Pump (SRP) is the simplest artificial method known and most widely used choice of artificial methods. Basically a SRP system consists of a tube divided into chambers by plunger and a simple surface unit including power plant (Fig. 1.21). Operating principle is depending on the two valves work with plunger, transferring fluid from bottom chamber to top. Although sucker rod pumping considered as a simple system, installation has to be properly designed by the engineer.

有杆泵(SRP)是一种最简单的也是应用最广泛的人工举升方法。有杆泵系统由井下抽油泵、抽油杆和地面抽油机三部分组成(图 1.21)。工作原理是两个阀门与柱塞相互配合,将流体从底部腔室转移到顶部。虽然有杆泵系统被认为是一个简单的系统,但必须由工程师设计安装。

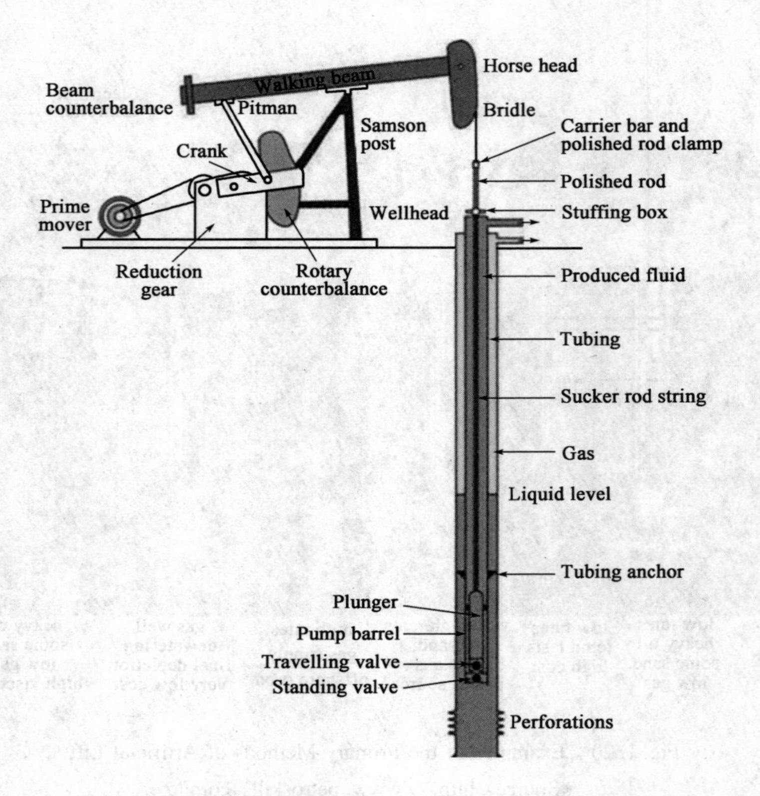

Fig. 1.21 Sucker Rod Pumping System

(Source：http://www.dynamicsystems.asmedigitalcollection.asme.org)

1.2.4.2 Gas Lift

After SRP, gas lift is the second most widely used artificial method in the world, generally offshore. Gas lift system depends on the principle of lightening the gradient of fluid by injecting high pressure gas down the annulus. Specially designed gas lift valves installed on the tubing string. Gas under pressure is injected down through the space between casing and the tubing. Gas enters from valves and the fluid standing above the gas inlet point displaced. A complete gas lift system consists of a gas compression station, a gas injection manifold with injection chokes and time cycle surface controllers, a tubing string with installations of unloading valves and operating valve, and a down-hole chamber. Fig. 1.22 and 1.23 depict the gas lift system.

1.2.4.2 气举

气举应用的广泛程度仅次于有杆泵，尤其在海上。气举系统的工作原理是通过沿油套环空注入高压气体来减小流体压力梯度。将设计的特殊气举阀安装在油管柱上，气体通过油套环空向下注入。气体从气举阀进入，入口点上方的流体被举升。一个完整的气举系统包括气体压缩站、有注入控制阀的气体注入管柱、装有卸荷阀和操作阀的管柱，以及井筒。气举系统示意图详见图1.22和图1.23。

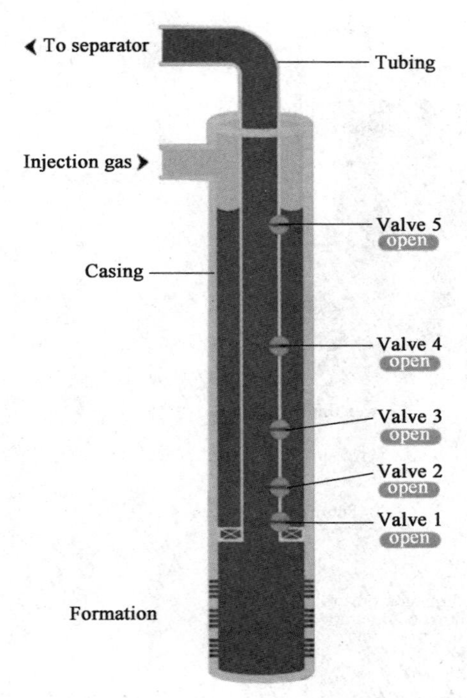

Fig. 1.22 Configuration of a Typical Gas Lift Well
(Source: http://www.epiclift.com/gas-lift/)

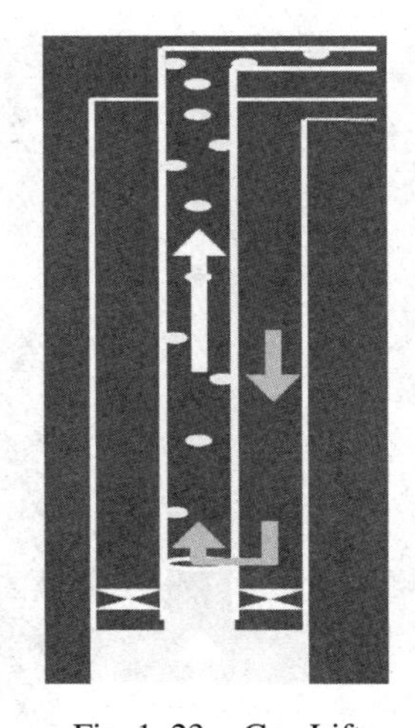

Fig. 1.23 Gas Lift
(Source: http://www.core.ac.uk)

1.2.4.3 Hydraulic Pumps

Hydraulically powered downhole pumps are powered by a stream of high-pressure water or oil (power fluid) supplied by a power-fluid pump at the surface. Hydraulic pumps are of two types:

(1) Piston pumps, which are similar to beam pumps.

(2) Jet pumps that operate by power fluid passing through a venturi, exposing the formation to low pressure.

The use of hydraulic system applications is becoming more common. One of the most common ways to produce by the use of a hydraulic jet pump. A hydraulic jet pump operates by pumping a high pressure power fluid produced by a surface unit into the tubing then down to the bottom of the well where it is passed through a nozzle and finally through a diffuser when this occurs the result is a low pressure point where the formation can provide inflow and the production and power fluids can be returned together up the annulus. Fig. 1.24 depicts the hydraulic pumping.

1.2.4.3 水力泵

水力泵由地面流体泵提供的高压水或油(动力液)提供动力。水力泵有两种类型:

(1)水力活塞泵,类似于有杆泵。

(2)水力射流泵,通过管内的动力液,使井筒压力降低。

水力泵在气井中的应用变得越来越普遍,其中水力射流泵比较常用。水力射流泵将地面装置产生的高压动力液泵送到管道,然后通过喷嘴、喉管和扩散管,这时地层流体和动力液可以一起返回到油套环空。图1.24为水力泵系统。

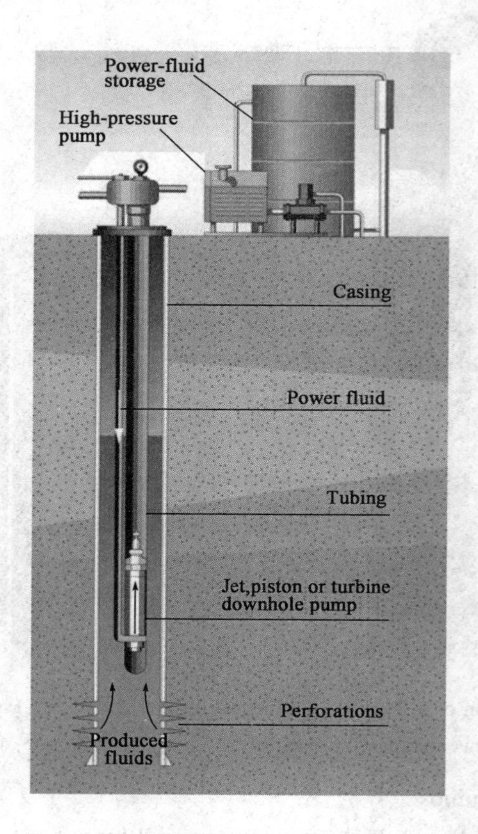

Fig. 1. 24　Hydraulic Pumping
(Source：http://www. . rigzone. com)

1.2.4.4　Progressing Cavity Pumping

Audio 1. 3

Progressive cavity pumps (PCPs) are a special type of rotary positive displacement pumps, and were first introduced in petroleum engineering as an artificial lift method in 1970's. Fig. 1. 25 and 1. 26 depicts a PCP system. In a PCP, the flow through the pump is almost axial, while in all other rotary pumps, pumping fluid is forced to travel circumferentially. This gives PCP unique axial flow pattern and low internal velocity, which reduces fluid agitation and churning and therefore reduces fluids solids erosion.

1.2.4.4　螺杆泵

螺杆泵(PCP)是一种特殊类型的回转容积泵,在20世纪70年代被引入石油工程中作为一种人工举升工具。螺杆泵系统如图1. 25和图1. 26所示。在螺杆泵中,流体的流动几乎是轴向的,而在所有其他回转泵中,流体是被迫沿周向流动的。这使得螺杆泵具有独特的轴向流动模式和低的流动速度,从而减少了流体搅动,因此减少了固体侵蚀。

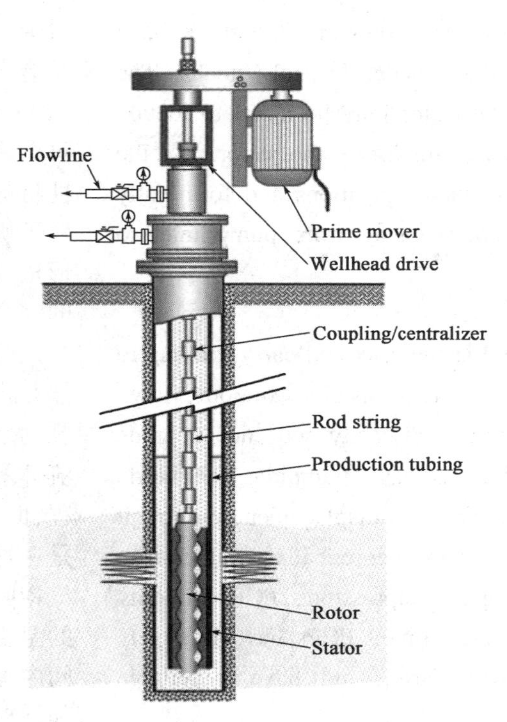

Fig. 1.25　PCP System
(Source: http://www. . higp. hawaii. edu).

Fig. 1.26　Progressive Cavity Pump and Motor Drive
(Source: http://www. petroskills. com)

PCP has advantages of lower investment, broader applications to fluid mixtures, less maintenance, and higher efficiency to other artificial lift methods. It is becoming a popular lift tool and, for same wells, the best choice in artificial lift methods.

螺杆泵具有投资少、应用范围广、维护成本低、效率高的优点,已经成为一种流行的人工举升方式,对于同样的井,它是最佳选择。

In petroleum industry, the most commonly used progressive cavity pump is a single lobe pump that consists of a single external helical rotor turning eccentrically inside a

在石油工业中,最常用的螺杆泵是单吸泵,它由一个能转动的单螺杆(转子)和一个固定的

double internal helical stator. The rotor and the stator have the same minor diameter and are made of metal (steel). The fits between the rotor and the stator may be metal to metal, or metal to elastomer which is set inside the stator. As the rotor rotates, the seal lines change positions and form fully enclosed cavities moving continuously from pump inlet to outlet.

衬套(定子)组成,转子和定子具有相同的内径并且由金属(钢)制成。转子和定子的配套可以是金属对金属,或金属对弹性材料(设置在定子内部)。当转子旋转时,密封线改变位置并形成从泵入口到出口连续移动的完全封闭的腔。

These cavities trap fluid at the inlet and carry it along to the outlet, thus providing a non-pulsation smooth flow. Unlike centrifugal pump, fluid viscosity will not degrade pump head of a PCP, but increase pumping volumetric efficiency. Since PCP is a positive displacement pump, it doesn't have gas lock problem theoretically, but due to temperature increase from gas compression, PCP can only handle high gas slug in a short time. PCPs have relatively low inertia of their rotating parts, and have a reliable working life.

这些空腔在入口处捕获流体并将其带到出口,从而提供非脉动平稳流动。与离心泵不同,流体黏度不会降低螺杆泵的扬程,但会提高泵效。由于螺杆泵是一种正排量泵,理论上它没有气锁问题,但由于气体压缩引起的温度升高,螺杆泵只能在短时间内输送气体。螺杆泵的旋转部件惯性相对较低,并具有可靠的工作寿命。

Although PCPs have been used for a few decades as an artificial lift method in petroleum engineering, no one has discussed the design of PCP wells in production system. Fluid production from a PCP pumping well is not only determined by the PCP, but also controlled by reservoir inflow performance, fluid outflow performance and surface condition. This section fills the gap by using nodal analysis method to design a PCP.

虽然螺杆泵在石油工程中用于人工举升已有几十年,但生产系统中螺杆泵的设计还没有研究过。在井中,螺杆泵的产液量不仅取决于螺杆泵,还受油藏流入动态、流体流出动态和地面控制的影响。本部分介绍通过节点分析方法来设计螺杆泵。

The design of a PCP pumping well is to determine the rotational speed of the pump to produce the well at desired liquid rate, or calculate the production rate of the well at a given rotational speed. Also system analysis is helpful in analyzing and improving the pumping efficiencies of existing PCP wells. Similar to other artificial lift designs, PCP design also associates with many factors like mechanical efficiency, reservoir temperature, working life, abrasion resistance, elastomer tolerance and so on. This section focuses on production rate and rotational speed.

螺杆泵的设计主要是确定泵的转速使井在预期产量下进行生产,或者在给定的转速下计算井的产量。螺杆泵系统分析有助于分析和提高螺杆泵的泵效。与其他人工举升设计类似,螺杆泵设计还涉及许多因素,如机械效率、储层温度、工作寿命、耐磨性、弹性体公差等。本部分重点介绍产量和转速。

1. Basic Correlations

A PCP rotor is a single external helical gear. Its cross section is a circle with the diameter of d at any place as shown in Fig. 1. 27. The centers of all the cross sections are on a helical line which has an eccentricity, e, with the rotor's axis. A stator is a double internal helical gear, and has the same minor diameter as the rotor. The stator has twice pitch length as that of the rotor. Generally, one and a half stator pitches are called a stage in a PCP.

1. 基本等式关系

螺杆泵转子是单个外部斜齿轮。它的横截面是一个直径为 d 的圆形,如图 1. 27 所示。所有横截面的中心位于螺旋线上,该螺旋线与转子轴线的距离称为偏心距 e。定子是双内螺旋齿轮,并且具有与转子相同的内径。定子导程是转子导程的两倍。通常,一个半定子的导程称为螺杆泵的一级。

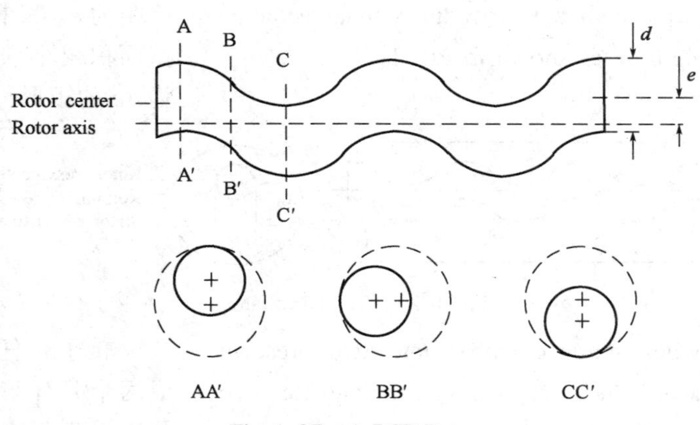

Fig. 1. 27 A PCP Rotor

A stator's cross section at any place is a long circle and can be described as two half circles of diameters d departed a distance of $4e$ as shown in Fig. 1. 28.

定子在任何位置的横截面都是一个椭圆,可以描述为直径为 d 的两个半圆,其距离为 $4e$,如图 1. 28 所示。

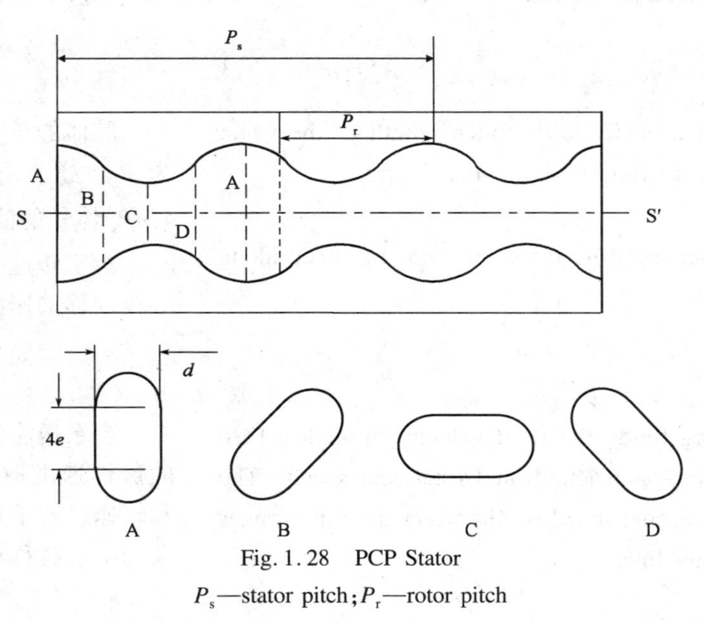

Fig. 1. 28 PCP Stator

P_s—stator pitch; P_r—rotor pitch

The stator has the same cross-sectional shapes along it axis but with different angles. The cross-sectional areas of the rotor and the stator are:

定子沿轴线具有相同的横截面形状,但角度不同。转子和定子的横截面积分别为

$$A_{\text{rotor}} = \frac{1}{4}\pi d^2, \quad A_{\text{stator}} = \frac{1}{4}\pi d^2 + 4ed \tag{1.14}$$

After setting a rotor in a stator, the rotor axis is not coincidence with stator center line. In addition to rotating around its axis, the rotor rotates eccentrically around the stator center line with the same eccentricity of e. The cross-sectional area at any place of a PCP reduces to a rectangle with width d and length $4e$ as shown in Fig. 1.29.

在定子中设置转子后,转子的轴线与定子的中心线不重合。除了围绕其轴线旋转之外,转子还围绕定子中心线偏心旋转,偏心距为 e。螺杆泵任何位置的横截面积减小为宽度为 d 且长度为 $4e$ 的矩形,如图 1.29 所示。

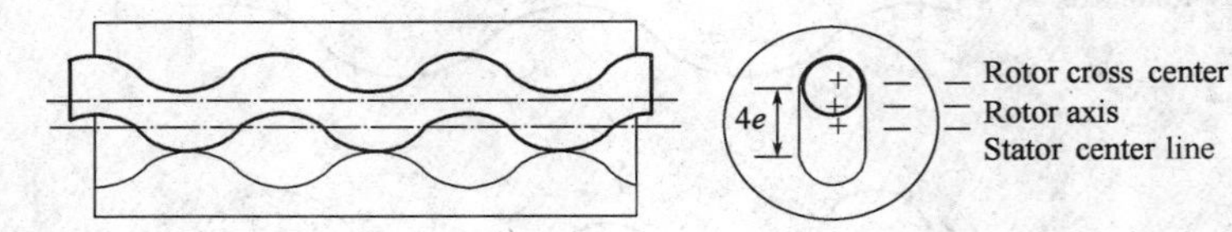

Fig. 1.29 Rotor in Stator

The rotor divides stator chamber into two crescent sections at any place. The two sections belong to two cavities and their areas change with the rotation of the rotor. The total area of the two sections is constant at any place along the pump, and it is the fluid flow area.

转子在任何位置将定子分成两个月牙形部分。这两个部分属于两个腔体,它们的面积随着转子的旋转而变化。两个部分的总面积在沿泵的任何位置是恒定的,并且它是流体的流动区域,总面积为

$$A_{\text{f}} = 4ed \tag{1.15}$$

The length of a cavity is the pitch length of the stator P_s. A PCP will move fluid of $4edP_s$ per rotation.

腔体的长度是定子的导程,为 P_s。螺杆泵每次旋转移动 $4edP_s$ 体积的流体。

For a rotational speed n, the cavity moving speed along stator center line is

对于转速 n,腔沿着定子中心线的移动速度是

$$v = nP_s \tag{1.16}$$

When pumping fluid, the fluid velocity through a PCP depends on its stator pitch length and rotational speed. The higher the rotor rotational speed is, the faster the fluid travels along pump's center line.

当泵送流体时,通过螺杆泵的流体速度取决于其定子导程和转速。转子转速越高,流体沿泵的中心线移动得越快。

Since the cross-sectional area for fluid flow is constant at any place, the flow rate in a PCP is

由于流体流动的横截面积在任何地方都是恒定的,因此螺杆泵中的流速为

$$q_t = A_f v = 4ednP_s \tag{1.17}$$

The correlation implies that the cavity fluid is displaced completely. The rate is the theoretical rate of the pump at the rotational speed n. However, internal leakage from outlet to inlet is always there whenever a differential pressure applies. This is because that the seals have small clearances for metal to metal fits or the seals are pushed apart by the pressures in cavities for metal to elastomer. The leakage, also called slip, depends upon pump type, the clearance between rotor and stator, fluid viscosity, and differential pressure. Pump size also affects slip. Larger pump has longer clearance length and thus has larger slip.

等式关系表示腔内流体完全排出。流速 q_t 是泵在转速 n 下的理论流速。但是,每当施加压差时,从出口到入口的内部泄漏始终存在。这是因为密封件有用于金属与金属配合的小间隙,或者密封件被金属与弹性体的腔体中的压力推开。泄漏也称为滑移,取决于泵的类型、转子和定子之间的间隙、流体黏度和压差,泵的尺寸也会影响滑移。较大的泵具有较大的间隙,因此具有较大的滑动。

Taking into account the slip rate, q_s, the actual discharge rate of a PCP, q_a, is

考虑到滑移量 q_s,螺杆泵的实际流速 q_a 可表示为

$$q_a = q_t - q_s = 4ednP_s - q_s \tag{1.18}$$

Volumetric efficiency of a PCP is defined as the ratio of the actual rate to the theoretical rate. The higher the slip is, the lower the volumetric efficiency. The efficiency of one hundred percent represents no slip.

螺杆泵的容积效率定义为实际流速与理论流速的比值。滑移越高,体积效率越低。若效率为百分之百则代表没有滑动。

$$E_v = \frac{q_a}{q_t} = 1 - \frac{q_s}{q_t} \tag{1.19}$$

2. Basic PCP Design

As show in Eq. 1.17, the theoretical flow rate varies with pump, and is proportional to pump rotational speed. For a higher production rate, selecting a large size pump will avoid high rotational speed. Also noted is that the rate has nothing to do with stage number or pump length. Unlike for an electric submersible pump, selecting a PCP with more stages will not increase production rate.

2. 螺杆泵设计基础

由式(1.17)可知,不同的泵其理论流量不同,并且与泵的转速成比例。为了获得更高的生产率,选择大型泵可以避免过高转速。还要注意的是,产量与级数或泵长度无关。与电潜泵不同,选择具有更多级数的螺杆泵不会提高产量。

More stages represent more seals and longer length. If each seal takes the same differential pressure, a PCP with longer length or more stages can take higher total differential pressure. When a high lift head is needed, a pump with more stages should be selected.

更多级数代表密封更好、长度更长。如果每个密封件具有相同的压差,则具有更长或更多级的螺杆泵可以承受更高的总压差。当需要高扬程时,应选择级数更多的泵。

In short, to get a higher production rate, one can use a larger pump or make a pump rotate faster. To overcome higher differential pressure, one can use longer pump or add more stages to a pump.

简而言之,为了获得更高的产量,可以使用更大的泵或使泵更快地旋转。为了克服更高的压差,可以使用更长的泵或增加泵的级数。

Eq. 1.17 gives the basic design correlation. The term $4edP_s$ is the theoretical volume or nominal displacement per rotor revolution. It represents a pump's volumetric capacity. If all the parameters are known, the theoretical rate can be calculated easily. Normally, PCP manufacturers provide the theoretical displacement directly. If Q_t is the theoretical displacement per revolution, $Q_t = 4edP_s$.

式(1.17)给出了基本关系式。$4edP_s$ 是每转子旋转的理论容积或额定排量,代表泵的容量。如果所有参数都已知,则可以计算理论产量。通常,螺杆泵制造商直接提供理论排量。如果 Q_t 是每转的理论排量,则 $Q_t = 4edP_s$。

Substituting Q_t into Eq. 1.17 and solving for the pump speed give the correlation of flow rate and rotational speed. For PCP design, replacing the theoretical rate q_t, by the total fluid rate at the pump intake q_{tL}, yields the correlation to calculate the required rotational speed from total flow rate.

将 Q_t 代入式(1.17),得出流速与转速的关系。对于螺杆泵,将理论流速 q_t 替换为泵入口处的总流速 q_{tL},可利用总流速计算出所需的转速。

$$n = \frac{q_{tL}}{Q_t} \tag{1.20}$$

The theoretical displacement tells the pump's volumetric capacity, and larger pump has larger Q_t. For US field units, the Q_t is in bbl/d/r/min, the flow rate is in bbl/d, and the rotational speed n is in r/min.

从理论排量可知泵的容积,较大的泵具有较大的 Q_t。美国的油田,Q_t 单位为 bbl/d/r/min,流速 q_{tL} 单位为 bbl/d,转速 n 单位为 r/min。

q_{tL} in the equation is the total fluid (the sum of liquid and gas) rate flowing into a PCP at pump intake condition, which excludes the gas separated and vented through the annular of tubing and casing. The in-situ fluid rate can be calculated from desired production liquid rate q_d, formation

等式中的 q_{tL} 是在泵吸入条件下流入泵的总流体(液体和气体的总和)流速,不包括通过油套环空分离和排出的气体。原流体产量可以根据所需的产量

volume factor at pump intake, and in-situ free gas amount into the pump.

q_d、泵入口处的地层体积因子和进入泵的原自由气体量来计算。

The design of PCP speed is to find a speed to produce the well at a given desired liquid rate. The desired liquid rate is the rate at standard condition, bbl/d. Many engineers like to use it directly in Eq. (1.17) to replace the total fluid rate at pump intake. This is correct when pumping water. For most oil well production, the formation volume factor at pump intake is greater than one. Therefore, using desired liquid rate as the total fluid rate at pump intake will underestimate the rotational speed. For a well with very low gas oil ratio, using the desired liquid rate will give similar results.

泵速的设计是计算给定标准条件产量下的泵的转速,产量的单位为 bbl/d。许多工程师喜欢用式(1.17)中的参数计算泵入口处的总流量。气井排液时这是正确的。对于大多数油井,泵入口处的体积系数大于1。因此,使用所需的液体流速率作为泵入口处的总体量会导致低估泵的转速。对于具有较低气油比的井,使用所需的产量将产生类似的结果。

For a given rotational speed, the production rate can be calculated by rearranging Eq. 1.20

对于给定的转速,可以通过重新使用式(1.20)来计算产量。

$$q_{tL} = nQ_t \tag{1.21}$$

Again, the calculated rate is the in-situ rate at pump intake. It can be converted to the desired liquid rate at surface by calculating in-situ formation volume factor and free gas fraction.

同样,计算的产量是泵吸入口对应的地层产量。通过计算地层体积系数和游离气体分数,可以将其转化为地面所需的液体产量。

In addition to PCP's theoretical displacement rate, pump lifting capacity is also a major parameter for PCP design and selection. The lifting capacity is generally a head and gives the maximum head limit of the pump.

除理论排量外,泵的举升能力也是螺杆泵设计和选择的主要参数。举升能力通常指扬程,泵的额定扬程通常直接给出。

Many manufacturers use theoretical displacement and lift capacity to name their pumps, like 60ABC200. The first number (60) represents one percent of the pump's lifting capacity, ABC is the pump type, and the last number (200) is the theoretical displacement rate at the speed of 100 r/min. The two parameters are generally measured at laboratory condition with pure water.

许多制造商使用理论排量和举升能力来命名他们的泵,如60ABC200。第一个数字(60)表示泵扬程的百分之一,ABC 是泵类型,最后一个数字(200)是转速为 100 r/min 时的理论排量。这两个参数通常在实验室利用纯水测量。

For instance, a PCP has a name of 60ABC200 in field units. The number 60 tells the pump has 6000 (60 × 100) feet of water lifting capacity. And the 200 tells the pump has a theoretical displacement rate of 200 bbl/d at 100 r/min (200bbl/d/100r/min)

例如,螺杆泵60ABC200,60表示泵的扬程为6000(60 × 100)ft 水柱;200表明泵在100 r/min下理论排量为200 bbl/d。

The design of a PCP pump is simple when using the desired liquid rate directly. For example, for a desired liquid rate of 600 bbl/d and using the pump 60ABC200, the required rotational speed is 600bbl/d(200bbl/d ÷ 100r/min) = 300r/min. If one wants the production rate at 500r/min, the production rate is 500r/min × 200bbl/d ÷ 100r/min = 1000r/min.

当直接使用所需的产量时,螺杆泵的设计很简单。例如,所需产量为600bbl/d时使用螺杆泵60ABC200,所需转速为600bbl/d ÷ (200bbl/d ÷ 100r/min) = 300r/min。如果泵的转速为500r/min,则产量为500r/min × 200bbl/d ÷ 100r/min = 1000bbl/d。

As shown in Eq. 1.18, the basic PCP design ignores volumetric slip rate, therefore is only correct for pumping water, and may also be used in the situations of no or very low slip pumping, or small differential pressure. In most cases, in petroleum production, it is only an approximation and accurate calculation should take into account the effect of slip.

如式(1.18)所示,螺杆泵的基本设计忽略了体积滑移率,因此仅对泵送水的情况是正确的,同时也可以用于没有或非常低的滑移或压差小的泵。在大多数情况下,在石油生产中,它只是一个近似值,准确的计算应考虑滑移的影响。

3. Rotational Speed Design

3. 转速设计

Substituting the given theoretical rate Q_t, into Eq. 1.18 yields

将给定的理论排量 Q_t 代入式(1.18)得

$$q_a = nQ_t - q_s \tag{1.22}$$

The slip q_s, varies with the structure of a PCP and the differential pressure on it. It doesn't change with pump rotational speed. Usually, the slip of a PCP comes from extensive test data provided by PCP manufacturers.

滑移量 q_s 随着螺杆泵的结构及压差而变化,不会随泵转速而变化。通常,螺杆泵的滑移量来自螺杆泵制造商提供的大量测试数据。

Fig. 1.30 is a typical graph of the flow capacity of a PCP versus lift capacity from manufacturers. It is generally called catalog performance of a PCP. The horizontal axis is the head across the pump. It represents the differential pressure by taking away the effect of fluid gravity. The ideal non-slip line (horizontal dash line) is added in Fig. 1.30 to describe the slip. The slip is the difference between the non-

图1.30是螺杆泵排量与举升能力的典型关系图,通常被称为螺杆泵的性能曲线。横轴是泵的扬程,通过举升高度来表示压差。在图1.30中添加了理想的非滑移线(水平虚线)来描述滑移。滑移量是防滑线和性能

slip line and the performance curve. As shown in Fig. 1.30, the slip increases with increasing differential pressure.

曲线之间的距离。如图 1.30 所示,滑移量随着压差的增加而增加。

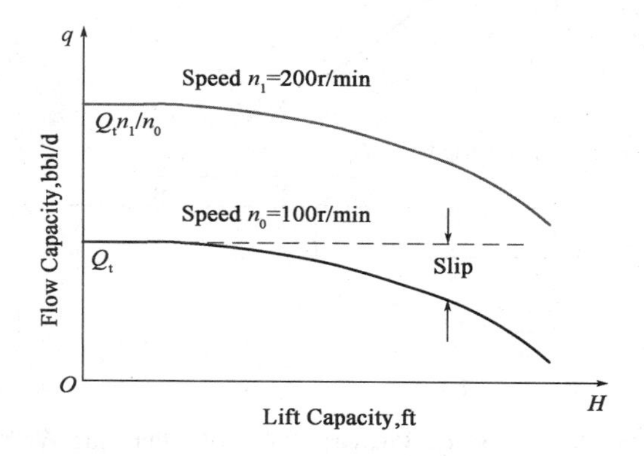

Fig. 1.30 Flow Capacity vs. Lift Capacity of a PCP

From Fig. 1.30, at zero head, the slip is zero and the actual flow rate equals the theoretical rate at the rotational speed. Thus, Eq. 1.22 reduces to Eq. 1.21 at this point.

从图 1.30 可以看出,在零扬程时,滑移量为零,实际流量等于转速下的理论速率。因此,在这时式(1.22)简化为式(1.21)。

Performance curves depend on pump rotational speed. The performance curves at different speeds are parallel to each other. To get the performance curve at a given rotational speed, one may use the point at zero head (differential pressure) to determine the position of the performance curve at the new speed. For instance, if the catalog performance curve is at the speed of n_0 (usually 100 r/min) and its flow capacity is Q_t at zero head, the performance curve at the speed of n_1 is parallel to the catalog curve with a flow capacity of $Q_t\,n_1/n_0$ at zero head.

动态曲线取决于泵的转速。不同速度下的动态曲线相互平行。为了获得给定转速下的动态曲线,可以使用零水头(压差)来确定新转速下动态曲线的位置。例如,如果标准性能曲线的转速为 n_0(通常为 100 r/min)且其流量为 Q_t,则转速为 n_1 的动态曲线与标准性能曲线平行,流量为 $Q_t\,n_1/n_0$。

During the design of a PCP well, Eq. 1.22 needs to be solved iteratively. As shown in Eq. 1.22, the actual production rate depends on the slip, the slip is a function of differential pressure across the PCP, and the differential pressure is determined by the actual production rate. Fig. 1.31 is a typical pressure profile of a pumping well at a production rate.

在设计螺杆泵时,式(1.22)需要进行迭代计算。如式(1.22)所示,实际产量取决于滑移,滑移量是压差的函数,并且压差由实际产量确定。图 1.31 是特定产量下抽油井的典型压力曲线。

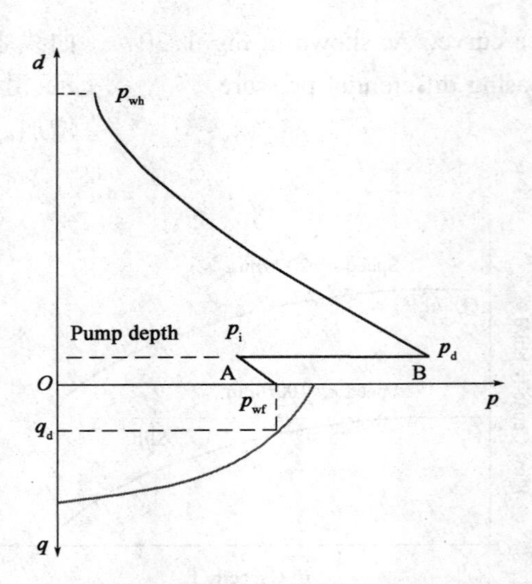

Fig. 1. 31　Typical Pressure Profile of a Pumping Well

As shown in Fig. 1. 31, the pressure profile consists of the pressures at well perforation place (p_{wf}), at pump intake (p_i), at pump discharge (p_d) and at well surface (p_{wh}). The differential pressure across a PCP is the pressure difference of points B and A. For different production rate, the profile and differential pressure are different.

如图 1. 31 所示,压力分布包括井底压力(p_{wf})、泵入口处的压力(p_i)、泵排出口压力(p_d)和井口压力(p_{wh})。螺杆泵上的压差是 B 点和 A 点的压力差。对于不同的产量,压力分布和压差是不同的。

The pressure curve in Fig. 1. 31 can be determined before selecting a PCP for a desired production rate. Therefore, the design of a pumping well is to find a proper pump to satisfy the required differential pressure. The method is also the basis of all artificial lift design.

在特定产量下可以确定图 1. 31 中的压力曲线。因此,泵的设计是找到合适的泵来满足所需的压差。该方法也是所有人工举升设计的基础。

To obtain the pressure profile curve before selecting a pump, nodal analysis method should be used. The node is set at the pump setting depth. The inflow to the node is from reservoir to well bottom hole, and then from bottom hole to the pump intake. The calculation direction is from reservoir to pump intake. The outflow of the node is from the pump discharge to wellhead. The calculation direction is from wellhead down to the pump discharge.

要获得压力曲线,可使用节点分析方法。节点设置在下泵深度处。节点的流入是从储层到井底,然后从井底到泵入口,计算方向是从储层到泵的进液口。节点的流出是从泵到井口,计算方向是从井口到泵出口。

As illustrated in Fig. 1. 31, for a desired production rate (liquid rate at standard condition) q_d, one can get the flowing bottom hole pressure p_{wf}, from the inflow performance relationship of the reservoir. Then from the pressure p_{wf} and temperature (reservoir temperature) at the

如图 1. 31 所示,对于特定的产量(标准条件下的液体流速)q_d,可以从储层的流入动态关系得到井底压力 p_{wf}。然后根据井底压力 p_{wf}和温度(储层温

bottom hole, one can calculate the pressure and temperature at any place from bottom hole to the pump intake by using multiphase flow correlations. The fluid properties, such as water cut, gas oil ratio, bubble point pressure, the densities of the oil, gas, and water, are known variables for the calculation. In addition to the pressure and temperature, in-situ fluid density, liquid rate, and free gas rate are also calculated. Many commercial programs can provide the calculation.

度),通过多相流基本模型来计算从井底到泵入口任何位置的压力和温度。流体性质,如含水率、气油比、泡点压力、油气水的密度,是用于计算的已知变量。除了压力和温度之外,还可以计算地下流体密度、液体流量和游离气体流量。许多商业程序可以提供该计算。

The outflow performance curve in Fig.1.31 is calculated from wellhead (known pressure and temperature) down to the pump discharge by using multiphase flow correlation. The pressures, temperatures, fluid rates (liquid and gas) and fluid densities along the well are calculated. It should be noted that the fluid properties may be different from those upstream of pump intake since some gas may be separated and vented through the annular of tubing and casing.

图 1. 31 所示的流出动态曲线是通过多相流关系式从井口(已知压力和温度)到泵出口计算得到的,也计算了沿井筒的压力、温度、流体(液体和气体)速率和流体密度。应该注意的是,流体性质可能与泵入口上游的流体性质不同,因为一些气体可以析出并通过油套环空分离。

Once the differential pressure across the PCP has been determined, one can transfer it to head by dividing it by the average fluid density between pump intake and pump discharge. The calculated head and the total fluid rate at pump intake are used to design the rotational speed of the PCP.

一旦确定了螺杆泵的压差,就可以用泵入口和出口之间的流体平均密度计算液柱高度。计算的扬程和泵入口处的总流体速率可用于设计螺杆泵的转速。

The graphical method to design the rotational speed of a selected PCP is shown in Fig. 1. 32. Draw a vertical line from the calculated head H_a, on the horizontal (lift) axis, draw a horizontal line from the total fluid rate (liquid + gas into pump) q_a, on the vertical (flow) axis, then find the intersection of the two lines, point A. Move the performance curve at speed 100r/min up or down vertically until point A on the curve.

可以用图版设计螺杆泵转速,如图 1. 32 所示,从水平轴 H_a 处绘一条垂线,从总流量 q_a(液体+气体)处绘一条水平线,在纵轴上找到两条线的交点 A,将动态曲线以 100r/min 的速度垂直向上或向下移动,直到曲线上的 A 点。

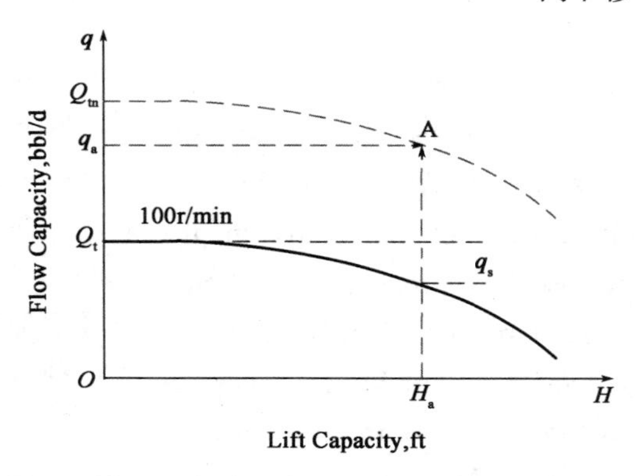

Fig. 1. 32 Designing the Rotational Speed of a PCP

Find the theoretical rate Q_{tn}, from the new curve at zero differential pressure. The solution speed of the pump is

在零压差下从新曲线中找出理论速率 Q_{tn},则泵的转速为

$$n = \frac{100Q_{tn}}{Q_t} \tag{1.23}$$

Another method is using the slip from the catalog performance curve. Since PCP slip doesn't vary with pump speed, one may use it directly to calculate the design speed:

另一种方法是使用标准性能曲线中的滑移量。由于螺杆泵的滑移量不随泵速变化,可以直接用它来计算设计转速:

$$n = \frac{100(q_a + q_s)}{Q_t} \tag{1.24}$$

Fig. 1.32 also shows the method graphically. The slip at head H_a is q_s from the catalog performance curve. The slip at any speed is the same. Therefore $Q_{tn} = q_a + q_s$.

图1.32也以图形方式展示了该方法。H_a 的滑移量是标准性能曲线的 q_s。任何转速的滑移量都是相同的,因此 $Q_{tn} = q_a + q_s$。

A computer program can be used to design the rotational speed. The catalog performance curve can be expressed by a polynomial correlation as

计算机程序可用于设计转速。目录动态曲线可以用多项式关系表示

$$q_a = C_0 + C_1H + C_2H^2 + C_3H^3 + C_4H^4 + C_5H^5 \tag{1.25}$$

where C_0 to C_5 are polynomial coefficients. The theoretical rate Q_t of the PCP equals the first coefficient C_0.

其中 C_0 至 C_5 是多项式系数。螺杆泵的理论速率 Q_t 等于系数 C_0。

The slip at any head H is

任意处的滑移量为

$$q_s = -(C_1H + C_2H^2 + C_3H^3 + C_4H^4 + C_5H^5) \tag{1.26}$$

1.2.4.5 Electrical Submersible Pumping

1.2.4.5 电潜泵

1. Introduction

1. 引言

As the fluids from a reservoir are produced, with time, the pressure in the reservoir decreases and causes the production to also decrease and eventually stop. In such cases, energy is artificially added to the fluids in the wellbore to sustain or increase oil production. Electric submersible pumps (ESPs) are a popular form of artificial lift. ESPs (Fig. 1. 33) consist of multiple stages of centrifugal pumps stacked on top of each other in a narrow

随着生产时间的推移,储层不断产生流体,使储层中的压力下降并且导致产量降低直到停产。在这种情况下,可以人为地增加井筒中的流体能量以维持或增加产量。电潜泵(ESP)是一种应用广泛的人工举升方式,它(图1.33)由多级离心泵组成,

housing that allows the system to be lowered into restrictive wellbores.

这些离心泵彼此串联,可以下放到井筒中。

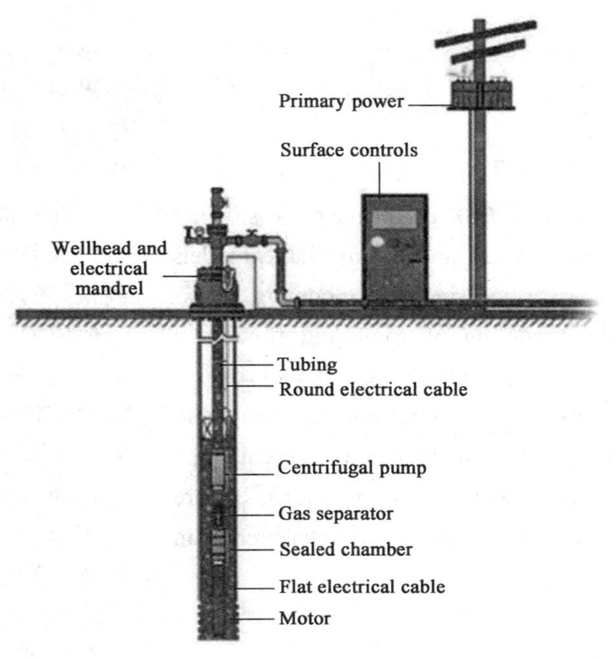

Fig. 1. 33 ESP System

(Source: http://www. narolouisiana. org)

As a centrifugal pump, ESP is good at pumping almost incompressible fluids such as oil and water, but its performance suffers in the presence of compressible gas. Once free gas is ingested into an ESP, the pump can experience gas lock. A gas lock can stop oil production.

由于电潜泵由多级离心泵组成,所以它主要用于泵送不可压缩的流体,如油和水。当存在可压缩的气体时,其性能会受到影响。一旦有游离气体进入泵中,就会出现气锁现象。气锁会阻止石油生产。

In many cases, it is impossible to avoid the presence of free gas at the pump inlet. Free gas may come from reservoir gas cap, be the solution gas liberated from oil due to pressure reduction from reservoir pressure to flowing bottom hole pressure. Engineers can increase the pump intake pressure (to reduce the free gas), but the intersection of the reservoir inflow curve and tubing outflow curve shows that this leads to reduced production. Another way to reduce free gas at pump intake is to divert it into the annulus but such natural separation is very inefficient. The best way to reduce free gas is to use gas separators below the pump intake. In many cases, despite various efforts of the engineers, free gas

在许多情况下,泵入口处不可能避免地存在游离气体。游离气体可能来自储层气顶,或是由于储层压力降低为井底压力后从原油中释放的溶解气体。可以增加泵的进气压力以减少游离气体,但储层的流入动态曲线和流出动态曲线的交点表明这会导致产量减少。减少泵入口处的游离气体的另一种方法是将其转移到油套环空中,但是这种天然分离效率非常低。减

does enter the pump. Thus there is the need to better understand the negative impact of free gas on ESP performance.

少游离气体的最佳方法是在泵入口下方使用气体分离器。在许多情况下,尽管工程师做了各种努力,但是仍有游离气体进入电潜泵。因此,需要更好地了解游离气对电潜泵性能的负面影响。

The impact of free gas on ESP performance is still not well understood in the oil industry and the available models are approximate. This is expected since the modeling of gas interference involves the modeling of gas-liquid flow in a curved impeller rotating at very high speeds. It took over a couple of decades for researchers to reasonably accurately model multiphase flow in far simpler symmetrical pipe (tubular) geometries that involved no rotations. More research is required to accurately model gas interference in ESPs.

在石油工业中,游离气体对电潜泵性能的影响仍然没有得到很好的认识,可用的模型都是近似模型。这是预料之中的,因为要对高速旋转的弯曲叶轮中的气液流动状态进行建模。研究人员花费了几十年的时间才能合理准确地模拟多相流动,证明了这种流动在更简单的对称管道(管状)几何形状中,不涉及旋转。因此需要更多的研究来准确模拟电潜泵中的气体干扰。

There were some studies done in the nuclear industry. However, all these studies, summarized by Sachdeva (1988), involved single stage centrifugal pumps that were vastly larger than the small-diameter multistage ESPs in the oil industry. The first serious effort to study the qualitative effect of free gas on pump performance in the oil industry was made by Lea and Bearden (1982). They gathered laboratory data on three different ESPs using water-air and diesel – CO_2 mixtures. The experiments showed that the free gas was more harmful at lower intake volumes at lower pump intake pressures. Also, the diesel – CO_2 mixture seemed to perform better than the air-water mixture. Sachdeva (1988) used the Lea-Bearden data and the qualitative findings of the studies in the nuclear industry to develop a dynamic model to simulate multiphase performance in ESPs. It is the first comprehensive analysis of pumping gas-liquid mixture in petroleum industry. Though thorough, the Sachdeva (1988) dynamic model is complicated and not easy to use in the field. So, as a gross simplification of the earlier dynamic model, Sachdeva et al. (1992) presented a far simpler empirical

核工业进行了一些相关研究。这些研究都由 Sachdeva (1988)总结,涉及单级离心泵,它比石油工业中的小直径多级电潜泵大得多。Lea 和 Bearden (1982)首次研究了游离气体对石油工业中电潜泵性能的影响。他们使用水—空气和柴油—二氧化碳混合物、收集了三种不同电潜泵的实验数据。实验表明,在较低的泵吸入压力下,游离气体在较低的进气量下更有害。此外,柴油—二氧化碳混合物比空气—水混合物的危害更小。Sachdeva (1988) 使用 Lea – Bearden 数据和核工业研究的定性研究结果开发动态模型来模拟电潜泵中的多相性能。这是对石油工业中泵送气液混合物的第一次综合分析。虽然全面,

correlation to estimate the pressure increase of an ESP stage in multiphase condition:

但 Sachdeva(1988)的动态模型很复杂，且不易现场使用。因此，作为早期动态模型的粗略简化，Sachdeva 等人在 1992 年提出了一个更简单的经验关系式来估计电潜泵各级在多相条件下的压力增加：

$$\Delta p_{\mathrm{m}} = (1/C_1) K (C_1 p_{\mathrm{in}})^{E_1} (\alpha)^{E_2} (C_2 q_{\mathrm{L}})^{E_3} \tag{1.27}$$

here, Δp_{m} is pressure increase across an ESP stage, kPa/stage; p_{in} is pressure at pump intake, kPa; α is free gas void ratio (free gas by volume) at stage intake, fraction; q_{L} is liquid flow rate, $\mathrm{m^3/d}$; C_1 and C_2 are unit conversion factors and added in Eq. 1.27 to change the original correlation from US field units to SI units. $C_1 = 0.145$ and $C_2 = 0.183453$.

式中，Δp_{m} 是电潜泵的压力增量，kPa/级；p_{in} 是泵入口处的压力，kPa；α 是游离气体分数；q_{L} 是液体流量，$\mathrm{m^3/d}$；C_1 和 C_2 是单位转换系数，将其带入式(1.27)中将原始关系式从美国油田现场常用单位改变为 SI 单位，$C_1 = 0.145$ 和 $C_2 = 0.183453$。

The present study does not look into the details of the movement of gas and liquid within the impeller. Nevertheless, it is pertinent to summarize the key reasons for the head degradation. The degradation in two-phase flow occurs when the gas tends to flow at a much lower velocity than the liquid. These studies also observed that in bubble flow, the gas tends to dampen the acceleration of the liquid phase in the impeller. This reduces the energy lost in the form of accelerated liquid phase velocities within the impeller. With increasing gas at the inlet, the bubble flow turns into a separated slug/churn flow where the liquid phase is accelerated even more, causing higher head degradation. With increasing liquid velocities, the gas phase tends to move slower and even stall, causing surging, and eventually, pump lock when the gas collects towards the impeller inlet. The Sachdeva model (1988) explains this behavior via the dynamic multiphase model that also indicates that the pump performance is dominated by the inter-physic fluid dynamics in the impeller and not the diffuser.

虽然现有的研究没有涉及叶轮内气体和液体流动的细节，但还是有必要总结扬程损失的主要原因。当气体流速比液体低得多时，会发生两相流的分层。同时还观察到，在泡流中，气体倾向于抑制叶轮中液相的加速，这减少了叶轮内加速液相的能量损失。随着入口处气体的增加，泡流变成分离的段塞流，其中液相逐渐加速，导致更高的扬程损失。随着液体速度的增加，当气体向叶轮入口聚集时，气相倾向于移动得更慢甚至失速，从而引起喘振，并最终导致泵锁。Sachdeva 模型(1988)通过动态多相模型解释了这种现象，该模型还表明泵性能主要由叶轮中的物理流体动力学而不是扩散器决定。

The simplified Sachdeva correlation [Eq. (1.27)] shed no predictive light on the onset of surging or on the applicability of affinity laws in multiphase conditions.

简化的 Sachdeva 关系[式(1.27)]没有预测喘振发生或多相条件下亲和力定律的适用性。

2. Empirical Model

2. 经验模型

As discussed in the Introduction, multiphase pumping behavior can be simulated by dynamic model and simplified empirical model. Dynamic model studies the mechanism of multiphase pumping and therefore should be more accurate theoretically. However, the dynamic model requires solving the equations of mass and momentum of the gas-liquid mixture in an ESP, and needs the detail design data of pump impeller and diffuser. Many researchers and engineers are apt to empirical model for its simplified form and easy to use.

如引言中所说的,可以通过动态模型和简化的经验模型来模拟多相混输行为。动态模型研究了多相泵的机理,因此理论上应该更准确。然而,动态模型需要求解电潜泵中气液混合物的质量和动量方程,需要泵叶轮和扩散器的详细设计数据。许多研究人员和工程师都倾向于采用经验模型来简化其形式使其易于使用。

In addition to the Sachdeva simplified model [Eq. (1.27)], many authors have developed simplified models to simulate the multiphase pumping behavior.

除了 Sachdeva 简化模型[式(1.27)]之外,许多研究者还开发了简化模型来模拟多相混输行为。

Turpin et al. (1986) presented the first empirical model to predict pump performance in gassy wells. The model was based on the data of Lea and Bearden, and has two different forms for different pump types. For pumps I-42B and K-70, the model is

Turpin 等人(1986)提出了第一个预测气井泵性能的经验模型。该模型基于 Lea 和 Bearden 的数据,并且针对不同类型的泵有两种不同的形式。对于 I-42B 和 K-70 型泵,模型为

$$H_{\mathrm{m}} = H\mathrm{e}^{-\alpha_1 (q_{\mathrm{g}}/q_{\mathrm{L}})} \tag{1.28}$$

$$\alpha_1 = \frac{346430\, q_{\mathrm{g}}}{(C_1 p_{\mathrm{in}}^2)\, q_{\mathrm{L}}} - \frac{410}{C_1 p_{\mathrm{in}}} \tag{1.29}$$

where, H_{m} is the head of gas-liquid mixture, is the head from manufacture catalog curve, q_{g} and q_{L} are the flow rates of gas and liquid at pump inlet, p_{in} is pump intake pressure, C_1 is a unit conversion factor and equals 0.145.

其中,H_{m} 是气液混合物的水头,是制造目录曲线的扬程;q_{g} 和 q_{L} 是泵入口处的气体和液体的流速;p_{in}是泵的进气压力;C_1 是单位换算系数,取0.145。

The model was suggested to be used in stable situation where pump surging didn't start. The critical model of judging pump surging is

该模型可以用于泵涌未开始的稳定情况。判断泵涌的关键模型是

$$\varphi = 2000\left(\frac{q_g/q_L}{3C_1 p_{in}}\right) \tag{1.30}$$

when $\varphi < 1$, pump worked in a stable condition and Eq. 1.30 could be used to calculate the head performance. If $\varphi > 1$, pump suffered surging.

当 $\varphi < 1$ 时，泵工作在稳定状态，并且均匀。式(1.30)可用于计算扬程。如果 $\varphi > 1$，泵会出现泵涌。

Used two mixed flow pumps, GN4000 and GN7000, and a radial flow pump, GN2100, Cirilo (1998) tested pump degradation behaviors using a mixture of air and water. Based on the test data, Cirilo presented a critical model of maximum gas void ratio to judge pump surging:

使用混流泵 GN4000 和 GN7000，以及径向流泵 GN2100，Cirilo(1998)用空气和水的混合物测试泵损失行为。根据测试数据，Cirilo 提出了最大气体空隙比的关键模型来判断泵涌：

$$\lambda_g = 0.0187(C_1 p_{in})^{0.4342} \tag{1.31}$$

where λ_g is the no-slip free gas void ratio.

其中 λ_g 是无滑移气体空隙率。

As shown in Eq. 1.31, the maximum gas void ratio is only a function of pump intake pressure. According to the correlation, to obtain a stable pumping behavior, the free gas void ratio at pump intake should be smaller than the maximum value.

如式(1.31)所示，最大气体空隙率仅是泵进气压力的函数。根据公式，为了获得稳定的泵送行为，泵入口处的自由气体空隙率应小于其最大值。

Based on Cirilo's test data, Romero (1999) established a multiphase head model for the type of mixed flow pumps:

根据 Cirilo 的测试数据，Romero(1999)建立了混流泵类型的多相模型：

$$\frac{H_m}{H_{max}} = \left(1 - \frac{q_{L-dL}}{q_{damx}}\right)\left[\alpha\left(\frac{q_{L-dL}}{q_{dmax}}\right)^2 + \frac{q_{L-dL}}{q_{dmax}} + 1\right] \tag{1.32}$$

where q_{L-dL} is the dimensionless liquid flow rate, which was defined as the in-situ liquid rate, q_L divided by the maximum liquid rate from manufacturer's catalog curve. And

其中 q_{L-dL} 是无量纲液体流速，定义为原液体速率 q_L 除以最大液体速率。且

$$\alpha = 2.092\lambda_g + 0.2751 \tag{1.33}$$

$$q_{dmax} = 1 - 2.0235\lambda_g \tag{1.34}$$

To judge pump surging, the author presented a critical model of minimum dimensionless liquid rate, and a critical model of maximum gas void ratio as

为了判断泵涌,研究者提出了最小无量纲液体速率的临界模型,以及最大气体空隙率的临界模型:

$$\left(\frac{q_L}{q_{max}}\right)_c = -6.6465\lambda_g^2 + 3.5775\lambda_g + 0.0054, \lambda_g = 0.004(C_1 p_{in})^{0.6801} \quad (1.35)$$

By using pumping data of air-water mixture, Duran and Prado (2004) presented models for mild and severe head degradations, which correspond to bubbly and elongated bubble flow regions. The bubbly flow region occurred before pump surging and the elongated bubble flow came after dramatic surging.

通过使用空气—水混合物的泵送数据,Duran 和 Prado (2004)提出了轻度和严重的扬程损失,对应泡流和长气泡流动区域。泡流在泵涌之前发生,并且在剧烈波动之后出现长气泡流。

For bubbly flow, the pressure performance correlation is

对于泡流,压力关系式为

$$\Delta p_m = (1-\alpha)\rho_L H\left(\frac{q_L}{1-\alpha}\right) + \alpha\rho_g H\frac{q_g}{\alpha} \quad (1.36)$$

The closure correlation for Eq. 1.36 is

式(1.36)的闭合相关性为

$$q_{g-n} = \left(\alpha\frac{\rho_m}{\rho_l} + b\right)q_{L-n}^c \quad (1.37)$$

where coefficients $\alpha = -0.843$, $b = 0.850$ and $c = 1.622$ in the correlation. ρ_m and ρ_L are the densities of the gas-liquid mixture and liquid respectively. q_{g-n} and q_{L-n} are in-situ normalized gas and liquid flow rates. The normalized gas and liquid flow rates were defined as

其中系数 $\alpha = -0.843$, $b = 0.850$, $c = 1.622$;ρ_m 和 ρ_L 分别是气液混合物和液体的密度;q_{g-n}和 q_{L-n}是标准化的气体和液体流速。标准化的气体和液体流速定义为

$$q_{g-n} = \frac{q_g}{q_{max}(1-\alpha)} \quad (1.38)$$

$$q_{L-n} = \frac{q_L}{q_{max}(1-\alpha)} \quad (1.39)$$

One may calculate the mixture density from the closure correlation from known in-situ liquid and gas flow rate, and then, solve for actual gas void ratio, from the correlation of $\rho_m = \rho_L(1-\alpha) + \rho_g\alpha$. The mixture pressure increase, Δp_m, will be calculated out by substituting the actual void ratio into Eq. 1.36.

可以从已知的原液体和气体流速的闭合相关性计算混合物密度,然后通过 $\rho_m = \rho_L(1-\alpha) + \rho_g\alpha$ 求出实际气体空隙率。混合压力的增加可以通过将实际空隙率代入式(1.36)来计算。

Duran and Prado also presented a critical model for the boundary between stable and unstable pumping by using of normalized gas flow rate:

Duran 和 Prado 还通过使用归一化气体流速提出了稳定和不稳定泵送之间边界的关键模型：

$$q_{g-n} = \left(a\frac{\rho_g}{\rho_L} + b\right)q_{L-n}^c \tag{1.40}$$

where ρ_g is in-situ gas density and coefficients $a = 5.580$, $b = 0.098$ and $c = 1.421$.

其中 ρ_g 是原气体密度；$a = 5.580$，$b = 0.098$，$c = 1.421$。

For elongated bubble flow, Duran and Prado (2004) presented

对于长气泡流，Duran 和 Prado(2004)提出

$$\Delta p_m = \frac{[a + b\ln(q_{g-n})]}{C_1} \tag{1.41}$$

where constants $a = -0.47075$, $b = -0.21626$.

其中 $a = -0.47075$，$b = -0.21626$。

The critical model of surging prediction for elongated bubble flow was given as

推导出长气泡流波动预测的临界模型为

$$q_{g-L} = aq_{g-n}^b \tag{1.42}$$

where constant $a = 1.6213$ and $b = 0.435$.

其中 $a = 1.6213$，$b = 0.435$。

1.2.5 Intermittent Producing

1.2.5 间歇生产

When a well first begins to show signs of liquid loading, this may be addressed by temporarily stopping and re-starting (cycling or intermitting) production. During the shut-in period the well pressure recovers (increases) around the wellbore and some liquids may be forced back into the formation, thus reducing the liquid load (level) in the wellbore. Then the well is returned to production and allowed to flow until the loading again occurs.

当气井刚开始有积液的迹象时，可以通过暂时停止和重新启动(循环或间歇)生产来解决这个问题。在关井期间，压力在井筒周围恢复(增加)，并且一些液体可能被迫返回地层，从而减少井筒中的液体积聚，然后在井再次发生积液之前恢复生产。

This method is sometimes used if the well has just started to experience liquid loading and is waiting for some type of artificial lift system to be installed. It may also be used if the well has a packer. The packer may reduce the effectiveness of some artificial lift methods. It may need to be removed, if possible.

当井刚开始出现积液并且正在等待安装某种类型的人工举升系统时，可以使用上述方法。如果有封隔器，封隔器可能会降低某些人工举升方法的效果。在可能的情况下，需要将封隔器移走。

This method may have a serious drawback. In some formations, the presence of water can damage or reduce the permeability for gas. So, in this case, it may damage the well to push (re-inject) water back into the formation.

该方法可能存在缺点:在一些地层中,水的存在会损害或降低气体的渗透性。因此,在这种情况下,将水重新注入地层可能对井造成伤害。

However the method is inexpensive and it can work with a packer in place. Therefore, it is commonly used as an interim method of production before an artificial lift system can be installed.

然而,该方法成本低,并且可以与封隔器一起工作。因此,在安装人工举升系统之前,它通常可作为临时的方法。

1.2.6 Compression

1.2.6 井口增压法

As the wellbore pressure of the gas well decreases, the gas well liquid loading occurs. The pressure in the wellbore is not high enough to discharge the slug flow formed by the liquid to the wellbore. This leads to decrease or stop the gas production. If the wellbore pressure can be increased by the wellhead pressurization device and the wellhead pressure can be reduced, and the gas well production capacity can be improved.

气井压力降低会导致气井积液,大部分气井积液的主要原因是气井产量低于临界携液流量,同时井筒内压力较低,不足以将积液形成的段塞流排至地面,从而导致气井产量下降甚至无气量。如果能通过井口增压装置提高井筒压力、降低井口压力,则可以有效发挥积液井生产潜力,提高气井生产能力。

Compression is crucial to all gas well production as it is the primary means to transport gas to market. Compression is also vital to deliquification, lowering wellhead pressure and increasing gas velocity. The lower bottom hole producing pressure from deliquifying wells and lowering surface pressures with compression can result in substantial production and reserves increases. These increases can range from a few percent to many times the current production. This uplift requires investment for the compressor and associated equipment as well as operating costs for the maintenance and power to continue running the compressor. However, many times compression can be the most economical way to keep wells deliquified, providing higher production rates at lower pressures.

井口增压对于所有气井都至关重要,因为它是将天然气输送到市场的主要手段。它对于排液、降低井口压力和提高气体速度也至关重要。通过排液采气降低井底压力和降低表面压力可使其可采储量大幅增加,增加的范围可以是当前产量的几个百分点到很多倍,这需要对压缩机和相关设备进行投资,需要维护运行成本和连续运行压缩机的动力。然而,多次进行井口增压是气井排液采气的最经济的方式,在较低压力下就可以提高生产率。

Compressing associated gas in oil wells often is seen as a simple "rate acceleration" project that seldom has good economics. The argument has been made successfully that compressing gas reservoirs exposes a significantly larger portion of the original gas in place (OGIP) to production and it actually adds significant reserves. This phenomenon is very pronounced in CBM and other adsorption reservoirs.

油井中有伴生气时的井口增压通常被视为一种简单的"提高产量"的方式，却很少取得良好的经济效益。对天然气储层进行井口增压使得更多原始天然气地质储量(OGIP)可以被开采，实际上大量增加了可采储量。这种效果在煤层气储层和其他吸附储层中非常明显。

The process of choosing how to apply compression and the proper equipment to achieve the desired pressures and rates is important in optimizing results. Fortunately Systems Nodal Analysis can be used effectively to help in the process of evaluating wells and compression equipment.

选择井口增压的方法及适当的设备以达到所需的压力和速率对优化结果非常重要。系统节点分析可以有效地评估井和压缩设备。

Compression and reduced surface pressure is usually the first tool used in the life of a gas well to keep it deliquified and sometimes the only artificial lift method used, but compression can also be used to increase the effectiveness of other artificial lift deliquification methods including foamers, gas lift, beam pumping, ESPs, and velocity strings. When applying compression or any deliquification method it is important to insure that downstream equipment has sufficient capacity to insure uplift to the overall production.

井口增压和降低表面压力通常是气井生产周期中排液采气的第一个方法，有时是唯一可使用的人工举升方法，也可用于提高其他人工举升方法的效率，包括泡沫、气举、射流泵、电潜泵和速度管柱。当采用井口增压或任何排液方法时，重要的是确保下游设备具有足够的容量以保证在整个生产过程中可以进行有效举升。

1.3 Unloading Method Selection

Audio 1.4

The industry uses a number of methods to remove liquid from (deliquify) gas wells. Research and development are constantly being performed to make these methods more effective and to develop new methods.

1.3 排液采气方法的选择

石油行业中有多种方法从气井中除去液体，且不断进行研究以使这些方法更有效，并开发新方法。

1.3.1 Factors Considered in the Selection of Artificial Lift Method

The selection of an appropriate lift option is critical to the project profitability. The proper artificial lift can improve productivity by removing liquid effectively and eventually raising project economics. On the contrary, a bad choice

1.3.1 人工举升方法的选择

选择合适的人工举升方法对项目的经济效益至关重要。采用适当的人工举升方法可以有效去除井底积液并提高经济

cannot be expected to improve production and adversely affects the economics of project. There are many factors that need to be considered when selecting an artificial lift option. These factors include site information like location, well characteristics, producing characteristics, fluid properties, power availability, surface facilities, reservoir characteristics, operation concerns, service availability, and economic points.

效益。相反,一个不恰当的选择会对项目的经济效益产生不利影响。在选择人工举升方式时需要考虑诸多因素,包括井场位置、井特征、生产特征、流体性质、电力供应、地面设施、储层特征、操作问题、服务可用性和经济效益。

1.3.1.1 Site Information (Location)

1.3.1.1 井场位置

The production facilities are installed differently depending on the well's location, especially the surrounding environment. The offshore well needs to install all surface facilities on the platform with limited aerial extent. Thus, special considerations are required to utilize concentrated area. For this reason, some artificial lifts requiring spacious area may not be used. Also, the logistics on equipment and power is difficult to establish compared to onshore.

生产设施的安装方式取决于井场位置,尤其是周围环境。海上油井需要在空间范围有限的平台上安装所有地面设施,因此需要特别考虑空间问题,可能无法使用一些需要宽敞区域的人工举升方法。此外,与陆上油井相比,海上油井设备运输和电力传输比较困难。

Regardless of well location, either onshore or offshore, we are given the regulations or approval conditions to observe, which are imposed by authorized organizations. Such conditions generally include safety rule, environmental protection rules, and pollution treatment plans, etc. For instance, some artificial options making high noise should be excluded if it is not able to meet the condition.

无论在陆上还是海上,都必须遵守相关制度及条件,包括安全规则、环境保护规则和污染治理方案等。例如,不能采用噪声不符合条件的人工举升方法。

1.3.1.2 Well Characteristics

1.3.1.2 井特征

Well characteristics like depth, deviation, and size of tubular are the most significant criteria in the selection process for an appropriate artificial lift. It is said that the depth may have little effect in determining the method of artificial lift. However, if we need to design an artificial lift for a very deep well below 15,000 ft, depth may indeed be a factor. In this case, there are limited options available to be applied due to their maximum depth of operation. Depending on the amount of deviation of hole, the efficiency of each lifting option is different, with efficiency usually reduced. For a highly deviated well or horizontal well, some options are not recommended to be used.

井深、井斜和油管尺寸是选择人工举升方法时最需要考虑的因素。有研究称井深对人工举升方法的选择影响不大,但是如果为15000ft以下的深井选择人工举升方法,深度可能确实是一个需要考虑的因素。在这种情况下,由于受其最大深度限制,可选择的人工举升方法有限。井斜不同,每种举升方法的效率是不同的,井斜通常会降低效率。对于斜井或水平井,一些举升方法则不能采用。

1.3.1.3 Producing Characteristics

According to how much liquid is being produced or how much gas is being produced, the production scheme should be differently made. The production rate in total volume basis is one of the criteria that screen artificial lift options, and the GLR is one of the most important factors in the selection process. For instance, in extremely high rates of production like above 15,000 bbl/d, ESP and gas lift systems can only be considered. On the other hand, with a low rate of production, we can consider all possible artificial lift techniques. Specifically, the pumping systems are inefficient if GLR exceeds 500 ft^3/bbl.

1.3.1.3 生产特征

产量不同,则其生产机制也就不同。产量是选择人工举升方法的依据之一,气液比是选择过程中最重要的因素之一。例如,当产量极高时(例如15000 bbl/d以上),只能选择电潜泵或气举。若产量低,则可以选择任意一种人工举升方法。具体而言,如果气液比超过500ft^3/bbl,则泵送系统效率很低。

1.3.1.4 Fluid Properties

The consideration points in fluid properties are viscosity, density, and composition. In general, viscosities less than 10 cP (above 30° API) are not a factor in determining the lift method. For instance, sucker rods do not fall down easily for highly viscous fluid, and such a phenomenon results in inefficient production. The fluid composition is important to see if the fluid results in a corrosion problem.

1.3.1.4 流体性质

流体性质包括黏度、密度和组成。通常,黏度小于10cP(高于30°API)时不影响人工举升方法的选择。例如,对于高黏流体,抽油杆不易下落,这种现象造成产量降低。此外,流体组成对于流体是否会导致腐蚀问题非常关键。

1.3.1.5 Power Availability

The power sources for artificial lift prime movers are usually electricity or pressurized gas. Most of these wells are located in isolated places far from residency areas. In some situations, it is not useful for a well to be supplied stable electricity from power generation stations. To use pumping systems like ESP and hydraulic pumps, stable electricity should be obtained, but for gas lift system, high pressurized gas is required to supply securely.

1.3.1.5 电力供应

人工举升电动机的动力源通常是电力或加压气体,这些井大部分位于远离居住地的偏远地区,在某些情况下,没有可用发电站向井供应稳定电力,要使用电潜泵和液压泵等泵送系统需要稳定的电力。但对于气举系统,需要高压气体才能安全供电。

1.3.1.6 Surface Facilities

Surface facilities like flow lines, choke valve, and separators are to be considered when choosing a proper lifting method. In general, they need to be designed in such a way that one can accommodate new equipment.

1.3.1.6 地面设施

在选择人工举升方法时应考虑地面设施,如管道、节流阀和分离器。通常,它们的设计需要满足新设备。

1.3.1.7 Reservoir Characteristics

Reservoir characteristics is a factor that should be considered in order to make a production plan that figures out

1.3.1.7 储层特征

储层特征是应该考虑的因素,以便制定生产计划,计算每

how much it will be produced monthly and how further time it will be produced. Based on production remaining time and production rate, the selection of lifting method can be made.

月生产多少及生产多久。根据生产剩余时间和生产率,可以选择举升方法。

1.3.1.8 Operation Concerns

1.3.1.8 操作问题

The operation concerns include, but are not limited to scale, corrosion and erosion. In addition, bottom-hole temperature and surface climate are also consideration points in operation. Sand causes erosion problems for all types of artificial lift methods. Downhole corrosion may be caused by electrolysis between different metal types, H_2S or CO_2 content in the produced fluid, highly saline or saturated brine water, or oxygenation of metals. For paraffin, once it is accumulated in the upper tubing string or flow-line, it will cause pressure drop. Scale deposition will decrease flow efficiency by reducing ID of tubing.

操作问题包括水垢、腐蚀、侵蚀,但又不仅仅局限于这些问题。此外,井底温度和地面气候也是操作过程中需考虑的问题。对于每一种人工举升方法,若出砂则会产生侵蚀问题。井下腐蚀可能是由不同金属类型之间的电解产生的流体中的H_2S或CO_2、高盐度或饱和盐水,或金属氧化引起的。石蜡一旦积聚在管柱上或流线中,就会引起压力损失。水垢沉积会因为减少油管的内径而降低流动效率。

Bottom hole temperature is one of factors that should be considered before final selection, because very high temperatures in bottom hole will damage equipment like pump motor and cable. Lifting equipment capable of operating over certain high temperatures should be selected.

井底温度是最终选择举升方法之前应考虑的因素之一,因为井底温度过高则会损坏电动机和电缆等设备。故应选择能够在某些高温下运行的举升设备。

1.3.1.9 Service Availability

1.3.1.9 服务可用性

Some types of lifting methods require work-over or pulling units in time of service or replacement, while other types of methods can be serviced by using wire-line. Sometimes the methods may be sensitive in terms of operation cost. Basically, checking points about service availability are to investigate which service personnel, replacement parts, and service rigs or equipment is available.

某些类型的举升方法在维修或更换时需要翻转或拉动设备,而其他类型的举升方法则可以通过使用电线来维修。有时这些方法操作成本相差较大。基本上检查服务可用性的要点是检查服务人员、更换部件和服务设备是否可用。

1.3.1.10 Economic Point

1.3.1.10 经济效益

Generally, we consider economic point of view in the last stage of selection process after technical evaluation is done. If several options are verified for application from a technical point of view, we will evaluate their influence on economics of project. The capital expenditure, operation expenditure, expected income, and other economic factors (e. g. equipment life of time, etc.) will be considered and

一般来说,在技术评价完成之后,在选择过程的最后阶段需要考虑经济因素。如果从技术角度验证所选的方式,将评估它们对项目经济性的影响,考虑资本支出、经营支出、预期收入和其他经济因素(如设备寿命等),

the result of economic evaluation will be used eventually when the final decision is made.

1.3.2 Adaptability of Various Unloading Methods

1.3.2.1 Plungers

1. Applications

A plunger system is good at removing liquid in gas wells if the well has sufficient GLR and the pressure is enough to lift the plunger and liquid slugs. Common applications for plunger lift are as followings:

(1) Gas wells with liquid loading problems;

(2) Intermittent gas lift wells with fallback problems;

(3) Wells with scale and paraffin problems;

(4) Oil production with associated gas.

In order to apply the plunger system, there are additional specific requirements or limitations. Wells must produce at least 400 ft^3/bbl per 1,000 ft of depth, meaning that high gas-liquid ratio is required to apply plunger system. The other limitation is that the wells should have shut-in pressure that is 1.5 times of sales line pressure.

2. Advantages

The greatest advantage of the plunger system is the cost. It is very cost effective method: lower installation and lower operation cost. This system requires no outside energy source to operate because it uses the well's natural energy. The plunger enables the well to clean off paraffin deposits so it is useful for the wells experiencing paraffin and scale problems.

1.3.2.2 Foaming

1. Applications

The best applications for this technology usually occur in higher GLR applications where the agitation necessary exists and in higher water cut applications where the surfactant acts mostly on water. Normally GLR of 1000 to 10000 ft^3/bbl will allow for the agitation necessary and the higher the water cut the more likely suitable foam will be determined. A lower limit of 50% water cut is usually a practical limitation, below this the chemical costs can get

并在最终决定时使用经济评估结果来做决定。

1.3.2 各种排液采气方法的适用性

1.3.2.1 柱塞法

1. 应用

如果井具有足够高的气液比且压力足以提升柱塞和液体段塞，则柱塞系统能够很好地去除气井中的液体。柱塞法的常见应用如下：

(1)存在积液问题的气井；

(2)间歇气举效率低的井；

(3)具有水垢和石蜡问题的井；

(4)具有伴生气的油井。

为了应用柱塞系统，还有一些特定要求或限制：井每1000ft深度产量至少为400ft^3/bbl，即应用柱塞系统需要高气液比；关井压力应为销售管线压力的1.5倍。

2. 优点

柱塞法的最大优点是成本较低，安装成本和运营成本低，是一种非常经济的方法。柱塞系统不需要外在能源供应，因为它使用了井的天然能量。柱塞系统可对井进行清蜡，这对出蜡和结垢井是有用的。

1.3.2.2 泡沫法

1. 应用

该技术最适合用在气液比较高的井中及加入表面活性剂作用于较高含水率的井中。通常，气液比为1000～10000ft^3/bbl，并且含水率越高，将更可能确定合适的泡沫。含水率为50%通常是实际的下限，若低于50%，则成本可能会非常高，并且可能

quite high and foaming may be impossible. A high condensate production particularly can be an issue as the condensate can act as a natural defoamer. In the case the other produced fluid is oil then many times it can be lifted in conjunction with the foamed water however in limited quantities.

2. Advantages

Foaming agents are very simple and inexpensive means of unloading low productivity gas and gas-condensate wells. There are no downhole modifications required and the surface equipment depends on the type of treatment.

1.3.2.3 Velocity Strings

1. Applications

The design for the velocity string depends on well conditions. The gas velocity must meet or exceed a minimum or critical velocity to prevent a well from loading up. There are two popular methods for determining the minimum gas velocity: a rule of thumb widely accepted in the petroleum industry, and a theoretical correlation presented by Turner et al. (1969).

The rule of thumb sets the minimum gas velocity at 10 ft/s. Thus, a well can be restored to flowing production if the gas velocity at the bottom of the tubing remains above 10 ft/s. However, the actual critical velocity depends on the well conditions.

The correlation presented by Turner et al. (1969) uses a theoretical analysis of the flow regime. In order to prevent liquid loading of the well, the liquid in the tubing must be suspended as a mist (qualities above 95%) or the flow regime in the tubing must be in annular-mist flow. In these flow regimes, as long as the gas velocities exceed the settling velocity of liquid droplets, high gas velocities force the liquid out of the tubing.

无法发泡。由于冷凝物可以充当天然消泡剂,因此高凝析油产量成为问题。在产生的其他流体是油的情况下,可以多次将其与泡沫水一起进行举升,但是举升数量有限。

2. 优点

泡沫法是低产井和凝析井排液采气简单且廉价的方式。排出低生产率气体时不需要井下改造,地面设备取决于其处理类型。

1.3.2.3 速度管柱法

1. 应用

速度管柱的设计取决于井况。气体速度必须达到或超过临界速度,以防止产生积液。确定最小气体流速有两种常用的方法:石油工业中广泛接受的经验法则,以及 Turner 等人提出的理论公式(1969)。

经验法则规定最小气体速度为 10ft/s。因此,若油管底部的气体流速保持在 10ft/s 以上,则可以恢复生产。但是,实际临界速度取决于井况。

Turner 等人提出的理论公式使用流动状态的理论分析。为了防止液体积聚,管道中的液体必须悬浮成雾状(质量分数高于 95%),或管道中的流动状态必须是环形雾流。在这些流动状态下,只要气体速度超过液滴的沉降速度,高气体速度就会迫使液体流出管道。

2. Advantages / Disadvantages

The velocity string is one of the most attractive options since it is low cost and requires no further maintenance after installation.

Apart from mechanical considerations, such as interference with the SSSV, the main drawback of the velocity string is that the introduction of the string increases the frictional flow resistance in the well. This inevitably leads to a reduction of the productivity of the well. Hence, the result for the suppression of liquid loading is decreased production. This makes selection of the optimum size of the velocity string critical. It has to be selected such that liquid loading is avoided or at least delayed over a considerable period of time, whilst maintaining the highest possible production.

1.3.2.4 Sucker Rod Pumping

1. Consideration

Sucker rod pumping systems should be considered for new, low volume stripper wells because operating personal are usually familiar with these mechanically simple systems and can operate them more efficiently. Inexperienced operating personnel operate this type of equipment with greater effectiveness than other types of artificial lift. Sucker rod pumping systems can operate efficiently over a wide range of well producing characteristics. Most of these systems have a high salvage value.

The sucker rod string, parts of the pump and unanchored tubing are continuously subjected to fatigue. Therefore, the system must be more effectively protected against corrosion than any-other lift system to insure long equipment life. Sucker rod pumping systems and crooked holes are often incompatible.

2. Applications

Sucker rod pump has very broad applications:

2. 优缺点

速度管柱法是最有吸引力的选择之一，因为它成本低，并且在安装后不需要进一步维护。

除了机械方面的考虑因素，如井下安全阀的干扰外，速度管柱法的主要缺点是引入管柱增加了井中的摩擦流动阻力，这不可避免地导致井的产量降低。因此，排液造成了产量减少。这使得选择速度管柱的最佳尺寸变得至关重要。必须保证在相当长的一段时间内避免或减少积液，同时保持最高可能的产量。

1.3.2.4 有杆泵

1. 考虑因素

有杆泵系统适用于低产井，因为操作人员通常对这些简单系统的操作很熟悉，并能有效地操作。与其他类型的人工举升方法相比，没有经验的操作人员可以更有效率地操作这种设备。有杆泵系统可以在油井生产中大范围地高效运行。这些系统大部分具有很高的残余价值。

抽油杆、泵的部件和油管会连续受到磨损。因此，该系统必须比任何其他举升系统更有效地防止腐蚀，以确保其使用寿命。有杆泵系统通常与斜井眼不匹配。

2. 应用

有杆泵系统具有非常广泛的应用：

(1) Applicable to sandy fluid, gaseous, high viscosity;

(2) All types of wells: horizontal, slant, directional and vertical well.

3. Limitation

Limitations of the system should be considered while choosing the equipment. Strength of the rods determines the maximum performing depth which is up to 12000 ft. Metallurgy of the component should be in compliance with the well environment. Corrosives, contaminants and salinity play important role in equipment life. The amount of fluid past through the interval between barrel and plunger called plunger slippage. Overall pump efficiency can decrease dramatically if plunger slippage increases because of improper design of barrel inner diameter (ID) and plunger outer diameter (OD).

And the ability of sucker rod pumping systems to lift sand is limited. Paraffin and scale can interfere with the efficient operation of sucker rod pumping systems. If the gas-liquid separation capacity of the tubing-casing annulus is too low, or if the annulus is not used efficiently, and the pump is not designed and operated properly, the pump will operate inefficiently and tend to gas lock.

1.3.2.5 Gas Lift

Of all artificial lift methods, gas lift most closely resembles natural flow and has long been recognized as one of the most versatile artificial lift methods. Because of its versatility, gas lift is a good candidate for removing liquids from gas wells under certain conditions.

1. Advantages / Disadvantages

The most important advantages of gas lift over pumping lift methods are:

(1) Most pumping systems become inefficient when the GLR exceeds some critical value, typically about 500 ft^3/bbl (90 m^3/m^3), due to severe gas interference. Although remedial measures are possible for conventional lift systems,

(1)适用于含砂、含气和高黏度流体;

(2)适用于所有类型的井,包括水平井、斜井、定向井和垂直井。

3. 局限性

选择设备时应考虑系统的局限性。杆的强度决定了最大操作深度,其最高可达12000ft。组件的材料应符合井环境。腐蚀性、污染物和盐度对设备寿命的影响很大。通过泵筒和柱塞之间的流量称为漏失量。如果由于桶内径和柱塞外径设计不当导致漏失量增加,整体泵效率会显著降低。

有杆泵系统携砂能力有限。石蜡和水垢会影响有杆泵系统的有效运行。如果油套环空的气液分离能力太低,或者如果没能有效地使用环空,并且泵的设计和操作不正确,则泵将无法运行且会产生气锁。

1.3.2.5 气举

在所有人工举升方法中,气举时的流体流动最接近自然流动,并且长期以来被认为是最通用的人工举升方法之一。由于其多功能性,气举是在某些条件下从气井中去除液体的良好方法。

1. 优缺点

气举相对于泵送系统最重要的优点是:

(1)当气液比超过临界值时,大多数泵送系统效率低,通常约为500ft^3/bbl($90m^3/m^3$),这是由严重的气体干扰造成的。

gas lift systems can be applied directly to high GLR wells because the high formation GLR reduces the need for additional gas to lower the formation flowing pressure.

尽管传统举升系统可采取补救措施,但气举系统可直接应用于气液比较高的井,因为高气液比减少了对额外气体的需求,降低了地层流动压力。

(2) Production of solids will reduce the life of any device that is placed within the produced fluid flow stream, such as a rod pump or ESP. Gas lift systems generally are not susceptible to erosion due to sand production and can handle a higher solids production than conventional pumping systems.

(2)固体颗粒产出会缩短任何装置的寿命,例如杆式泵或电潜泵。气举系统通常不会因产生砂而易受侵蚀,并且可以处理比常规泵送系统更高的固体含量。

(3) In highly deviated wells, it is difficult to deploy some pumping systems due to the potential for mechanical damage to deploying electric cables or rod and tubing wear for beam pumps. Gas lift systems can be employed in deviated wells without mechanical problems.

(3)在高度偏斜的井中,由于可能对电缆造成机械损坏或者对有杆泵造成杆和管道磨损,因此难以部署一些举升系统。气举系统可用于斜井,且不会产生机械问题。

(4) Another advantage that gas lift has over other types of artificial lift is its adaptability to changes in reservoir conditions. It is a relatively simple matter to alter a gas lift design to account for reservoir decline or an increase in fluid (water) production that generally occurs in the latter stages of the life of the field. Changes to the gas lift installation can be made from the surface without pulling tubing by replacing the gas lift valves via wireline and reusing the original downhole components. However, many gas lift installations in onshore lower volume gas well may choose to use conventional mandrels where the tubing must be pulled to access gas lift valves and to replace valves.

(4)气举相对于其他人工举升方法的另一个优点是其对储层条件变化的良好适应性。改变气举设计以解决储层下降或流体(水)产量增加问题(通常发生在油田寿命的后期阶段)相对简单。通过电缆更换气举阀并重新使用原有的井下部件,可以从地面进行气举装置的改变而不需要下放管柱。然而,许多陆上低产量气井的气举装置可以选择使用传统心轴,必须拉出油管以接近气举阀并更换阀门。

The two fundamental types of gas lift used in the industry today are continuous flow and intermittent flow. Depending on the production capacity gas injection can be continuous or intermittent. If gas injected into the well by intervals because of the need for the build up in the tubing it is the intermittent gas lift. A well which is able to maintain a column of fluid above gas injection point called under continuous flow gas lift.

气举可以分为连续气举和间歇气举。根据生产能力,气体注入可以是连续的或间歇的。如果需要给井中间歇注入气体则为间歇气举,在注入点以上保持液柱则为连续气举。

2. Applications

A gas lift system applies to continuous or intermittent flow. When the well is not as economic with continuous gas lift, as occurs when bottom hole pressure declines, the well is converted to intermittent gas lift. The converting time is when production rate is about 200 bbl/d. The applications for gas lift are summarized below:

(1) Tubing and casing flows;

(2) Wells available to get supply pressurized gas for injection;

(3) Well with insufficient bottom hole pressure;

(4) Relatively high GLR wells.

1.3.2.6 Hydraulic Pumping

There are two kinds of hydraulic pumps currently on the market-positive displacement pumps and jet pumps.

1. Strength

(1) Free pump. Being able to circulate the pump in and out of the well is the most obvious and significant feature of hydraulic pumps. It is especially attractive on offshore platforms, remote locations, populated areas and in agricultural areas.

(2) Deep wells. Positive displacement pumps are capable of pumping depths to 17,000 feet, and deeper. Working fluid levels for jet pumps are limited to around 9000 feet.

(3) Speed control. By changing the power fluid rate to pumps, production can be varied from 10 percent to 100 percent of pump capacity. The optimum speed range is 20 to 85 percent of rated speed.

(4) Crooked wells. Deviated wells typically present no problem to hydraulic free pumps. Jet pumps can even be used in TFL installations. Jet pumps can be used for sandy fluid, because they have no moving parts, can handle sand and other solids very well.

(5) Viscous oils. Positive displacements pumps can handle viscous oils very well. The power fluid can be heated or it can be diluent to further aid getting the oil to the surface.

2. 应用

气举系统适用于连续或间歇流动。当井底压力下降,使用连续气举不具有经济效益时,应使用间歇气举。转换时间是当生产率约为 200 bbl/d 时。气举的应用条件有以下几点:

(1)油管套管连通;

(2)具有充足的高压气体源;

(3)井底压力不足;

(4)气液比相对较高。

1.3.2.6 水力泵

目前市场上有两种水力泵——活塞泵和射流泵。

1. 优点

(1)活动式井下泵。水力泵最显著的特点是能够使泵进出井内,在海上平台、偏远地区、人口稠密地区和农村地区具有很大吸引力。

(2)深井。活塞泵的泵深可至 17000ft 处。喷射泵的工作液面限制在 9000ft 左右。

(3)速度控制。通过改变动力液的速率,其生产能力可以从泵容量的 10% 变化到 100%。最佳速度范围是额定转速的 20% ~85%。

(4)斜井。对于活动式水力泵,斜井通常没有问题。射流泵甚至可以油管外组装。对于出砂油井可以采用射流泵,因为他们没有移动部件,而且可以很好地处理砂子或其他固体。

(5)高黏油。活塞泵可以有效处理高黏油,通过加热动力液或者稀释剂将油采出。

(6) Corrosion. Corrosion inhibitors can be injected into the power fluid for corrosion control.

(6) 腐蚀。可以在动力液中注入腐蚀抑制剂来防止腐蚀。

2. Limitation

2. 局限性

(1) Power fluid cleaning. Removing solids from the power fluid is very important for positive displacement pumps. Maintenance of surface plunger pumps is also affected by solids in the power fluid. Jet Pumps, on the other hand, are very tolerant of poor power fluid quality.

(1) 动力液净化。从动力液中去除固体对活塞泵非常重要。一方面,动力液中的固体会对柱塞泵的维护造成影响;另一方面,射流泵对于传动液有严格的质量要求。

(2) Pump life. Positive displacement pumps, on average, have shorter life between repairs than jet, sucker rod and electric submersible pumps. Mostly, this is a function of the quality of power fluid, but also, on average, they are pumping from greater depths. Jet pumps, on the other hand, have very long pump life between repairs.

(2) 泵的使用寿命。活塞泵的平均寿命比射流泵、有杆泵和电潜泵的短。一方面,其寿命与动力液的性能和工作深度有关。另一方面,射流泵的维修年限长。

(3) Bottom hole pressure. Whereas positive displacement pumps can pump to practically zero bottom hole pressure, Jet Pumps cannot. Jet Pumps require approximately 1000 psi bottom hole pressure when set at 10,000 feet and approximately 500 psi when set at 5000 feet.

(3) 井底压力。活塞泵需要的井底压力可以几乎为零,而射流泵不能。在 10000ft 时,射流泵大约需要 1000psi 的井底压力,而在 5000ft 时,大约需要 500psi 的井底压力。

(4) Skilled personnel. Positive displacement pumps generally require more highly skilled operating personnel, or perhaps, just more attention, than Jet Pumps and other types of artificial lift. There are two reasons for this. First, pump speed needs to be monitored daily and not allowed to become excessive. Secondly, power fluid cleaning systems need frequent checking, and action taken, to keep them operating at their optimum effectiveness.

(4) 熟练的操作人员。活塞泵与射流泵和其他人工举升方式相比通常需要更熟练的操作人员。这主要有两个原因,首先,需要每天监测泵速,不能使泵速过大;其次,动力液净化系统需要经常检查使其保持最佳效果。

1.3.2.7 Progressing Cavity Pumping

1.3.2.7 螺杆泵

1. Applications

1. 应用

PCP can be applied to the wells producing sand-laden heavy oil and bitumen, high water-cut wells, and in the gas wells that require dewatering. Operating depth is somewhat limited, as it is believed that the maximum depth of operation is 6,000 ft. Well's deviation is not a factor, so PCP is applicable regardless of hole deviation.

螺杆泵可以应用于产生含砂重油和沥青的井、高含水井、需要排水的气井。其操作深度有限,因为它最大操作深度为 6000ft。不用考虑井斜因素,因此无论井斜多少,螺杆泵都适用。

2. Advantages / Disadvantages

Solid and gas handling is good or excellent while corrosion handling is just fair. This system can be installed and operated economically due to low capital investment and power consumption. Compared to other pumping methods, it is able to operate more quietly.

2. 优缺点

螺杆泵处理固体和气体的效果好，耐腐蚀相对较好。由于投资少、能耗低，该系统的安装和操作成本低。与其他泵送方法相比，它运行时的噪声小。

1.3.2.8 Electrical Submersible Pumping

1.3.2.8 电潜泵

In ESP system the entire unit is lowered to the bottom of the well with an insulated cable from the surface. Basic elements are a centrifugal pump, the shaft and an electric motor. High production rates from deeper depths considered to be accomplished by using ESP. Improving technology increases the usage of ESP making it flexible for different rates.

在电潜泵系统中，整个装置通过地面绝缘电缆下放到井底，基本元件是离心泵、轴和电动机。使用电潜泵可以实现较深储层更高的产量。技术进步增加了电潜泵的使用范围，使其适用于不同的排量。

1. Applications

1. 应用

The ESP system is typically reserved for application of the well producing primarily liquid. In gas wells, ESP can be applied when it is necessary to handle large liquid volume. The other considerations for application are summarized below:

电潜泵系统通常用于气井排液采气。在气井中，当需要处理大量液体时，可以应用电潜泵。应用时的其他考虑因素总结如下：

(1) Adaptable to highly deviated wells-up to 800 m;

(1)适用于斜井:高达800m;

(2) High volume lift requirement (100 ~ 30,000 bbl/d);

(2)大排量举升要求(100 ~ 30000bbl/d);

(3) Deep wells / deviated wells;

(3)深井和斜井;

(4) Waterflood or high water-cut wells;

(4)注水井或含水率高的井;

(5) ESP's operate from shallow depths to as deep as 10, 000 and deeper.

(5)操作范围从浅到深达10000ft及更深。

2. Strength

2. 优点

Factors leading to the selection of the submersible pump as the most economical method of lift are as follows:

选择电潜泵作为最经济的举升方式的原因如下：

(1) Adaptable to highly deviated wells-up to 800 m.

(1)适用于高达800m的大斜度井。

(2) Permit use of minimum space for subsurface controls and associated producing facilities.

(2)井下控制器和相关生产设施占地空间小。

(3) Method is quiet, safe and sanitary for acceptable operations in an offshore and environmentally conscious area.

(3)安全、清洁、环保，适合在海上采油。

(4) Generally considered a high volume pump provides for increased volumes and water cuts brought on by pressure maintenance and secondary recovery operations.

(4)通常认为大容量泵可以通过维持压力和二次采油增加容量及含水量。

(5) Permits placing well on production immediately after drilling and completion.

(5)钻井和完井完成后,可以立即进行生产。

3. Limitation

3. 局限性

Some of the weaknesses of the submersible system are as follows:

电潜泵举升系统的局限性如下:

(1) Will tolerate only minimal percent of solids (sand) production.

(1)允许的出砂非常少。

(2) Costly pulling operations to correct downhole failures.

(2)处理井下故障的成本较高。

(3) While on DHF there is a loss of production during time well is covered by drilling operations in immediate vicinity.

(3)当井下设备发生故障时,损失的生产时间和邻井钻井作业时间差不多。

(4) They are installed in deviated wells, but the unit must be landed such that it is straight even if the wellbore is deviated.

(4)可以安装在斜井中,但组件必须保持垂直,即使井眼是倾斜的。

(5) Power must be available and is transmitted down a three phase cable to the motor.

(5)必须有电力供应,并通过三相电缆传输到电动机。

(6) High solids concentrations may cause the unit to fail if they are allowed to be pumped, although special abrasion resistant units can be used.

(6)如果泵送高固相含量的液体,可能导致设备失效,但可以使用特殊的抗磨损装置。

1.3.2.9 Compression

1.3.2.9 井口增压法

The adaptability of compression are as follows:

井口增压法的应用情况如下:

(1) Compression is used for single wells or for multiple wells.

(1)用于单井或丛式井。

(2) Nodal analysis will help predict the expected results to be achieved.

(2)节点分析有助于预测要实现的结果。

(3) Lower well head pressure has many beneficial effects.

(3)降低井口压力会产生有利影响。

(4) Lower pressure keeps water in vapor state so this is artificial lift method in itself.

(4)较低的压力使水保持蒸汽状态,因此这本身就是一种人工举升方法。

(5) Biggest percentage gains for low pressure wells.

(5)低压井的增产幅度最大。

(6) Lower wellhead pressure improves artificial lift methods in general.

(6)较低的井口压力改善了人工举升方法。

(7) Once wells die completely, compression may not restart them without swabbing and other trouble.

(7)一旦死井,可能无法在没有抽吸的情况下重新启动。

(8) Compression is good combination method with plunger.

(8)是与柱塞组合的好方法。

(9) Compression is needed to lower the annulus for pumping wells to achieve best results.

(9)需要压缩来降低抽油井的环空以达到最佳效果。

The above various methods are suitable for different developmental stages of gas reservoir. The velocity strings, foaming and the plungers are mainly applied for the gas wells with the ability of natural flow. The gas lift and machine pumping are mainly for unloading.

以上各种工艺适用于不同的气藏开发阶段,其中适用于气藏自喷末期、具有自喷或间喷能力产水气井的工艺有速度管柱法、泡沫法、柱塞法,适用于气井强排水或水淹气井复产的工艺有气举、机抽排液采气工艺。

When selecting the unloading methods, the following principles should be followed: the selected gas well must have a certain productivity and recoverable reserve; in the selection of the process type, the unloading methods without the moving tubing string are preferred, and followed by the methods to move the string; the selected methods should remove the bottom hole liquid as soon as possible to restore the gas well productivity; the selected methods should be considered in the long term, and the application period of the process should be relatively long to avoid the gas well reloading in a short period of time; the choice of unloading methods should be economical with less investment, simple operation and easy management. Fig. 1. 34 shows the lift selection guidelines.

在选择排液采气工艺时,要遵循以下原则:所选气井必须具有一定的产能,具有一定的可采储量;在工艺类型的选择上,优先选择不用动管柱的排液采气工艺,然后再选择动管柱的排液采气工艺;优选出的排液采气工艺要能尽快排出气井井底积液,恢复气井产能;所选的排液采气工艺要从长远考虑,工艺的应用期要相对较长,尽量避免气井在短期内再次水淹;排液采气工艺的选择要从经济投入角度出发,尽量选用投资较低、作业较简单、易于管理的排液采气工艺。图 1.34 为人工举升方式选择指南。

	Rod Lift	Progressing Cavity	Gas Lift	Plunger Lift	Hydraulic Piston	Hydraulic Jet	Electric Submersible
Operating Depth	To 16,000′ TVD	To 6,000′ TVD	To15,000’ TVD	To 19,000′ TVD	To 17,000′ TVD	To15,000’TVD	To 15,000′ TVD
Operating Volume	To 5,000 BPD	To 4,500 BPD	To 30,000 BPD	To 50 BPD	50～4,000BPD	300～15,000 BPD	200～30,000 BPD
Operating Temperature	100～550°F	75～250°F	100～400°F	120～500°F	100～500°F	100～500°F	100～400°F
Corrosion Handling	Good to Excellent	Fair	Good to Excellent	Excellent	Good	Excellent	Good
Gas Handling	Fair to Good	Fair	Excellent	Excellent	Fair	Good	Poor to Fair
Solids Handling	Fair to Good	Excellent	Good	Poor to Fair	Poor	Good	Poor to Fair
Fluid Gravity	>8°API	<35°API	>15°APl	GLR Required-300 SCF/BBL/1000′Depth	>8°APl	>8°APl	>10°APl
Servicing	Workover or pulling Rig	Workover or pulling Rig	Wireline or Workover Rig	Wellhrad catcher or Wireline	Hydraulic or Wireline	Hydraulic or Wireline	Workover or pulling Rig
Prime Mover	Gas Engine or Electric	Gas Engine or Electric	Compressor	Wells'Natural Energy	Gas Engine or Electric	Gas Engline or Electrlc	Electric Motor
Offshore Application	Limited	Good	Excellent	N/A	Good	Excellent	Excellent
Overall System Efficilncy	45%～60%	40%～70%	15%～30%	N/A	45%～55%	10%～30%	35%～60%

Fig. 1. 34 Lift Selection Guidelines

(Source：http://www. petroskills. com)

1.3.3 Unloading Selector

1.3.3.1 Introduction

The unloading selector is a logical artificial-lift application selection process for gas well unloading. The unloading selector works by assigning a high or low value to each of only four readily available surface-gathered data points; liquid rate, flowing tubing pressure, water cut percentage and gas liquid ratio. Start in the middle of the selection tool and match your high or low answers to these four variables. Then view inside the outer most ring of the tool to see an artificial-lift-type selection has been made. Once the lift selection has been identified, move to the corresponding outer four quadrants of the tool for further analysis of that lift selection.

1.3.3 排液方法选择器

1.3.3.1 引言

排液方法选择器用于选择合适的气井排液方式。排液方法选择器的工作原理是只为四个现有的地面数据采集点(液体流量、油管压力、含水率和气液比)分配一个高值或低值;从选择工具的中间开始,将这四个变量的高值或低值进行匹配;然后查看该工具的最外环的内部,看到已经选择了一种人工举升方式;一旦确定了举升方式,就移动到工具对应外部的四个象限,来进一步选择举升方式。

1.3.3.2 Lift Systems

1. Positive-Displacement Lift Systems

A gas well ultimately depletes to the extent that gas volume or pressure build rate is insufficient for intermittent plunger lift or foam lift to work and bottom hole pressure is inadequate to support gas-lift pressures. This is when positive-displacement lift systems are deployed. They may not be the least expensive lift systems to install and operate, but they are the only forms of lift that can deplete liquid-producing gas wells to their lowest possible abandonment pressure. Fig. 1.35 shows the positive-displacement lift analysis.

1.3.3.2 举升系统

1. 活塞举升系统

气井在气体流量或压力不足以使间歇柱塞举升或泡沫举升工作,或井底压力不足以提供气举压力时停产,这时需要使用活塞举升系统。它可能不是安装和操作最便宜的举升系统,但它是唯一能够将产液气井开采直到最低废弃压力的举升方式。图 1.35 为活塞举升选择分析图。

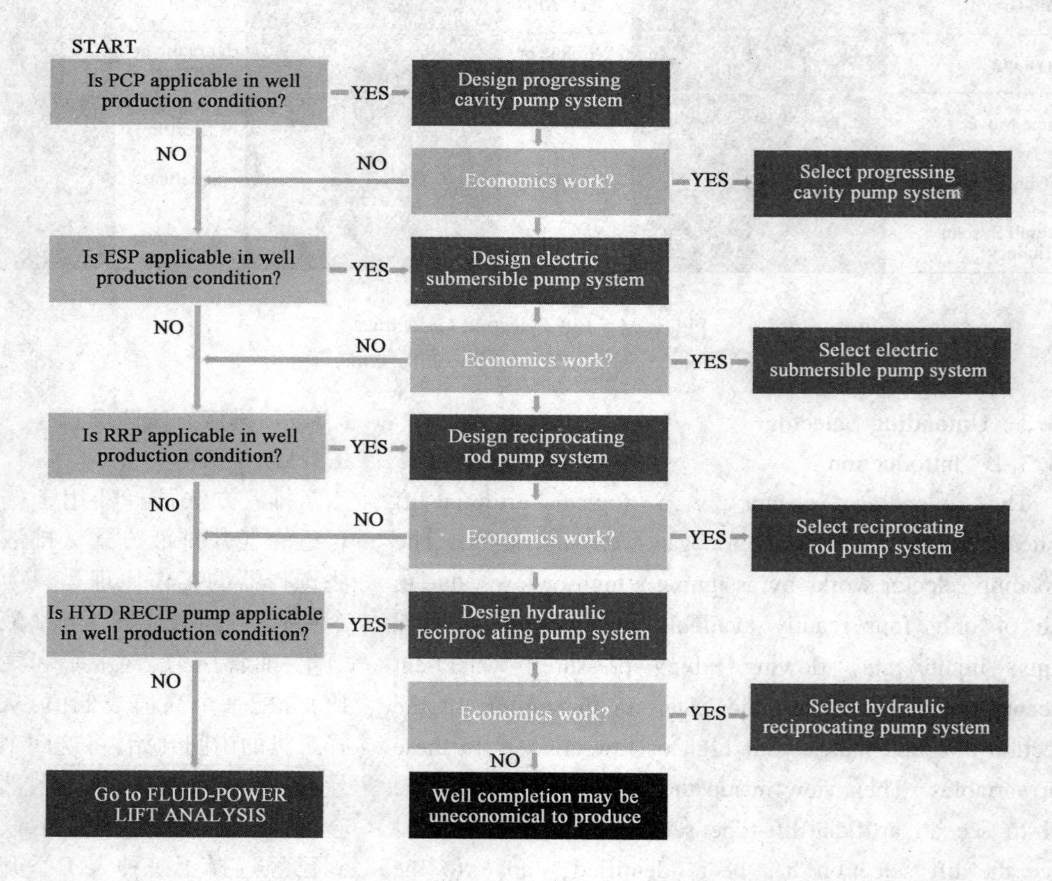

Fig. 1.35 Positive-Displacement Lift Analysis
(Source: http://apps.weatherford.com)

2. Fluid-Power Lift Systems

While not always an obvious choice, a fluid-power lift system can be an extremely economical solution for gas-well liquid loading problems. These systems use either liquid or gaseous fluids to generate lift capacity to carry liquids from a well. Fig. 1.36 shows the fluid-power lift analysis.

2. 动力液举升系统

虽然并不总是一个很好的选择，但动力液举升系统是气井积液问题极其经济的解决方案。该系统使用液体或气体将井中液体举升到地面。图 1.36 为动力液举升选择分析图。

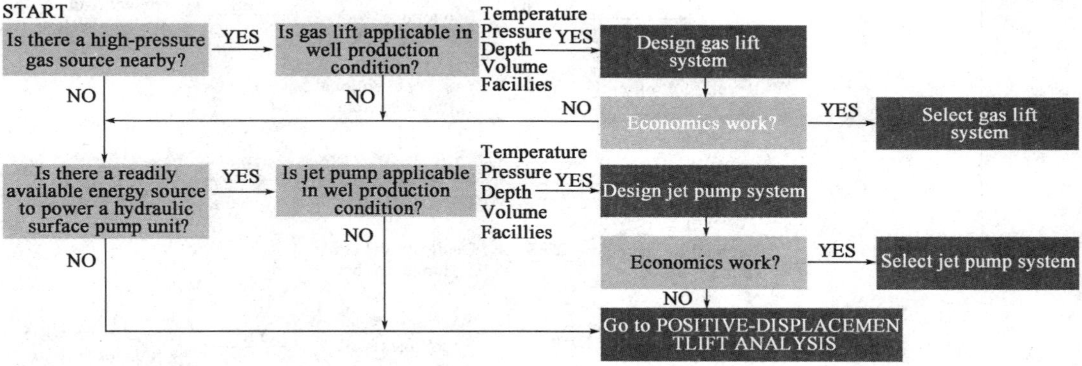

START
Is there a high-pressure gas source nearby?
YES
Is gas lift applicable in well production condition?
Temperature
Pressure
Depth
Volume
Facillies
YES
Design gas lift system
NO
NO
NO
Economics work?
YES
Select gas lift system
Is there a readily available energy source to power a hydraulic surface pump unit?
YES
Is jet pump applicable in wel production condition?
Temperature
Pressure
Depth
Volume
Facillies
YES
Design jet pump system
NO
NO
Economics work?
YES
Select jet pump system
NO
Go to POSITIVE-DISPLACEMEN TLIFT ANALYSIS

Fig. 1.36 Fluid-Power Lift Analysis

(Source: http://apps. weatherford. com)

3. Plunger-Lift Systems

As a gas well depletes, the gas velocity is insufficient to naturally carry liquid accumulations from the wellbore. However, energy in the system is still sufficient for the well to flow naturally. This is when plunger and foam lift systems should be installed, as both use the well's own internal energy to unload liquids. Fig. 1. 37 shows the plunger-lift analysis.

3. 柱塞举升系统

当气井能量耗尽时，气体速度不足以携带出井筒积液，系统中的能量仍然足以使气体自然流动。这时应该安装柱塞和泡沫举升系统，因为它们都使用井自身的内部能量来排出液体。图 1. 37 为柱塞举升选择分析图。

4. Foam-Lift Systems

Foam-lift systems alter the physical properties of the produced fluid by applying surfactant. Surface tension and apparent liquid density are altered to reduce the critical velocity needed to lift the water from the system. The surfactants react with water, so application is effectively limited to wells in which most of the liquid phase is water rather than hydrocarbon. Foam-lift systems excel in deliquification of low-reservoir-pressure wells. Fig. 1. 38 shows the foam-lift analysis.

4. 泡沫举升系统

泡沫举升系统通过使用表面活性剂改变生产流体的物理性质，如改变表面张力和液体密度以降低排液所需的临界速度。表面活性剂要与水反应，因此泡沫举升系统在大部分液相是水而不是烃的井中适用。泡沫举升系统在低储层压力井的排液方面表现优异。图 1.38 为泡沫举升选择分析图。

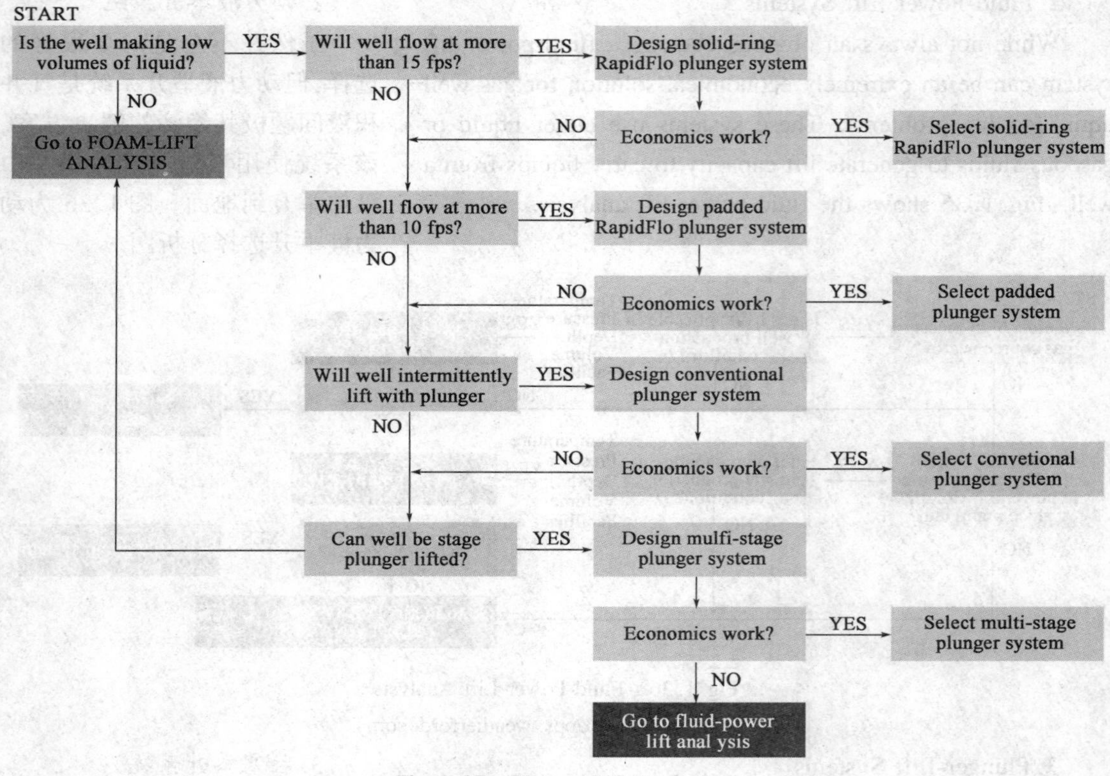

Fig. 1. 37 Plunger-Lift Analysis

(Source: http://apps. weatherford. com)

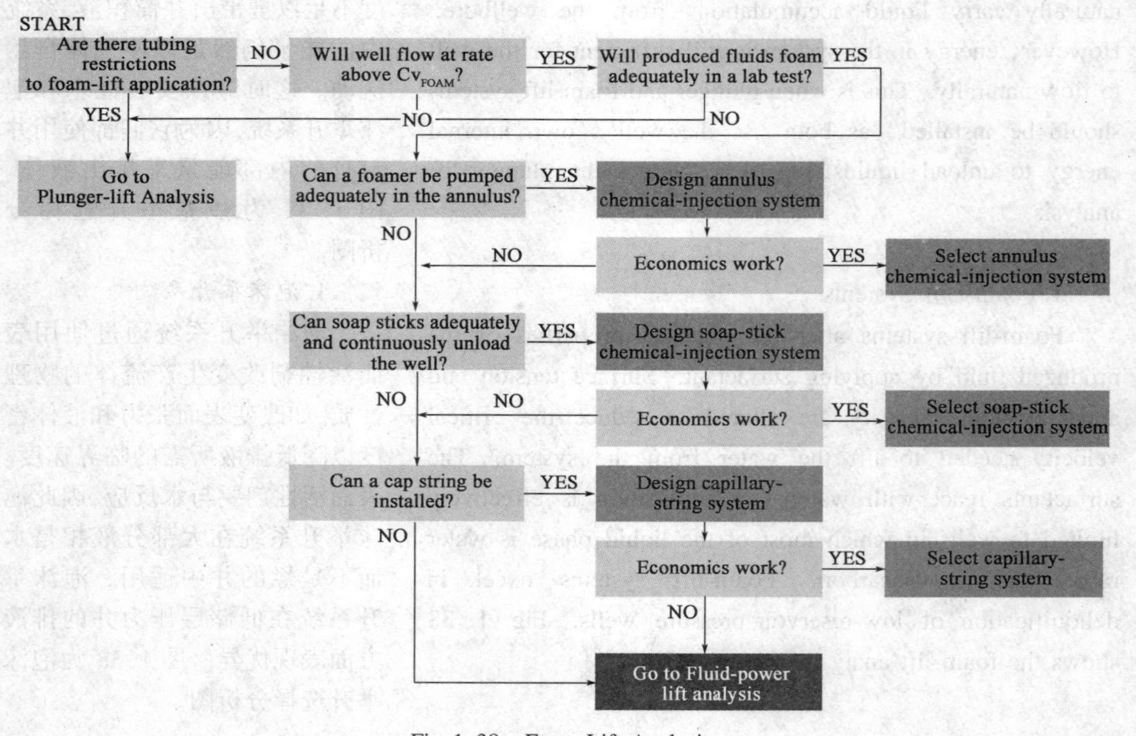

Fig. 1. 38 Foam-Lift Analysis

(Source: http://apps. weatherford. com)

Exercises/练习题

1. Describe flow patterns and their characteristics in gas wells. /简述气井的流态和特征。

2. What are the well symptoms indicating liquid loading?/气井积液的征兆有哪些?

3. What is critical gas velocity?/什么是气井临界流速?

4. Describe the main features of the models to estimate critical gas velocity. /简述气井临界流速计算模型的特点。

5. Describe a plunger lift cycle. /简述柱塞举升的循环过程。

6. How can foaming for unloading gas well?/为什么泡沫能排液采气?

References/参考文献

[1] Kermit E. Brown. The Technology of Artificial Lift Methods [M]. Tulsa: Petroleum Publishing, 1980, 7 – 193.

[2] Lea Jr J F, Rowlan L. Gas Well Deliquification[M]. Oxford: Gulf Professional Publishing, 2019, 23 – 66, 107 – 188.

[3] Neely B, Gipson F, Clegg J, et al. Selection of artificial lift method[C]. SPE Annual Technical Conference and Exhibition. Society of Petroleum Engineers, 1981.

[4] Henderson L J. Deep Sucker Rod Pumping for Gas Well Unloading[C]. SPE Annual Technical Conference and Exhibition. Society of Petroleum Engineers, 1984.

[5] Heinze L R, Winkler H W, Lea J F. Decision Tree for selection of Artificial Lift Method[C]. SPE Production Operations Symposium. Society of Petroleum Engineers, 1995.

[6] Park H Y, Falcone G, Teodoriu C. Decision matrix for liquid loading in gas wells for cost/benefit analyses of lifting options[J]. Journal of natural gas science and engineering, 2009, 1(3): 72 – 83.

[7] Lea J F, Nickens H V. Solving gas-well liquid-loading problems[J]. Journal of Petroleum Technology, 2004, 56(04): 30 – 36.

[8] Wang J Y. Well Completion for Effective Deliquification of Natural Gas Wells[J]. Journal of Energy Resources Technology, 2012, 134(1): 13102.

[9] Lea J F. Dynamic analysis of plunger lift operations[J]. Journal of Petroleum Technology, 1982, 34(11): 2617 – 2629.

[10] Ceylan S E. Design and Economical Evaluation OF Sucker Rod and Electrical Submersible Pumps: Oil Wells in a Field, Turkey[D]. Ankara: Middle East Technical University, 2004.

[11] Brown K E. Overview of artificial lift systems[J]. Journal of Petroleum Technology, 1982, 34(10): 2384 – 2396.

[12] Damola Fadipe. Candidate Selection and Assessment of the Benefits of Velocity Strings on the Dunbar Field [R]. Partial fulfilment of the requirements for the MSc and/or the DIC, 2012.

[13] Clegg J D, Bucaram S M, Hein N W. Recommendations and Comparisons for Selecting Artificial-Lift Methods (Includes Associated Papers 28645 and 29092)[J]. Journal of Petroleum Technology, 1993, 45 (12): 1128 – 1167.

[14] Soponsakulkaew N. Decision Matrix Screening Tool to Identify the Best Artificial Lift Method for Liquid-loaded Gas Wells[D]. College station: Texas A & M University, 2010.

[15] Carrascal J F. Multiphase flow application to ESP pump design program[D]. Lubbock: Texas Tech University, 1996.

[16] Turner R G, Hubbard M G, Dukler A E. Analysis and prediction of minimum flow rate for the continuous

removal of liquids from gas wells[J]. Journal of Petroleum Technology, 1969, 21(11): 1475 - 1482.

[17] Coleman S B, Clay H B, McCurdy D G, et al. A new look at predicting gas-well load-up[J]. Journal of Petroleum Technology, 1991, 43(3): 329 - 333.

[18] Jafarov T, Alnuaim S. Critical Review of the Existing Liquid Loading Prediction Models for Vertical Gas Wells[C]. Offshore Technology Conference Asia. Offshore Technology Conference, 2016.

2 Recent Advances in Hydraulic Fracturing Technology

水力压裂技术新进展

Since the 1950s, hydraulic fracturing has been proven to be a very robust technology, and applied to many different types of reservoirs. Although fracturing is a very complex process, it has been established as the most widely used production enhancement procedure in the petroleum industry.

Audio 2. 1

A lot of oil and gas wells worldwide is being stimulated, ranging from small skin-bypass fracs at less than $ 20,000 to massive fracturing treatments over $ 1 million per well. Commercial developments of unconventional resources would not be possible today without hydraulic fracturing. There is no question that the hydraulic fracturing will continue to overwhelmingly dominate the development of low-permeability reservoirs.

In this chapter, we focus on the hydraulic fracturing technique and its advances in past decades. To begin with, a brief introduction of the technique is surely necessary, which can provide an overview of this technology in the petroleum industry. The advanced topics are discussed in three sessions.

自20世纪50年代以来,水力压裂技术的现场实践证明它是一项有效的增产技术,适用于各类不同性质的地层。水力压裂尽管是一项非常复杂的工艺,但它依然是油气行业如今应用最为广泛的增产技术。

世界各地的许多口井正在进行水力压裂作业,从只花费不到两万美元的小型解堵型压裂到每口井花费超过一百万美元的大型水力压裂作业。非常规资源的商业开发也得益于水力压裂。毫无疑问,水力压裂技术将是低渗透油气藏开发的主要技术。

本章介绍水力压裂技术的新进展。首先,对水力压裂技术进行一个简要的介绍,以便总体把握其在整个石油工业中的作用。然后将水力压裂技术新进展划分为三个大部分,并逐一进行介绍。

2.1 Overview of Hydraulic Fracturing Technology

Hydraulic fracturing entails pumping a viscous fluid at a sufficiently high pressure into the formation. As the pressure continues to increase, eventually a point will be reached where the pressure becomes greater than the maximum stress that can be sustained by the formation to induce a parting of the formation. The fracture is then filled with granular materials called proppants to hold the fracture open after the treatment is finished. Eventually, the artificial fracture with certain length, width and height is created within the formation.

2.1 水力压裂技术概述

水力压裂,就是将黏性流体以较高的压力泵入地层,随着压力的不断增加,当压力超过地层可以承受的最大应力时,地层便被压开产生裂缝。之后向裂缝中填入支撑剂(颗粒状材料),以在压裂完成后保持裂缝的张开,从而在地层中形成具有一定长宽高的人工裂缝。

2.1.1 History of Hydraulic Fracturing

As shown in Table 2.1, the first attempts at fracturing formations for the purpose of improving production were not hydraulic in nature-they involved the use of high explosives to break the formation apart and provide "flow channels" from the reservoir to the wellbore. There are records indicating that this took place as early as 1890. This type of reservoir stimulation reached its ultimate conclusion with the experimental use of nuclear devices to fracture relatively shallow, low-permeability formations in the late 1950s and early 1960s (Fig. 2.1).

2.1.1 水力压裂发展历史

表2.1显示,实际上最早为了增产而进行的地层压裂并非水力压裂,而是使用爆炸物炸开地层形成地层与井筒之间的流动通道。记录表明,这种技术最早出现于1890年,当时在埋藏较浅的低渗储层中使用核装置进行了压裂实验(图2.1),在20世纪50年代末60年代初终结。

Table 2.1 History of Hydraulic Fracturing

Types	Overview
Explosive Fracturing	Records showed that this fracturing was first appeared in 1890, and liquid nitroglycerin was used in shallow, hard formations in Pennsylvania, West Virginia and other areas for oil well stimulation
Nuclear Fracturing	In 1957, nuclear device at the Nevada test site was used for reservoir stimulation, marking the ending of explosive fracturing
Acid Fracturing	Acid fracturing has been widely used in the late 1930s, and many operators have found that when above a certain pressure, the amount of acid injected into the formation will increase significantly
Hydraulic Fracturing	In 1947, Indiana Mobil Oil Company completed its first official vertical well hydraulic fracturing operation in the Hugoton Gas Field in western Kansas. Halliburton then purchased the patent from Indiana Mobil Oil Company and it developed into a mainstream fracturing method
Waterfree Fracturing	Waterfree fracturing refers to a new type of fracturing technology that does not use water, but use anhydrous methanol, anhydrous ethanol, carbon dioxide, nitrogen, or liquefied natural gas, etc. as fracturing fluid. The result is better, but the cost is higher and there is safety risk

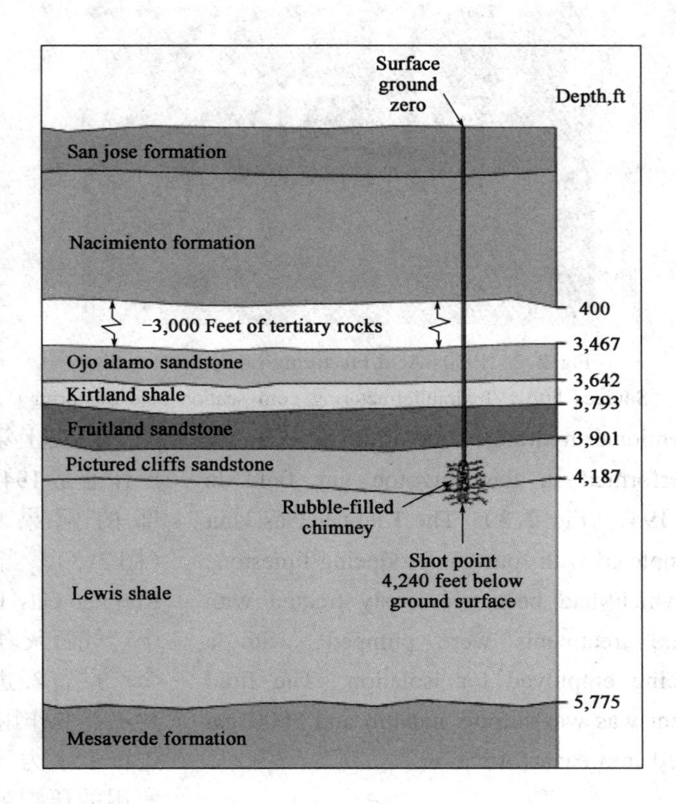

Fig. 2. 1　Project Gasbuggy of Nuclear Fracturing in New Mexico in 1967
(Source: http://dictionary. sensagent. com/ProjectGasbuggy)

By the late 1930s, acidizing had become an accepted well development technique (Fig. 2. 2). Several practitioners observed that above a certain "breakdown" pressure, injectivity would increase dramatically. It is probable that many of these early acid treatments were in fact acid fractures. In 1940, Torrey recognized the pressure-induced fracturing of formations. He presented data to show that the pressures generated during some operations could part the rocks along bedding planes or other lines of "sedimentary weakness".

在20世纪30年代末期,酸化技术成为主要的油气井增产技术(图2. 2)。一些从业者在酸化过程中发现,当注入酸液的压力超过破裂压力值时,注入酸液的量会急剧增加。因此,这些早期的酸化作业很有可能就是酸化压裂。1940年,Torrey发现了压力而引起的地层破裂。他提供的数据显示,过高的压力会使岩石沿着层面或其他"沉积弱点"处分开。

Fig. 2. 2　1950s Acid Fracturing Equipment on Site
(Source：http://hydraulicfracturing. com/section/acid-fracturing)

The first intentional hydraulic fracturing process for stimulation was performed in the Hugoton gas field in western Kansas, in 1947 (Fig. 2. 3). The Klepper Gas Unit No. 1 well was completed with four gas producing limestone intervals, one of which had been previously treated with acid. Four separate treatments were pumped, with a primitive packer being employed for isolation. The fluid used for the treatment was war-surplus napalm and 3000 gal of fluid were pumped into each formation.

历史上第一次水力压裂增产作业是1947年在堪萨斯州西部的雨果顿气田进行的(图2.3)。当时进行压裂的The Klepper Gas Unit No. 1井有四个产气石灰岩完井段,其中有一个产层曾经进行过酸处理。在封隔器的作用下,对每个产层都进行了水力压裂施工。施工所使用的压裂液是剩余的凝固汽油,每层注入量为3000gal。

Fig. 2. 3　Klepper Gas Unit No. 1, Hugoton Field, Kansas: The Very First Frac Job
(Source：http://en. wikipedia. org/wiki/fracturing)

Although post-treatment tests showed that the gas productivity of some zones had been increased relative to others, the overall productivity was not increased. It was therefore concluded that fracturing would not replace acidizing for limestone formations. However, an explosion of the practice was followed in the mid-1950s and a considerable

尽管压后结果表明一些层段的天然气产能相比其他层段得到了提高,但该井的整体产能并没有增加。因此,当时人们认为水力压裂不能替代酸化处理技术。然而,水力压裂作业规模在

surge in the mid-1980s. Since then, hydraulic fracturing grew to become a dominant stimulation technique, primarily for low permeability reservoirs in North America. By 2013, 40 percent of new oil wells and 80 percent of gas wells in the United States were fracturing treated.

20 世纪 50 年代中期和 80 年代中期经历了两次爆炸式增长,并在此之后成为北美低渗透油气藏的主要增产技术。到 2013 年,美国 40% 的新油井和 80% 的气井都进行过水力压裂,这一数字仍在不断上升。

With improved modern fracturing capabilities and the advent of high permeability fracturing (HPF) around 1993, fracturing has expanded further to become the choice of all types of wells in United States. Today, it is established as one of the major recent developments in petroleum industry.

随着现代压裂技术的不断改进和高渗透压裂技术的诞生,水力压裂技术的应用范围正在不断扩展,成为美国几乎所有井的增产选择。如今,水力压裂技术是石油工业近期的主要发展方向之一。

2.1.2 Role of Hydraulic Fracturing Technology

Since its introduction, hydraulic fracturing has been, and will remain, one of the primary engineering tools for improving well productivity. But why can hydraulic fracturing achieve such a goal almost without fail? How does the technology increase overall productivity? What on earth makes it the one in a million? In order to answer why we frac, we must figure out the production increasing mechanism of fracturing.

2.1.2 水力压裂技术作用

水力压裂技术已经是,并且未来也会是提高油气井产能的重要技术之一。为什么水力压裂技术能够提高产能?它是怎样提高产能的?为什么其他增产技术不能达到它的效果?为了回答这些问题,必须先了解水力压裂增产的机理。

2.1.2.1 Production Increasing Mechanism

The well-known radial well inflow equation:

2.1.2.1 增产机理

一口直井的产量计算公式为

$$Q=\frac{2\pi K_o h(p_e-p_{wf})}{\mu_o B_o\left(\ln\frac{r_e}{r_w}-\frac{1}{2}+S\right)}=\frac{2\pi K_o h(p_e-p_{wf})}{\mu_o B_o\left(\ln\frac{r_e}{r'_w}\right)} \tag{2.1}$$

where Q is the production rate, m^3/s; K_o is the effective permeability of reservoir, μm^2; h is the effective height of reservoir, m; p_e is the reservoir pressure, Pa; p_{wf} is the bottom hole pressure, Pa; μ_o is the oil viscosity, Pa · s; B_o is the oil volume factor; r_e is the drainage radius, m; r_w is the wellbore radius, m; S is the skin factor; r'_w is the effective wellbore radius, m, clearly shows that well production rate (Q) can be increased by:

其中 Q 是产量, m^3/s; K_o 是地层有效渗透率, μm^2; h 是地层有效高度, m; p_e 是油藏压力, Pa; p_{wf} 是井底压力, Pa; μ_o 是原油黏度, Pa · s; B_o 是原油体积系数; r_e 是供给半径, m; r_w 是井筒半径, m; S 是表皮系数; r'_w 是有效井筒半径, m。从公式(2.1)中可以看出,提高油井产量的方法有:

(1) Increasing the formation flow capacity (K_h). The fracture may increase the effective formation height h or connect with a formation zone with a higher permeability K.

(2) Decreasing the skin factor S. The fracture can eliminate near wellbore formation damage.

(3) Increasing effective wellbore radius r'_w. r'_w is a function of the conductive fracture length L_f. If the hydraulic fracture has infinite conductivity i. e. the pressure drop along its length due to flow is negligible, then: $r'_w = L_f/2$. Thus effective radius has been enlarged to a value equal to half the single wing fracture length. Alternatively, if the actual wellbore radius is used, this improved inflow can be expressed as a negative skin.

(1)提高储层流动能力(K_h)。水力压裂产生的人工裂缝能增加储层有效厚度 h,或将储层与高渗透能力的地层相连通。

(2)降低表皮系数 S。水力压裂能够消除近井污染。

(3)增加有效井筒半径 r'_w。有效井筒半径 r'_w 是裂缝长度的函数。如果水力裂缝有无限导流能力,即沿着裂缝长度方向的流体压降可以忽略不计,则$r'_w = L_f/2$。因此高导流能力的水力裂缝使得井筒的有效半径变为单翼裂缝长度的一半。另外,如果在计算过程中使用的是实际井筒半径,则水力压裂的效果可以用负的表皮因子来表示。

In addition, if we take the production system in a nodal way (Fig. 2. 4), it can also explain the mechanism. The wellbore flowing pressure (p_1) after treatment has been increased, compared to an unimpaired (p_2) or impaired well (p_3):

除此之外,如果用节点法来分析整个生产系统(图 2.4),同样能解释水力压裂增产的机理。在相同的流速下,与未损害井井底流动压力(p_2)和损害井井底流动压力(p_3)相比,压裂井井底流动压力(p_1)增加:

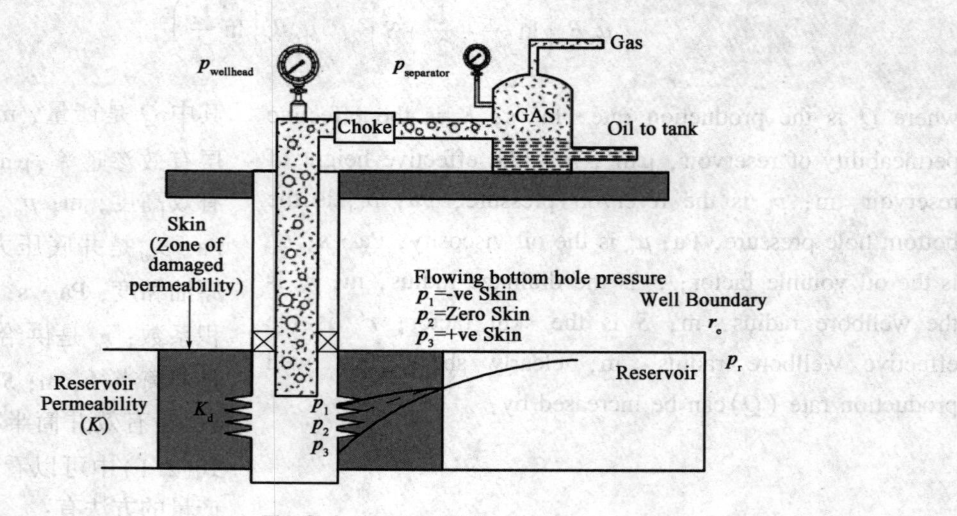

Fig. 2. 4 Production System

(1) The pressure (p_2) of a well with an ideal completion ($S = 0$);

(2) The even lower pressure (p_3) showing a positive skin due to formation damage.

Obviously, the hydraulically fractured well with the negative skin will have the greatest production rate.

(1)对于理想完井情况下($S=0$)的井底压力(p_2);

(2)储层伤害引起的正表皮系数降低了井底压力(p_3)。

很明显,具有负表皮因子的压裂井产量最高。

2.1.2.2 Why Do We Frac

2.1.2.2 进行水力压裂的原因

As the technology can increase overall productivity, hydraulic fracture operations may be performed on a well for one (or more) of mainly three reasons:

水力压裂技术可以增产,因此进行水力压裂施工的原因主要有以下三个方面:

1. Damage Bypass

1. 近井解堵

The fracture, filled with proppant, creates a very conductive flow path. This flow path has a very large permeability, frequently five to six orders of magnitude larger than the reservoir permeability. Typical intended propped widths in low permeability reservoirs are on the order of 0.25 cm (0.1 in), while the length can be several hundred meters (Fig. 2.5). In high permeability reservoirs, the targeted fracture width is much greater, perhaps as high as 5 cm (2 in), while the length might be as short as 10 meters (30 ft). In almost all cases, an overwhelming part of the production comes into the wellbore through the fracture; therefore, the original near wellbore damage is "bypassed", and the pre-treatment skin does not affect the post-treatment well performance (Fig. 2.6).

充填有支撑剂的人工裂缝为油气进入井筒提供了高效的导流通道。人工裂缝的渗透率通常都很大,一般是地层渗透率的5～6倍。人工裂缝通常很窄,约0.25cm,却有几百米长(图2.5)。在高渗储层中,人工裂缝的宽度(经过设计和控制)要大得多,接近5cm(2in),而长度可能只有10m(30ft)。无论是在哪种情况下,大部分的油气都从人工裂缝进入井筒,因此,近井地带的储层伤害被"越过",压裂前的表皮系数不再影响压裂后井的产能(图2.6)。

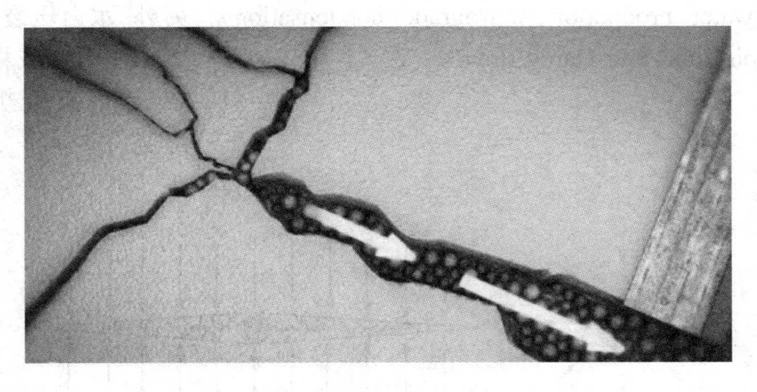

Fig. 2.5 Typical Hydraulic Fractures
(Source: http://cbu.ca/hfstudy/propped-fracture)

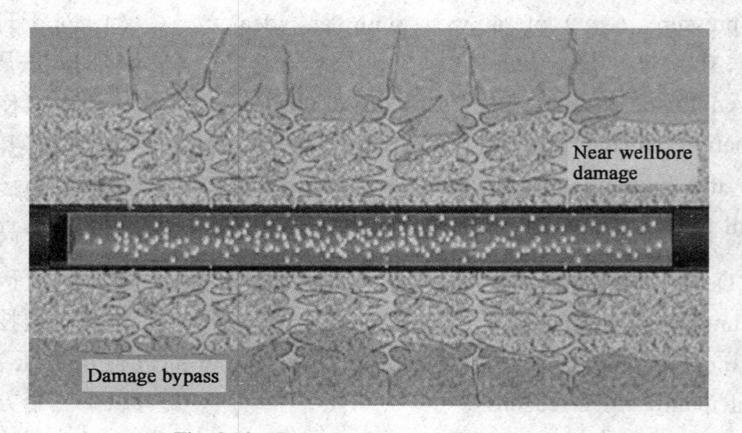

Fig. 2. 6 Damage Bypass of Fracturing

(Source: http://scienceaccess. com/technique/fracking)

2. Improved Productivity

When we drill a well, we are creating a region of low pressure at the wellbore, and the conductive path is provided by the formation's permeability. In many cases, however, the pressure stored in reservoir is all used for hydrocarbons to flow to the wellbore, thus leaves little energy for them to flow to the well head. But things are different after fracturing. As shown in Fig. 2. 7, fluid flow pattern has changed from radial flow to liner flow after fracturing. Thus hydraulic fracturing leaves more energy available for bringing the oil and gas to the surface. It also reduces the minimum energy (i. e. pressure) required in the reservoir to achieve economic flow to the wellbore, thereby extending production beyond reserve levels that might otherwise be considered depleted. Plus, it minimizes secondary pressure-dependent effects such as water production, retrograde condensation within the reservoir, and non-Darcy flow.

2. 提高产能

钻井时,在井筒附近会有一个低压区,并且流体以储层孔隙作为导流通道流入井筒。然而,在很多情况下,储存在地层中的压力都被油气从地层流入井筒这一过程所消耗,留给油气流向井口的能量很少。但压裂后这一情况发生了变化,如图 2. 7 所示,油气的渗流状态从径向流变为线性流,从而使得油气井筒流的能量增加。水力压裂还降低了流体从地层到井筒经济流动所需的最小能量(即压力),使得产油气范围扩展到储层深处。另外,水力压裂还最大限度地减少了压力衰减相关的影响,如储层产水、反凝析现象和非达西渗流。

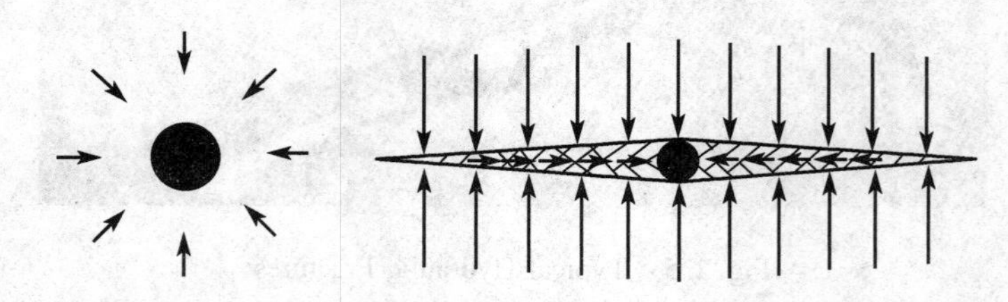

Fig. 2. 7 Change of Fluid Flow Pattern From Radial Flow to Liner Flow

3. Adoptable in Almost All Reservoirs

(1) Low Permeability Reservoirs. Originally suggested for low-permeability gas, hydraulic fracturing still plays a crucial role in developing low-permeability sandstone formations, and is increasingly used to produce from low-permeability carbonates, shales and coal seams. Low-permeability formations require stimulation because the permeability of the formation just isn't high enough for the well to produce naturally at economic rates. Although the reservoir may contain significant reserves, sufficient production cannot be obtained. In low-permeability formations, it is easy to produce a fracture that is many times more conductive than the formation. Fractures in low-permeability reservoirs are designed for length.

(2) High Permeability Reservoirs. High-permeability fracturing is primarily a near-wellbore flow enhancement technique. Usually, production is constrained by effects around the wellbore, caused by the phenomena of skin damage and the increasing constriction of flow as it approaches the wellbore. Fractures in high-permeability formations are designed to be short and highly conductive, which in turn usually means maximizing propped fracture width. And often its side effects-such as sanding prevention-might be the primary reason of application. In recent years high-permeability fracturing has become as significant in the economic sense as low-permeability fracturing.

3. 适用于几乎所有地层

(1)低渗储层。水力压裂技术最初被应用于低渗气藏增产，但现在也广泛应用于低渗砂岩地层，并且也越来越多地被应用于低渗碳酸盐岩、页岩和煤层增产。低渗储层由于渗透率较低，在不进行增产的情况下难以获得经济产能。尽管低渗储层可能拥有很大的储量，但却无法获得可观的产量。在低渗储层中，水力压裂很容易压裂出导流能力远远高于地层的人工裂缝。低渗储层水力压裂的重点在于控制缝长。

(2)高渗储层。对于高渗储层而言，水力压裂主要起增强近井地带流动的作用。通常，在高渗储层中，生产主要受到井眼附近的影响，这是由表皮效应和流体接近井眼时流量的急剧收缩引起的。一般要求高渗储层的人工裂缝短且具有高导流能力，也就是要扩大裂缝的宽度。通常水力压裂的附带作用(如防砂)可能是进行压裂的主要原因。近些年来，高渗储层压裂在经济效果上已经变得与低渗储层压裂一样重要。

2.1.2.3 Limitations to Hydraulic Fracturing

Just as the old proverb goes: every coin has two sides. Despite all the benefits hydraulic fracturing has, it still has its limitations. Hydraulic fracturing operations should only be considered when the:

(1) Well is connected to adequate producible reserves;

(2) Reservoir pressure is high enough to maintain flow;

2.1.2.3 水力压裂技术的局限性

万物皆有利弊，水力压裂技术也是如此。尽管水力压裂技术有其优越性，它同样也存在自身的局限性。水力压裂只应在以下情况下被使用：

(1)生产井附近拥有足够的可采储量；

(2)储层压力能保证生产；

(3) Production system can process the extra production.

(3)整个生产系统可以容纳水力压裂后的额外产量。

These minimum criteria are summarized in Table 2. 2. There is, however, one extra, unique requirement for propped hydraulic fracturing:

这些最基本的要求总结于表2.2。对于水力压裂技术的应用,还有一个额外的、独特的要求:

Professional, experienced personnel are available for treatment design, execution and supervision along together with high quality pumping, mixing and blending equipment.

有经验丰富的专业人员可以进行施工设计、执行和监督,以及现场必须拥有高品质的泵送、混合和搅拌设备。

Table 2. 2 Minimum Hydraulic Fracturing Candidate Well Selection Criteria

Parameter	Oil Reservoir	Gas Reservoir
Hydrocarbon Saturation	>40%	>50%
Water Cut	<30%	<200 bbl/10^6 ft^3
Permeability	1 ~ 50 mD	0.01 ~ 10 mD
Reservoir Pressure	<70% depleted	twice abandonment pressure
Gross Reservoir Height	>10 m	>10 m
Production System	20% spare capacity	

This latter requirement arises because a hydraulic fracturing treatment has a complexity and difficulty an order of magnitude greater than any other treatments the industry has ever experienced. And the complexity as well as difficulty arise from the inherent multidisciplinary nature of the fracturing process. Proper treatment design is tied to several disciplines:

(1) production engineering; (2) rock mechanics; (3) fluid mechanics; (4) material mechanics; (5) operations.

之所以有这个要求是因为与其他技术相比,水力压裂技术在各个方面都复杂得多、困难得多。这些复杂性和困难性主要来源于水力压裂技术自身的多学科交叉特性。良好的压裂设计与多个学科息息相关,包括:(1)油气开采工程;(2)岩石力学;(3)流体力学;(4)材料力学;(5)现场实施工程。

Also, the design must be consistent with the physical limits set by actual field and well environments. Because of this absolutely essential multidisciplinary approach, there is only one rule of thumb in fracturing: that there are no rules of thumb in fracturing. Thus the ability to complete the treatment to the specified design requires numerous, on-site adjustments during the treatment (Fig. 2. 8).

同时,水力压裂设计还必须与现场环境和井筒限制一致。也正是由于水力压裂的这种多学科特性,人们总结出了一个经验法则,即水力压裂没有经验法则可遵循。因此,必须经过现场的无数次调整才能达到设计所要求的水力压裂施工效果(图2.8)。

Fig. 2. 8　Fracturing Operations on Site
(Source: https://commons. wikimedia. org/w/index. php? curid = 27166623)

2.1.3　Hydraulic Fracturing Operation and Evaluation

After knowing what is hydraulic fracturing and its role in oil industry, it is time to understand how exactly a hydraulic fracturing operation is carried out. Afterwards, what concerns us the most is the evaluation of the operation.

2.1.3　水力压裂施工与评估

在了解了什么是水力压裂技术及其在石油工业中的作用之后，本部分介绍水力压裂施工，以及施工之后对压裂作业的评估。

2.1.3.1　Hydraulic Fracturing Operation

The major steps needed to carry out a hydraulic fracture treatment are summarized as follows:

(1) Pumping the fracturing fluid at a sufficiently high pressure to overcome the rock stresses i. e. initiate a fracture.

(2) Continue pumping to ensure efficient fracture propagation.

(3) The created fracture is then filled with proppant to hold it open and provide conductivity for fluid flow when fluid pumping is halted.

(4) The viscous fracturing fluid is degraded after the treatment to a viscosity similar to that of water by incorporation of a chemical breaker into the fracturing fluid formulation. This will allow it to be produced back after the treatment, followed by the initiation of hydrocarbon production.

Fig. 2. 9 summarizes the main stages in the process involved in creation of a propped hydraulic fracture. And the surface and well set up required to achieve the above is schematically illustrated in Fig. 2. 10.

2.1.3.1　水力压裂施工

水力压裂施工主要步骤有：

(1) 在足够高的压力下泵入压裂液以克服岩石应力产生裂缝，即裂缝起裂。

(2) 持续泵入压裂液以实现裂缝的扩展。

(3) 向人工裂缝中填入支撑剂以保证施工结束后裂缝张开并提供地层流体流动的通道。

(4) 将破胶剂加入压裂液中，压裂液逐渐降解。这样压裂完成后压裂液就能返排出地层并开始油气生产。

图 2.9 总结了水力压裂的主要阶段。图 2.10 为进行水力压裂所需的地面和井口安装的设备。

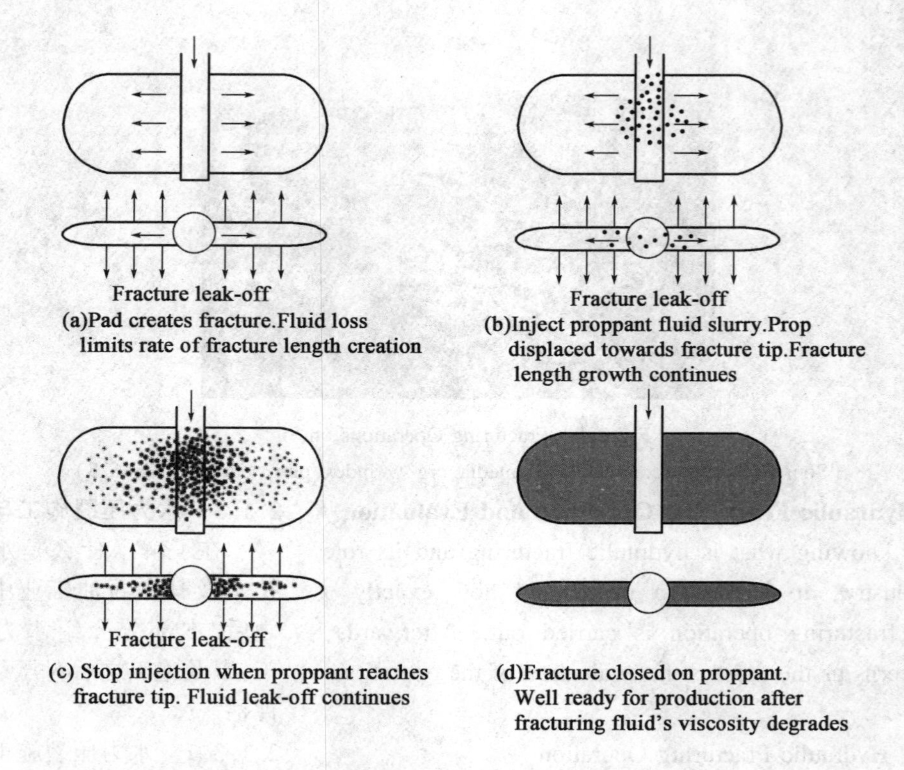

Fig. 2. 9 Creation of a Propped Hydraulic Fracture

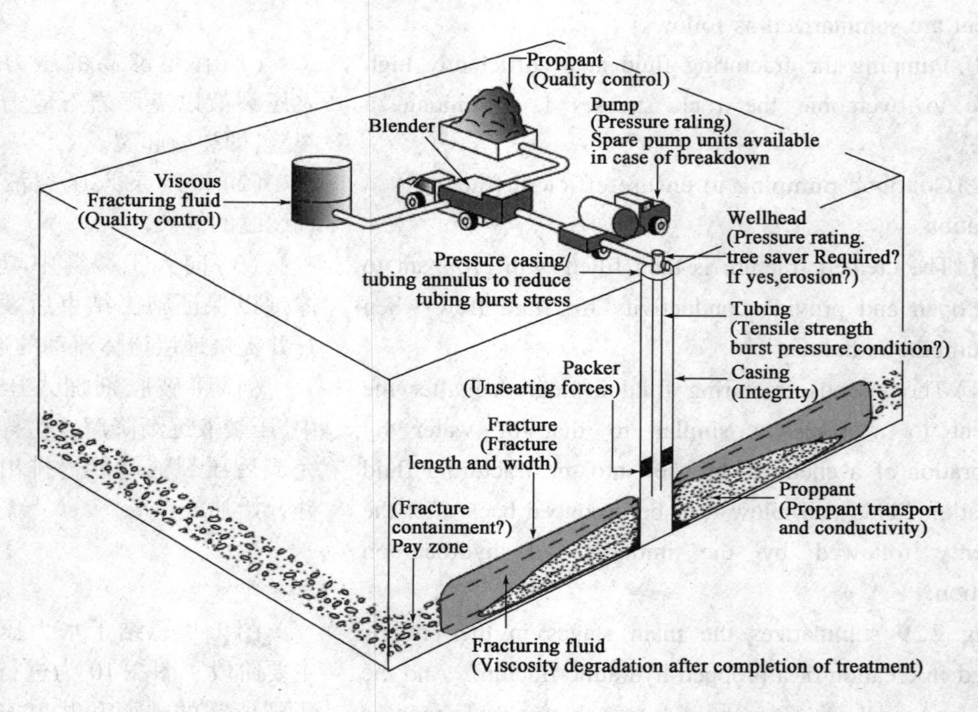

Fig. 2. 10 A Hydraulic Fracturing Operation Illustration

(Source: https://eogresources. com/responsibility/hydraulicfracing. html)

The created length of the fracture is considerably longer than the propped length since the fracturing treatment is still in progress. Issues to be evaluated during the design include:

压裂时的缝长要比支撑缝长(有效裂缝长度)更大,这是因为压裂结束后裂缝仍然会扩展一段距离。水力压裂过程中需要考虑的主要问题有:

(1) transport of the proppant to the fracture tip;

(1)支撑剂能运移至裂缝端部;

(2) settling of proppant due to inadequate fracturing fluid viscosity;

(2)压裂液黏度不足引起的支撑剂沉降;

(3) creation of the required proppant pack width and;

(3)压出设计的缝宽;

(4) degradation of the fracturing fluid to minimize permeability damage to the proppant pack and formation.

(4)压后压裂液的降解,使其对支撑剂层和地层渗透率的伤害最小。

2.1.3.2 Hydraulic Fracturing Evaluation

2.1.3.2 水力压裂评估

Before and after a hydraulic fracturing treatment, the most concerned issue is how much productivity can be achieved by fracturing. The main factors affecting the productivity of a fracturing well lay in the reservoir itself (e. g. permeability) and fracture parameters (i. e. Fracture length, fracture width, fracture height and fracture conductivity). By analyzing the inflow performance of a fractured well in the following text, the relationship between productivity versus formation parameters and fracture parameters is obtained.

无论是在水力压裂设计阶段还是评估阶段,人们最关心的问题就是压裂后能带来多少产能。影响压裂井增产幅度的主要因素是储层本身(储层渗透率)和裂缝参数(缝长、缝宽、缝高、裂缝导流能力)。本部分通过分析压裂井的流入动态,介绍产量和油层特性与裂缝参数之间的关系。

The inflow performance of a fracture stimulated well is controlled by the dimensionless fracture conductivity (F_{CD}):

压裂井的流入动态由无因次裂缝导流能力(F_{CD})所控制:

$$F_{CD} = \frac{K_f w}{K L_f} \tag{2.2}$$

where K_f is the fracture permeability, μm^2; w is the fracture width, m; $K_f w$ represents fracture conductivity, μm^2 · m; K is the reservoir permeability, μm^2; L_f is the half length of fracture, m.

其中 K_f 是水力裂缝渗透率,μm^2;w 是裂缝宽度,m;$K_f w$ 代表裂缝导流能力,μm^2 · m;K 是储层渗透率,μm^2;;L_f是裂缝半长,m。

These parameters are illustrated in Fig. 2. 11. Therefore, the fracture conductivity is increased by:

这些参数列在图 2. 11 中。因此,增加裂缝导流能力可以通过以下方式:

(1) an increased fracture width (w),

(2) an increased proppant permeability (large, more spherical, proppant grains have a higher permeability) and,

(3) minimizing the permeability damage to the proppant pack from the fracturing fluid.

(1)增加缝宽 w；

(2)增加支撑剂导流能力(粒度大、球度好的支撑剂能得到高的渗透率)；

(3)减少压裂液对支撑剂层渗透率的伤害。

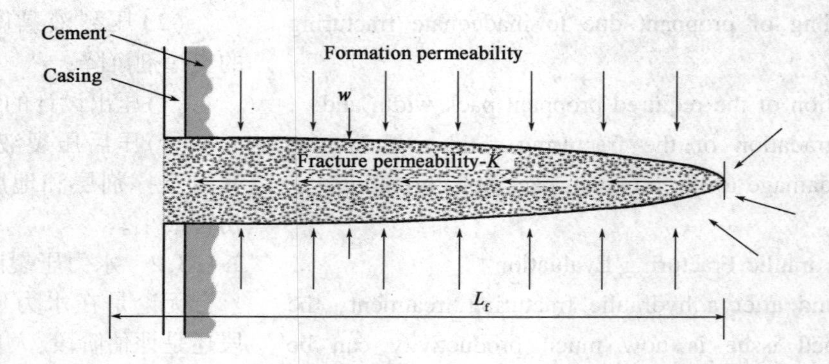

Fig. 2. 11 Factors Contributions to the Dimensionless Fracture Conductivity

Frequently the increased production achieved by carrying out a hydraulic fracturing treatment is represented by the "Folds of Increase" or FOI:

表征水力压裂后产量增加的参数称为增产倍数(FOI)：

$$FOI = \frac{J_f}{J_0}\left(\frac{7.13}{\ln 0.472 R_e / r_w}\right) \tag{2.3}$$

where J_0 and J_f are the well production before and after carrying out the hydraulic fracturing treatmeat.

式中 J_0 和 J_f 分别指压裂前后井的产量。

There have been several studies of the composite effect of fracture length, fracture conductivity and formation permeability on the well inflow performance. A widely used correlation is that published by Cinco-Ley and Samaniego in which r'_w/L_f, or effective wellbore radius divided by conductive fracture length, is plotted against the dimensionless fracture conductivity (F_{CD}). This is illustrated in Fig. 2. 12. This figure shows that an F_{CD} value of 15 is required to ensure that the well inflow is not being limited by the fracture conductivity.

已经有许多关于裂缝长度、裂缝导流能力和储层渗透率对压裂井流入动态影响的研究，最为人们所接受的是 Cinco-Ley 和 Samaniego 提出的，即有效井筒半径除以裂缝半长 r'_w/L_f 与无因次裂缝导流能力 F_{CD} 的关系。这一关系如图 2. 12 所示。该图表明，F_{CD}需要达到 15 以确保压裂井流入动态不受裂缝导流能力的限制。

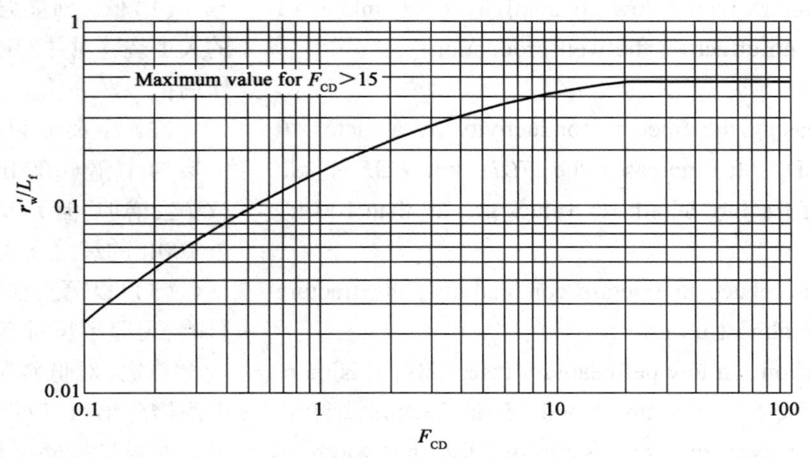

Fig. 2.12 Cinco-Ley and Samaniego's 1981 Correlation between Effective Wellbore Radius and Fracture Conductivity

The above correlations and equations can be used to quantify the relationship between the increased production (*FOI*) as a function of the fracture length L_f, formation permeability K and the fracture conductivity $K_f w$, see Fig. 2.13. Fig. 2.13 shows that for wells in low permeability (0.1 mD) formations:

(1) High values of the *FOI* are possible;

(2) *FOI* is related to fracture half length, while the fracture conductivity has a limited effect, providing its value is greater than a certain minimum.

上述的相关性和方程可用于量化增产倍数 *FOI* 与裂缝长度 L_f、地层渗透率 K 和裂缝导流能力 $K_f w$ 之间的关系，如图2.13所示。对低渗储层（渗透率小于0.1 mD）而言：

（1）高 *FOI* 是可能的；

（2）*FOI* 与裂缝半长相关，而裂缝导流能力对 *FOI* 的影响很小。

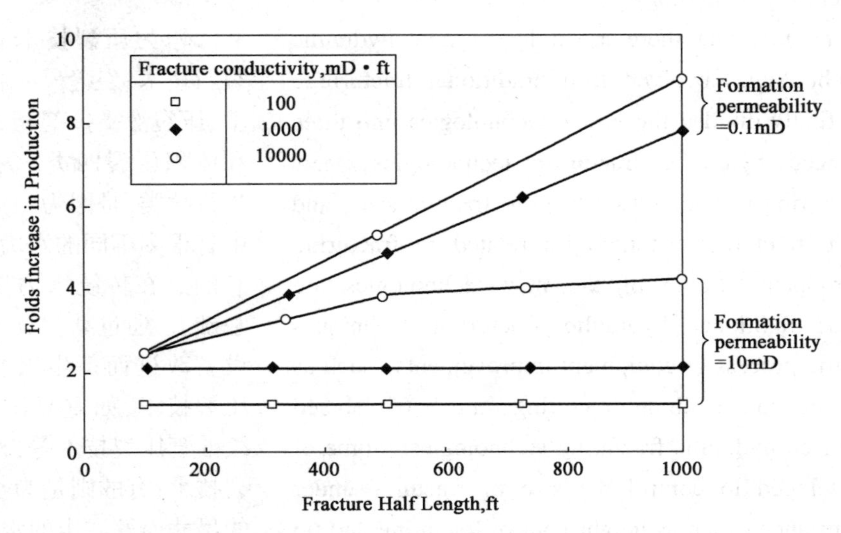

Fig. 2.13 Well Productivity Response to Hydraulic Fracturing

An increase in the formation permeability to 10 mD results in a different picture:

当地层渗透率上升到10mD时，情况则完全不同：

(1) A fracture with a low conductivity (100 mD · ft) has essentially no effect on the well production;

(2) Increasing the fracture conductivity by a factor 10 (to 1,000 mD · ft) increases the *FOI*; but *FOI* is still independent of fracture length for values greater than 100 ft;

(3) *FOI* is related to fracture conductivity, the fracture half length has almost no effect.

In conclusion, in low permeability reservoirs, it is more beneficial to increase fracture length than fracture width. While for high permeability reservoirs, fracture width is much preferred.

(1) 低导流能力(100mD · ft)的人工裂缝对井的产量没有任何作用;

(2) 若人工裂缝的导流能力增加十倍(1000mD · ft),则 *FOI* 会增加,但 *FOI* 仍与长度超过 100ft 的裂缝无关;

(3) *FOI* 受裂缝导流能力的影响,裂缝半长对其影响很小。

总之,对低渗储层而言,要提高增产倍数,应以增加缝长为主。而对高渗储层正好相反,应以增加裂缝导流能力为主。

2.2 Advanced Hydraulic Fracturing Technologies

From hydraulic fracturing to waterfree fracturing (i. e. non-aqueous fracturing), from vertical well fracturing to horizontal well staged multi-cluster fracturing, and from conventional fracturing to volume fracturing, hydraulic fracturing technology is in the process of continuous development. Starting from this section, the latter part of this chapter will introduce various types of hydraulic fracturing technology.

Audio 2.2

Given the fact that there are a lot of new hydraulic fracturing technologies evolved from traditional fracturing, the textbook firstly divided these new technologies into three types: advanced hydraulic fracturing technologies, new hydraulic fracturing technologies related to fracture size, and new hydraulic fracturing technologies related to fracturing fluid and proppant. Some of the new technologies are derived from traditional hydraulic fracturing techniques through specific process or equipment improvements, such as re-fracturing and coiled tubing fracturing, the book labeled them as advanced hydraulic fracturing technologies. Some of them are developed to control the size of fractures under certain circumstance, such as height-control fracturing and tip

2.2 常规压裂技术新进展

从水力压裂到无水压裂,从直井压裂到水平井分段多簇压裂,再从常规压裂到体积压裂,水力压裂技术正处于不断发展的阶段。从本节开始,将对各类水力压裂工艺技术进行介绍。

水力压裂技术分为常规压裂新技术、裂缝尺寸相关压裂技术、压裂液支撑剂相关压裂技术和体积压裂四类。之所以这样进行分类,是因为当前石油工业中有很多不同的水力压裂技术。有的是在传统水力压裂技术的基础上,通过某些具体的工艺或设备改进而衍生出的新的水力压裂技术,如重复压裂技术、连续油管压裂技术等,称为常规压裂技术;有的则是为了控制裂缝的尺寸而发展出的水力压裂新

screen out (TSO), the book labeled them as hydraulic fracturing technologies related to fracture size. While some other new technologies are formed through the development of fracturing fluids and proppants, such as waterfree fracturing and hiway fracturing, labeled as hydraulic fracturing technologies related to fracturing fluid and proppant. Ultimately, Chapter 3 introduced hydraulic fracturing technologies in horizontal wells, especially volume fracturing.

技术,如控缝高压裂技术、端部脱砂压裂技术等;还有的则是依靠压裂液和支撑剂的进展而发展出的水力压裂新技术,如无水压裂技术、高速通道压裂技术等。本书第3章将介绍水平井压裂技术,特别是体积压裂。

2.2.1 Multilayer Fracturing

Vertical wells usually penetrate a number of potential producing zones. The placement of the hydraulic fracture treatment in these instances is often accomplished by pumping multiple fracture treatments (Fig. 2.14). The perspective reservoir is divided into discrete layers and hydraulically fractured separately to place the desired treatment in the most productive intervals and lessen the potential to fracture nonproductive reservoir rock.

2.2.1 分层压裂

直井经常会穿过一系列的储层,因此针对这种情况就需要进行多次水力压裂(图2.14)。分层压裂就是将储层分为多个独立的储层,对最有潜力的储层进行较大规模的改造,而对于潜力较小的储层则进行小规模改造,或封堵某些易产水层而仅压裂具有潜力的储层。

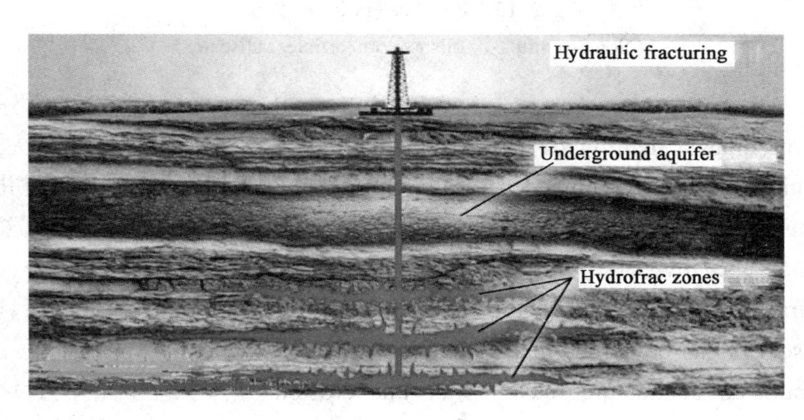

Fig. 2.14 Multilayer Fracturing in Vertical Wells
(Source: https://petrodomain.com)

While the number of stages that should be performed and the interval that can be fractured are sometimes in question and depend on a number of reservoir and rock characteristics, the ability to multilayer fracture is not in question. Multilayer fracturing has become a technically sound and economic method to stimulate large pay intervals.

当前,虽然需要进行压裂的层段数与地层和岩石参数有关,但分层压裂工艺所要达到的分段改造级数已不存在任何问题,分层压裂已成为在传统压裂技术基础上发展起来的较为成熟且经济的压裂技术。

2.2.1.1 Multilayer Fracturing with Mechanical Isolation

1. Ball Sealers

Ball sealers (Fig. 2.15) have been used successfully during hydraulic fracturing in cased-hole applications for many years. When the interval to be multilayer fractured is large enough that multiple perforations are scattered over a long producing interval, or downhole conditions prevent the use of plugs or other mechanical diversion techniques, ball sealers could be considered during hydraulic fracturing operations.

2.2.1.1 机械法分层压裂

1. 封堵球分层压裂

封堵球分层压裂(图2.15)已成功应用于套管井的分层压裂作业中。当压裂层段的间隔足够大、多个射孔段分散在较长的产层中时,或井下状况不允许使用桥塞或其他机械封隔设备,则建议使用封堵球进行分层压裂。

Fig. 2.15 Ball Sealers in Wellbore
(Source: https://oilsns.com/article/ballsealer)

Although ball sealers have been used successfully and economically at times, their wide use as a multilayer fracturing technique is not recommended. Ball sealers, like diverting agents, do not allow the determination of the interval at which the diversion occurred nor whether the subsequently fractured interval was the desired one. In addition, ball sealers can be eroded due to the erosive nature of proppants pumped at fracture rates. After this erosion occurs, the perforation initially plugged with the ball sealer may once again accept fracturing fluid.

尽管封堵球分层压裂是非常经济的一种方法,且有很多成功应用的实例,但并不推荐大规模使用该方法。这是因为封堵球跟暂堵剂一样,在具体施工时,不能确定封堵具体发生的层位,也不能确定下一个压开的具体层位。此外,封堵球还可能受到携砂液的冲蚀逐渐失效,使得压裂液滤失进被封堵层位导致封堵失败。

If ball sealers must be used, several guidelines have been established for their application. The ball sealer should be larger than the perforation they are to seal. The ball sealer should also be of adequate specific gravity that it will not float in the selected fracture fluid, and it should fall to the bottom of the well after fracturing. The ball sealer should also be sufficiently durable to minimize deformation during the fracture stimulation, thereby preventing the ball sealer from partially or completely entering the perforation and

如果必须得使用封堵球进行分层压裂,则必须遵循一定的原则。首先,封堵球直径要大于射孔孔眼直径。其次,封堵球必须具备一定的密度,这样就能保证在压裂过程中封堵球不漂浮在压裂液中,施工结束后可以沉到井底。最后,封堵球还必须具有一定的强度,这样就能减少在

causing a permanent block of the perforation and potentially restricting production after the treatment.

封堵过程中的变形,避免出现部分或整个进入射孔孔眼,影响压裂后的生产。

2. Bridge Plugs

The early days of multilayer fracturing saw the use of plugs and packers to isolate the zone of interest. Tools were developed with dual packers (Fig. 2. 16) or more, which permitted the treatment of a producing interval between them. This method employed the use of tubing and would therefore limit (1) the number of stages which could be effectively treated due to the limited number of packers which could be simultaneously deployed and (2) the rate at which the stimulation treatments could be pumped due to high friction associated with pumping fluids through small tubulars.

2. 桥塞分层压裂

早期的分层压裂技术都是使用桥塞和封隔器进行分层。人们设计了拥有多个封隔器(图2.16)的井下工具,以便在封隔器之间进行水力压裂。这种方法的局限性在于:(1)借助了油管,并且由于封隔器数目有限会导致能封隔的层段数有限;(2)由于压裂液通过较小的管道导致磨损压力增加,因此水力压裂的速度受到限制。

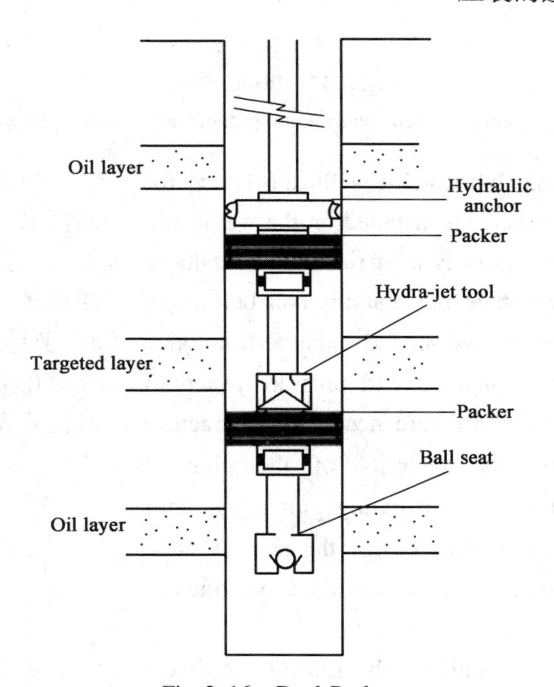

Fig. 2. 16 Dual Packers

(Source: https://51wendang. com/doc/4be4b522131e26a3e1803bd9/11)

The most prevalent method of multilayer fracturing today uses bridge plugs set inside the casing (Fig. 2. 17). The lowest or deepest interval (first stage) in the well is perforated first. The perforations in this zone may be pumped into and broken down with acid and tested for

目前分层压裂最常用的方法是使用套管内的桥塞封隔器(图2.17)。最下方(第一层)的层段首先进行射孔作业,注入酸液进行解堵,并在水力压裂之前

injectivity prior to mobilizing fracture equipment to the location. Then the designed fracture treatment is performed. Upon completion of the first stage, the treatment is flushed to the top perforation with acid and water or gelled water.

测试其注入能力。之后再在该层段进行水力压裂作业,作业完成之后,用酸和水(或凝胶水)进行冲洗。

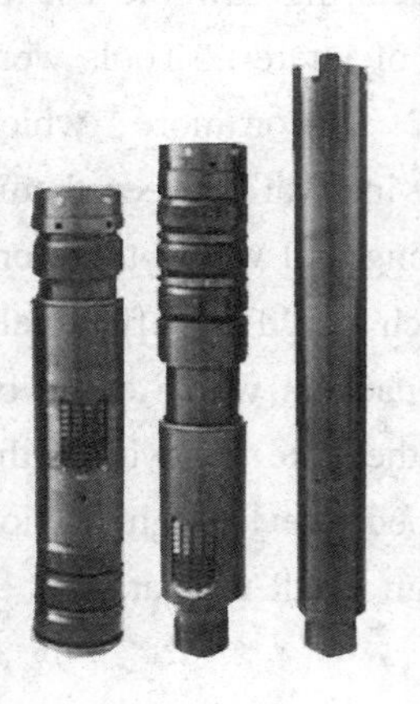

Fig. 2. 17 Bridge Plugs
(Source: https://ychuayuan. com/product. asp/Product_ID = 663)

Meanwhile, a wireline lubricator (wellhead device for the control of wellhead pressure) is installed on the wellhead. A wireline-conveyed bridge plug is then run in the wellbore and set just above the first stage perforations and below the interval selected for the successive second stage perforations. The wireline bridge plug is then set and pressure tested to verify that it will support the pressure necessary to fracture the ensuing stage. After the integrity of the plug is confirmed, a wireline-conveyed perforating gun is run into the wellbore and the casing perforated for the second stage interval. This second stage interval can then be fracture stimulated.

与此同时,在井口安装有线润滑设备(用于控制井口压力)。之后,电缆输送桥塞封隔器至第一层和上方第二层段之间,坐封,测压。测试完成后,下入射孔枪进行第二层段的射孔。射孔完成后进行压裂作业。

This process is repeated until all the selected reservoir intervals within the well are fracture stimulated. It is not uncommon to complete as many as six to seven stages using wireline-conveyed bridge plugs and perforating guns.

重复上述步骤直至压裂完所有层段。一般可以使用该方法进行六至七次分层压裂作业。

Many kinds of bridge plugs are available for use in multilayer fracturing. Most fracturing plugs are classified by the material with which they are made. Bridge plugs can be retrievable for reuse after the fracture stimulation is complete

能进行分层压裂的桥塞有很多种,一般通过制造材料对它们进行分类,分为可钻式桥塞和可取式桥塞,如图 2.18 所示。

or expendable (drillable), as shown in Fig. 2. 18. Recent technology has incorporated the use of flowthrough bridge plugs, fitted with ports or flappers that allow flow in one direction, thus allowing the retrieval or removal of the bridge plug to be delayed until the reservoir has been allowed to flow back. Drillable bridge plugs can be made of soft metal alloys or of composite material, enabling the drilling of the plug in minutes.

现在还多使用流通式桥塞,即在桥塞内部加入单流阀使得流体只能沿一个方向流动,这样桥塞在整个压裂完成后再被取出或钻穿。可钻式桥塞一般由软合金或复合材料制成,可在几分钟内被钻穿。

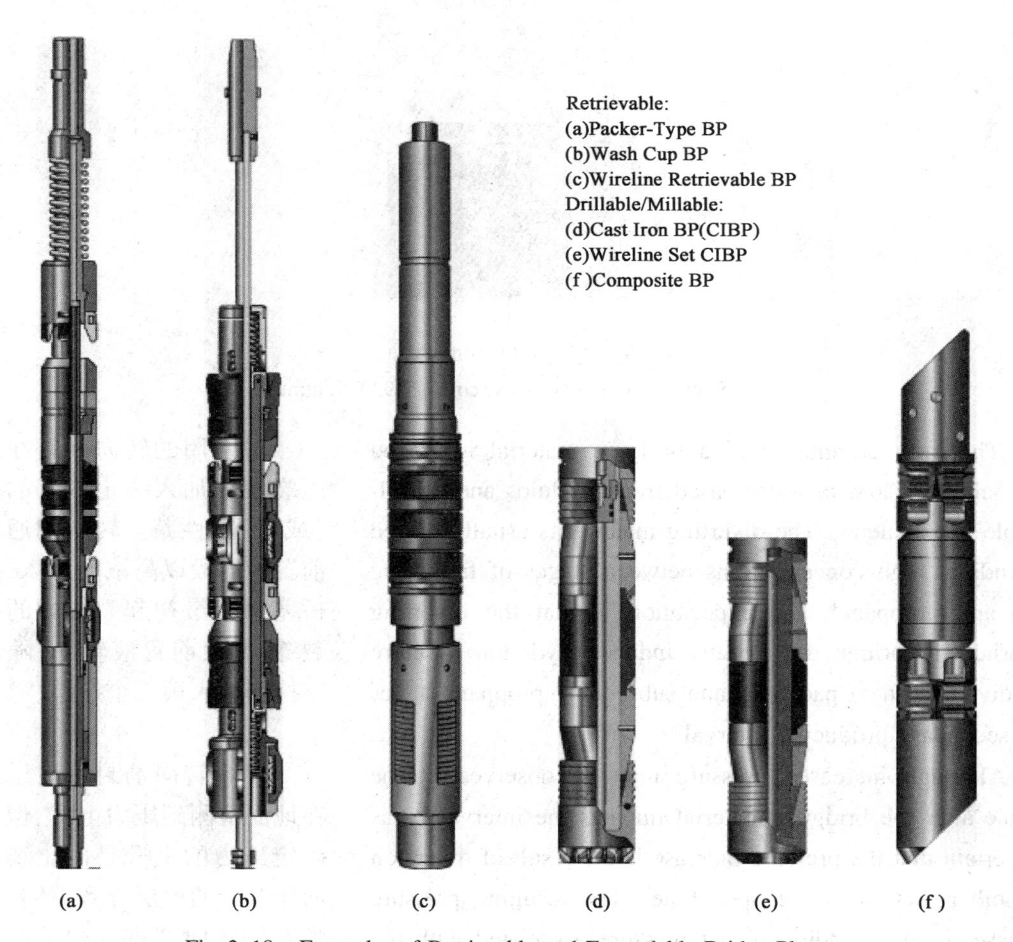

Fig. 2. 18 Examples of Retrievable and Expendable Bridge Plugs
(Source: https://onepetro. org/journal-paper/SPE – 2453 – PA)

2. 2. 1. 2 Multilayer Fracturing with Chemical Isolation

The use of particulate diverting agents (Fig. 2.19) during multilayer fracturing has been usually applied in openhole or uncased sections. The practice is not often recommended. It is only considered when the section is large enough that it would be impossible to mechanically isolate.

2. 2. 1. 2 化学法分层压裂

化学法分层压裂使用颗粒状转向剂(暂堵剂,图 2. 19)进行分层压裂,多应用于裸眼井或未射孔井段。只有当需要压裂的层段过大以至于无法进行机

When this scenario is encountered, particulate materials can be used to bridge the face of the open-hole section or the initially induced fracture, thus diverting treatment fluids to another section.

械封堵时才建议使用该方法。当进行化学法分层压裂时,转向剂封堵裸眼层段或最初压开的裂缝,这样压裂液就会转向流入别的地层。

Fig. 2. 19　Diverting Agents
Source: https://youboy. com/s5193075. html)

The most commonly used bridging material is graded rock salt (NaCl) with water-based fracture fluids and/or oil-soluble naphthalene. The diverting material is usually placed in fluids at high concentrations between stages of fracturing fluid and proppant. The expectation is that the diverting material will bridge the initially induced hydraulic fracture and divert the next pad fluid and subsequent proppant stages to a secondary producing interval.

最常用的转向剂是在水基压裂液中加入一定浓度的盐和(或)油溶性萘。转向剂通常在前置液后段以高浓度注入,从而桥塞封堵最初起裂形成的人工裂缝,导致前置液转向,随后的支撑剂进入第二个产层。

Although increased pressure may be observed at the surface after the bridging material impacts the interval, it is not certain that the pressure increase is the result of diversion to another section of the open hole. The resulting pressure increase could be additional net pressure associated with the larger particulate material entering the first induced fracture. The increased pressure could also be a temporary diversion that occurs until such time that the abrasive proppant erodes the bridging material.

尽管转向剂封堵地层后会在地面检测到压力上升,但无法确定压力的上升是由压裂液转向压开新的地层导致,还是由大颗粒的转向剂进入初始裂缝引起的额外净压力;压力的上升还可能是压裂液暂时的转向,最后转向剂可能会被支撑剂磨蚀而失去封堵效果。

2.2.1.3 Limited Entry

Limited entry is, as the name implies, a method by which the number of perforations is limited within a given wellbore interval to be fracture stimulated. Limited entry is a proven and technically sound method to divert fluids from one segment of an interval to another.

2.2.1.3 限流法分层压裂

顾名思义，限流法分层压裂就是通过限制特定层段的射孔孔眼数目，来达到分配进入各个层段内压裂液液量的目的。限流法分层压裂是一种技术上成熟的、可靠的分层压裂方法。

In order to treat a specific perforated interval in a well, the pressure must be increased such that it exceeds the fracturing pressure of the reservoir to be stimulated. By limiting the number of perforations in the wellbore and their diameter, the friction pressure across the perforations varies with the rate in which the fracture fluid is being pumped. When the fracture rate is increased, the perforation friction pressure also increases, causing a restricted flow or "choke effect" at the perforation and subsequently increasing the differential pressure between the wellbore and reservoir. This increased differential pressure can cause additional perforations or intervals to accept fluid or divert the fluid to another interval (Fig. 2. 20).

为了对某一特定的射孔层段进行压裂，压力必须超过该层段的破裂压力。在施工中，限制不同层段射孔孔眼数及其直径，这样流经这些射孔的摩阻是不同的。当压裂液排量增加时，射孔孔眼摩阻随之增加，导致射孔孔眼的限流作用（或者说节流效应），这样进入不同层段的压裂液液量就不同，从而达到限流分层压裂的目的（图 2. 20）。

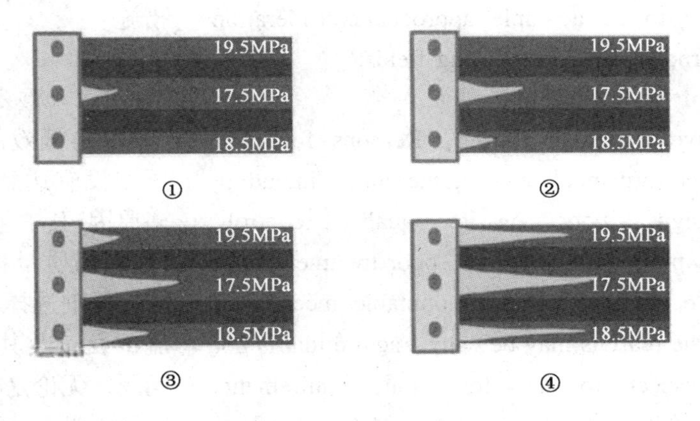

Fig. 2. 20 Limited Entry

The key to the limited entry is to design a reasonable perforation scheme, including the number, density and aperture of the perforating hole. The design of limited entry fracs is often an iterative process until the perforation scheme is achieved that satisfies the different intervals with different fracturing fluid distribution. For example, the zone with the lower fracture pressure would have the less perforations

限流法分层压裂的关键在于设计合理的射孔方案，包括射孔孔眼数、孔密和孔径。限流法分层压裂设计常常是一个重复验证的过程，需要经过不断的重复计算，直到获得不同层段具有不同量压裂液的射孔方案。例

compared with a higher fracture pressure zone with more volume of fracturing fluid.

如,在进行限流法分层压裂时,与地层压力较高的层段相比,地层压力较低层段的射孔数应较小,这样就会分配更多的压裂液进入较高地层压力的地层。

The Limited entry does have several limitations. First, the effects of erosion by proppant are difficult to quantify which may lead to alter the distribution of fracture fluid. Secondly, fracture pressure gradients of rock in different formation zones are usually estimated and cannot be measured until the perforations are shot. If the estimated values are substantially different from those encountered during pumping, the limited entry design can be grossly inaccurate.

限流法分层压裂同样具有局限性。首先,很难量化支撑剂对射孔孔眼的冲蚀程度,这种冲蚀将会导致压裂液分配的改变。其次,不同层段岩石破裂压力梯度值通常都是估测的,这些估测值若与实际不符,则整个限流法分层压裂设计就会存在很大的偏差。

2.2.2 Re-Fracturing

2.2.2 重复压裂

Given that a formation was suitable to be developed with a hydraulic fracture treatment during the initial completion phase, there is always the possibility that it will become a re-fracture candidate at a later date. Re-fracturing operations continue to be a staple approach/consideration when enhancing production from existing fields.

如果一个地层进行了初次水力压裂,那么它很有可能也非常适合进行重复压裂。重复压裂是油气田提高产量的主要措施。

2.2.2.1 Cause of Re-Fracturing

2.2.2.1 重复压裂的原因

(1) Ineffective Initial Fracturing: Reasons for poor initial fracture performance are numerous, including inappropriate design, poor on-site quality control, inappropriate fluid/proppant choice(s), poor treatment fluid recovery and the occurrence of insurmountable mechanical issues. Whatever the reasons may be, any fracture that is not optimum with respect to the formation requirements potentially presents itself as a candidate for re-fracture consideration.

(1)低效的初次压裂:初次压裂没有达到理想的效果有很多原因,其中包括压裂设计不合理、现场质量控制不达标、压裂液支撑剂选择不合理、压裂液回流不及时及其他不可控的机械问题。无论是哪种原因,如果初次压裂没有达到理想的效果,则该地层可以作为重复压裂的候选层。

(2) Transient Damage Effects (Fig. 2.21): Transient damage effects can reduce the reservoir performance with time. Scales and fines can invade and block the existing proppant pack and reduce the effective fracture conductivity, or long-term continuous proppant flowback may eventually negate the fracture potential. Once again, any fracture whose in situ properties have degraded to the point that it has become

(2)瞬态损伤效应(图2.21):随着油气藏的生产,瞬态损伤效应会减弱储层的生产能力。瞬态损伤指储层微粒运移并堵塞支撑剂之间的孔隙,降低支撑裂缝的导流能力;支撑剂回流进入井筒也会降低裂缝的生产能力。

the choke between the formation and the wellbore is a potential re-fracture candidate.

因此,当裂缝的性质降到使其成为地层和井筒之间的阻碍时,它也是一个潜在的重复压裂候选层。

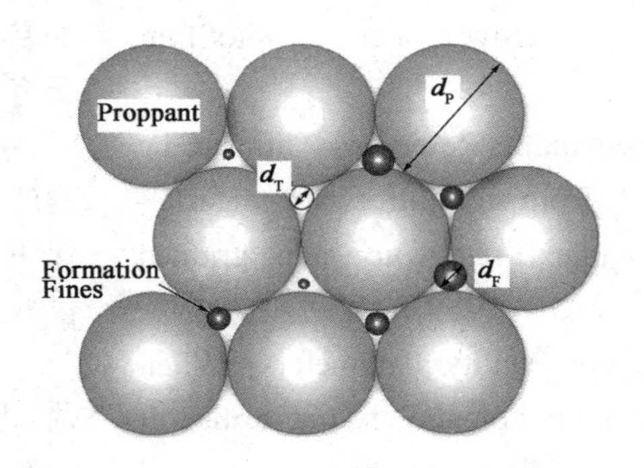

Fig. 2. 21 Particle Plugging within Proppant Pack

d_P—Proppant diameter; d_F—Formation fines diameter;

d_T—Maximum diameter of the inscribed circle in the unfilled pore space

(3) Transient Reservoir Properties (Fig. 2. 22): A reduction in pore pressure will create a change in stresses leading to increased effective stress on the proppant pack; it is quite feasible that the proppant pack may deteriorate under these conditions as proppant crushes and conductivity reduces. Another example would be in a condensate banking environment, where it is possible/probable that the original fracture treatment was not adequately designed to cope with the unusual flow conditions and fracture requirements to cope with two-phase flow. Under these conditions, the initial fracture (although potentially being optimal), does not provide an efficient solution for late-life field/well behavior.

(3)瞬态储层特性(图2.22):随着生产的不断进行,孔隙压力的降低会导致作用在支撑剂上的应力增加,从而使支撑剂受到挤压破坏乃至降低裂缝导流能力。另外,在凝析油地层中,初始裂缝难以应付生产后期流体的多相流动。在这些情况下,初始裂缝(一开始能够提供合理的生产)难以为油田或油井后期生产提供有效的油流通道。

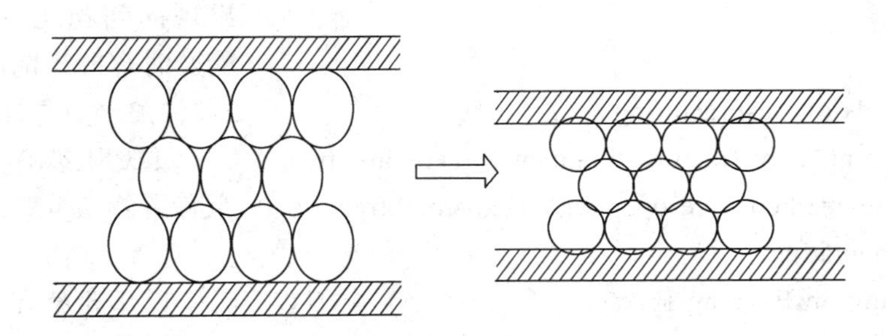

Fig. 2. 22 Proppant Crushes

2.2.2.2 Candidate Selection of Re-Fracturing

Candidate selection for re-fracturing is just as reliant on input data quality and availability as is candidate selection for an initial fracturing treatment. Conway et al. (1985) determined that candidate consideration for potential re-fracturing operations could be readily broken up into four simpler stages:

(1) Estimation of the Remaining Reserves: This can be achieved in a variety of ways, from straightforward material-balance approaches to complex reservoir and gridded simulations.

(2) Assessment of Fracture Quality/Parameters: When performing candidate selection for potential re-fracturing, it is most important to be able to differentiate between a good well with a bad fracture and a bad well. There are also a lot of methods that provide an assessment of the fracture parameters.

(3) Evaluate Fracturing Parameters: There have been numerous reported cases of fracture data-set manipulation aimed at determining either poorly fracture stimulated wells or common factors to success. The quality and availability of the data input must be carefully managed.

(4) Evaluate Economics based on Drivers and Costs: Clear understanding of the economics is essential for a re-fracturing campaign. All of the drivers and goals should be assessed before the program begins. Even more importantly, as the re-fracture program develops, the measured results must be quickly adjusted and compared to expectation. One of the recurring faults during re-fracturing campaigns is that they fail to learn from the present campaign rapidly enough.

2.2.2.2 重复压裂选井

重复压裂的选井和初次压裂一样,依赖地层数据的质量和可用性。Conway 等人(1985)在重复压裂选井所需考虑情况的基础上,提出重复压裂选井主要有以下四个步骤:

(1)剩余储量的估算:用准确的方法估算出剩余储量,以确保重复压裂后有足够的油气可供开采。

(2)裂缝质量(参数)的评估:当进行重复压裂选井时,能够区分具有不良水力裂缝的好井和不良井是最为重要的。有很多方法可以进行裂缝评估。

(3)评估压裂参数:已有许多案例显示地层数据对压裂的成功与否具有重要的影响。因此,必须谨慎地管理数据输入的质量和可用性。

(4)基于成本评估经济效益:清楚地了解成本和收益对重复压裂而言至关重要。在进行重复压裂之前必须了解所有的驱动因素和目标,更重要的是,随着重复压裂施工的进行,必须快速调整结果并将其与预期进行比较。在重复压裂期间反复出现的问题之一就是施工方未能及时掌握当前的压裂情况。

2.2.2.3 Methods of Re-Fracturing

When designing a re-fracture treatment, there are two distinct re-fracture methods: re-opening an existing fracture and fracture reorientation.

1. Re-Opening an Existing Fracture

If the initial fracture is simply being re-opened, then a close examination of the original fracture treatment must be

2.2.2.3 重复压裂方式

重复压裂有两种方式:压开原有裂缝和重新定向造出新缝。

1. 压开原有裂缝

如果重复压裂只是重新打开初始裂缝,则必须仔细检查原

made to see whether there is significant improvement can be achieved. If not, then this well may not be a suitable re-fracture candidate. However, if suitable improvements in the treatment design can be made, then the treatment may well be successful if sufficient recoverable reserves remain to justify the expense of the re-fracture stimulation.

有裂缝以确定重复压裂能否带来显著的增产效果。如果不能，则该井可能不适合的重复压裂。然而，如果剩余可采储量证明重复压裂的投入是合理的，那么重复压裂很有可能会很成功。

With this type of treatment, it is highly recommended to flush the existing proppant away from the wellbore, and subsequently perform a minifrac to evaluate tortuosity and fluid leak-off. Furthermore, in those particular cases where the initial treatment had experienced a premature screenout, it is recommend using acid soaks to remove polymer residues, caustic soda soaks to break up resin-coated proppant packs and a series of injection tests to ensure the near-wellbore region is free of proppant. After all reasonable steps have been taken to clean the near-wellbore region and flush the proppant deep into the formation, the re-fracture treatment can be designed based on the data generated by the injectivity testing, in exactly the same fashion as if the original fracture did not exist.

对于压开原有裂缝的情况，必须将现有的支撑剂完全冲离井筒，并在重复压裂前进行一次小型压裂以评估原有裂缝的曲折程度和压裂液漏失情况。此外，在初次压裂过早出现端部脱砂的情况下，建议使用酸液除去聚合物残留物，碱液浸泡以降解树脂涂层支撑剂，并进行一系列注入测试以确保近井眼区域没有支撑剂。在进行完上述所有措施后，清洁近井眼区域并将支撑剂冲入地层后，就可以依据注入测试的数据进行重复压裂设计。

2. Re-Fracturing Re-Orientation

2. 重新定向造出新缝

There are two principal reasons for a secondary fracture treatment to re-orient:

重复压裂之所以能重新定向造出新缝主要有两个原因：

1) Proppant Induced

1)支撑剂因素

Creating a hydraulic fracture will induce stresses within a formation perpendicular to the minimum principal stress direction. The magnitude of this induced stress will be equal to the additional net pressure on the proppant pack at closure but will decrease rapidly with distance from the fracture face.

水力压裂在地层中产生的应力总是垂直于最小地应力，该应力的大小等于裂缝闭合时作用在支撑剂上的净应力，但会随着与裂缝面距离的增加迅速减小。

This effect can cause a potential re-orientation of the re-fracturing treatment if the magnitude of the newly induced stress is sufficient to overcome the relative difference between the two original horizontal stresses (Fig. 2. 23).

此时，如果重复压裂新产生的应力足以克服原始地层水平主应力差，就会造成重复压裂裂缝的重新定向(图 2. 23)。

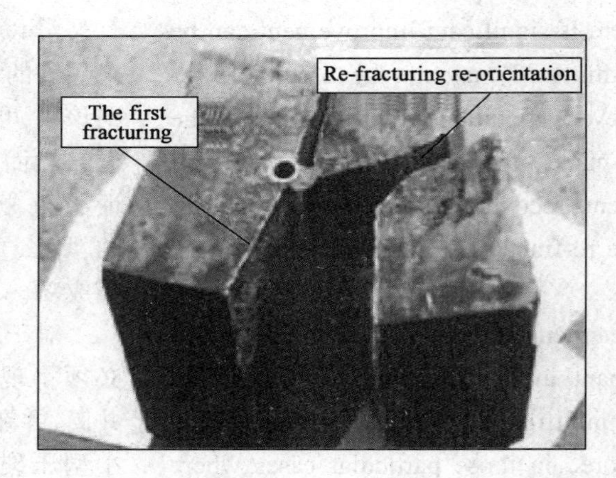

Fig. 2. 23 Re-fracture Reorientation

2) Production Induced

The second concept can be more readily understood with reference to Fig. 2. 24. The figure shows a horizontal cross-section through a vertical well containing an initial fracture of half-length x_f; continued production will result in a localized pore pressure change with an elliptical distribution. This pressure depletion will result in the horizontal stress component (parallel to the initial fracture) declining more quickly than the orthogonal stress. If this stress reduction exceeds the initial horizontal stress difference, then the direction of the minimum horizontal stress will become the direction of the maximum, and vice versa.

2)生产因素

图 2.24 显示了穿过一口垂直井、半缝长为 x_f 的初次压裂水力裂缝。连续生产将导致井筒附近孔隙压力的改变,使其最终呈椭圆形分布。需要注意的是,井筒附近压力的衰减将导致水平主应力(平行于初次压裂水力裂缝)比垂直主应力衰减更快。一旦应力衰减的程度超过了原始水平主应力差,那么最小水平主应力方向就变成了最大水平主应力方向,反之亦然。

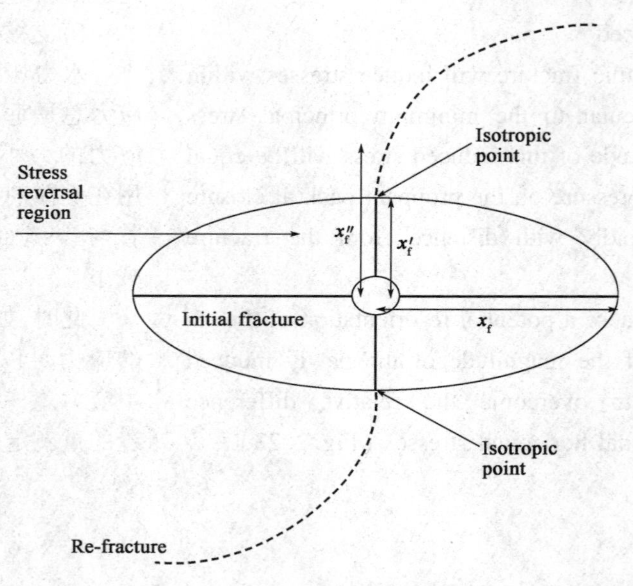

Fig. 2. 24 Re-Fracturing Re-Orientation Production Induced

In this case, the re-fracture treatment will now propagate orthogonally to the original fracture and grow to a half-length of x'_f at which point it will reach the isotropic or balance point. Once past this point the re-fracture treatment will begin to re-orient itself back to the far-field stress regime over the remaining half-length, x'_f. The distance to the isotropic point will be a function of the initial horizontal stress contrast, the initial fracture penetration, the production transient and the reservoir permeability.

在上述情况下，重复压裂作业就会产生一条垂直于原有裂缝且半缝长为 x'_f 的新裂缝，并且延伸到各向同性（平衡）点。一旦重复压裂裂缝超过这个点，它就会在地应力场的作用下重新定向回原有裂缝的方向。各向同性点距离井筒位置 x'_f 是初始水平主应力差、初始裂缝长度、生产时间和储层渗透率的函数。

2.2.3 Coiled Tubing Fracturing

In fact coiled tubing was invented to replace conventional tubing operations, and was mostly used for well drilling and completion technology (Fig. 2.25). The first application of the coiled tubing in hydraulic fracturing happened in 1992. Recently, the fracturing with coiled tubing has become a well-established technique of developing and stimulating multiple zone reservoirs.

2.2.3 连续油管压裂

实际上，最初的连续油管（图 2.25）主要是用于替代常规油管作业，多用于钻完井技术。连续油管在水力压裂中的应用最早出现于 1992 年。近年来，连续油管压裂技术已成为开发多层油藏的成熟方法。

Fig. 2.25 Coiled Tubing

(Source: https://en.wikipedia.org/wiki/Coiled_tubing)

The advantages of using coiled tubing are that multiple intervals can be rapidly treated and that the coiled tubing can then be used to circulate proppant from the wellbore and even gas lift the well back into production. This technique has also been successfully applied in coalbed methane formations.

使用连续油管进行压裂作业的优点在于能够快速处理多个层段，并且连续油管还能用于清洁井筒中的压裂液，甚至进行气举生产。同时，该技术也被成功应用于煤层气的开采。

After more than 20 years of development, on the basis of traditional coiled tubing fracturing, coiled tubing jet fracturing technology and coiled tubing stepless fracturing technology have been formed and applied. Nowadays, coiled tubing fracturing has become an essential method of production and stimulation.

经过二十余年的发展,在传统连续油管压裂的基础上,形成了连续油管水力喷射压裂技术和无限级压裂技术。目前连续油管压裂已经成为重要的开采和增产方法。

2.2.3.1 Coiled Tubing Conveyed Fracturing

2.2.3.1 连续油管传送压裂技术

The key to coiled tubing conveyed fracturing technology is the combination of bottom hole tools, as shown in Fig. 2.26. With this set of tools, coiled tubing can perform perforating, isolating, and fracturing (or without a perforating tool, coiled tubing is only used for fracturing).

连续油管传送压裂技术的关键是井底工具的组合,如图2.26所示。通过这套工具,使得连续油管可以进行射孔、分层、压裂(也可以没有射孔枪,仅使用连续油管进行分层压裂)。

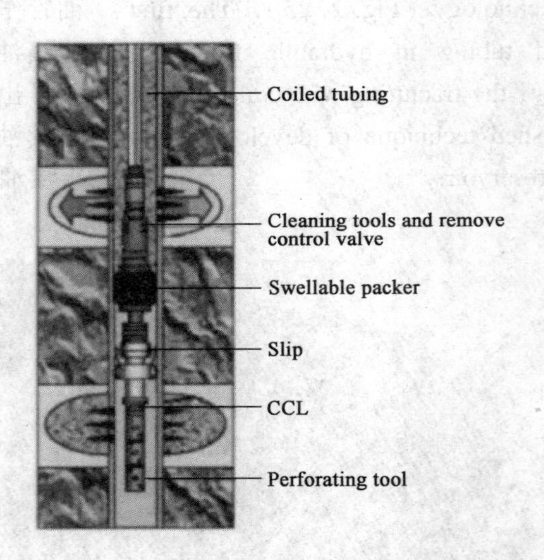

Fig. 2.26 Bottom Hole Tools of Coiled Tubing

Typical coiled tubing conveyed fracturing operation is shown in Fig. 2.27: (1) Lower the coiled tubing down to the lowest target layer; (2) Perforating the layer; (3) Lower the slips and packers below the target layer and anchor them; (4) The fracturing fluid is then injected into the formation through the annulus between the coiled tubing and the casing to perform hydraulic fracturing; (5) After the fracturing is completed, lift the coiled tubing to the upper target layer; (6) Repeat the above work until all target layers are fractured; (7) Cleaning the wellbore of proppant and prepare

典型的连续油管传送压裂作业流程如图2.27所示:(1)将连续油管下至最下方目标层位;(2)引爆射孔枪射孔;(3)将卡瓦和封隔器下至目标层位之下,坐封;(4)通过连续油管与套管之间的环空向地层注入压裂液,进行水力压裂;(5)压裂完成后,上提连续油管至上方的目标层;(6)重复上述作业直至所有

for production. In addition, if the coiled tubing is not equipped with a perforating tool, the perforating operation of each interval should be carried out in advance. Then the plugging and fracturing is performed on each layer from the bottom to the top by the coiled tubing.

目标层作业完成;(7)冲砂清理井筒,准备生产。如果连续油管没有射孔枪,则需提前进行各层段的射孔作业,之后用连续油管由下至上在每层进行封堵压裂。

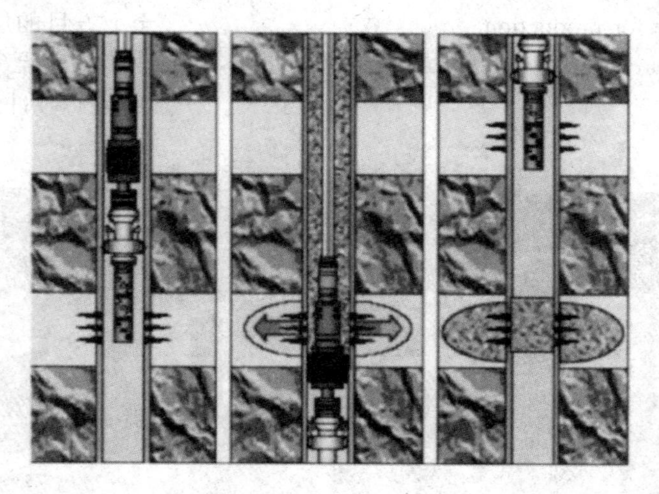

Fig. 2. 27　A Typical Coiled Tubing Fracturing Procedure

The advantages of coiled tubing conveyed fracturing is its ability to quickly treat multiple intervals, and the ability to transport proppants from coiled tubing, even gas lift to start production in oil and gas wells after treatments. It also saves a lot of work and time compared to conventional fracturing, thus the cost is greatly reduced. However, coiled tubing conveyed fracturing also has some limitations, such as stability (corrosion, internal and external pressure, tensile, vibration damage failure), the depth that coiled tubing can travel, and costs.

连续油管传送压裂的优势是能够快速作业多个层位,并且可以从连续油管中输送支撑剂,甚至可以气举,使油气井开始生产;与常规压裂技术相比,它能节约大量的作业时间,因此大幅度降低了成本。然而,连续油管传送压裂技术也存在一些问题,如自身的稳定性差(腐蚀、内外压、拉伸、震动破坏失效)、下入井内深度受限制,以及成本问题等。

2.2.3.2　Coiled Tubing Jet Fracturing

2.2.3.2　连续油管水力喷射压裂技术

Coiled tubing jet fracturing is a combination of hydrajet perforation and coiled tubing fracturing (Fig. 2. 28). It has the advantages of simpler down hole tools (without packer), small probability of complicated conditions in the well, and can shorten the well testing period as well as perforation cost. Specific operation procedures: (1) Lower the coiled tubing to the lowest target layer; (2) The nozzle carries out

连续油管水力喷射压裂技术是喷砂射孔技术和连续油管水力压裂技术的结合(图2.28),具有井下工具简单(不需要封隔器)、井下出现复杂情况概率小的优点,且能缩短试油周期、节约射孔费用。具体

hydrajet perforating; (3) Hydraulic fracturing is performed by hydrajeting pressure and the annular injection fracturing fluid pressure; (4) After the fracturing is completed, lift the tubing to the upper target layer; (5) Repeat the above steps until the fracturing of all target layers is completed; (6) Cleaning the wellbore and prepare for production.

施工工序:(1)下放连续油管至最下方目的层;(2)喷嘴进行喷砂射孔;(3)利用喷射压力在环空加注压裂液进行水力压裂;(4)该层压裂完成后,上提管柱至上方目的层;(5)重复上述作业步骤直至完成所有目标层的压裂;(6)冲砂清理井筒,准备生产。

Fig. 2. 28 Coiled Tubing Jet Fracturing

(Source: https://bhge. com/upstream/completions/stimulation-and-hydraulic-fracturing)

2. 2. 3. 3 Coiled Tubing Stepless Fracturing

The infinite-stage fracturing of coiled tubing (coiled tubing stepless fracturing) is a new technology that has emerged in the past ten years. Its technical tool is casing and sliding sleeve. After the cementing, the sliding sleeve is opened step by step by means of the coiled tubing tool. Thus multi-stage hydraulic fracturing is performed. Fig. 2. 29 is a schematic diagram of a sliding sleeve.

2. 2. 3. 3 连续油管无限级压裂技术

连续油管无限级压裂技术是近十多年来出现的新技术,它的技术原理是在下套管作业时,滑套与套管一起下入,固井后,依靠连续油管工具逐级打开滑套,进行多级水力压裂。图2. 29为无限级压裂滑套示意图。

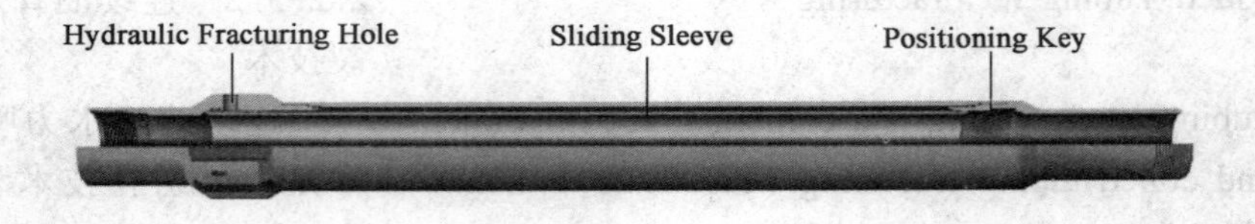

Fig. 2. 29 Sliding Sleeve

(Source: https://elyfrac. com/Tools/Sliding-Sleeve)

Specific operation procedures: (1) Lower the coiled tubing to the designed position; (2) Casing collar locator positioning and proofing depth; (3) Circulating base fluid positively; (4) Set the coiled tubing packer and perform pressure sealing test; (5) Open the sliding sleeve and carry out the hydraulic fracturing through annulus as designed previously; (6) Releasing the packer and lift the coiled tubing to the upper layer; (7) Repeat steps (1) – (6) to complete the treatment; (8) After the final fracturing, taking out the coiled tubing and discharging the casing.

具体作业步骤:(1)下放连续油管至设计位置;(2)套管接箍定位器定位、校深;(3)正循环基液;(4)坐封连续油管封隔器,并进行打压验封;(5)打开滑套,按设计进行环空加压裂液压裂施工;(6)解封封隔器,上提连续油管至上方层段;(7)重复上述步骤,完成施工;(8)施工结束后起出连续油管,套管放喷排液。

Just as the name of "infinite" or "stepless", the advanced technology can perform as many multilayer fracturing treatments as many layers there is in reservoir. Up to now, more than 120,000 hydraulic fracturing treatments have been completed in more than 6,000 wells through this technology. And the world record such as the number of layers that can be fractured in one well has been refreshed several times. At the same time, the technology has superior performance in fracture location control, fracturing efficiency, fracturing fluid dosage and real-time pressure monitoring. The fracturing time is shortened, the energy loss is small, and the water saving is more than 20% of total compared with conventional fracturing technology.

该技术可以无限制地进行分层压裂,至今已在6000多口井中完成了12万多次压裂,先后多次刷新单井压裂级数和一趟管柱压裂级数等世界纪录。同时该压裂技术在裂缝位置控制、压裂效率、压裂液用量和实时压力监测方面性能优越,压裂速度快、能量散耗小、节水20%以上。

Fracturing through coiled tubing has not been widely applied outside the Canadian operations and selected areas of the USA. This is because in order for the technique to be most cost-effective, the additional expense of using coiled tubing has to be competitive with the charges associated with the use of a workover rig. This has limited the application of coiled tubing fracturing technology to those areas with numerous shallow intervals and relatively high workover costs.

除加拿大和美国的部分区域外,连续油管压裂技术并没有得到广泛的应用。这是由于为了使该技术具有较高的性价比,使用连续油管时,修井设备的使用增加了额外费用。这限制了连续油管压裂技术在高的修井成本的多层油藏区域的应用。

2.2.4 Integral Fracturing

2.2.4 整体压裂

In contrast to jet fracturing that associated with drilling and completion engineering, integral fracturing (i. e. overall fracturing or systematic fracturing) is the product of combining hydraulic fracturing technology with reservoir engineering . At the end of 20th century, field engineers no longer focused only on the stimulation of a single well, but

相对于与钻完井工程相结合的喷射压裂而言,整体压裂(又称为开发压裂)是水力压裂与油藏工程相结合的成果。在20世纪末21世纪初,现场工程师不再仅将目光放在单井的压

the stimulation of the entire reservoir. The fracturing technology began to consider the whole reservoir as a unit, and took the well deployment, well spacing and fracture length into account, which led to the integral fracturing development in many fields.

裂增产上,而是着眼于整个油气藏的压裂开发。压裂技术开始以油气藏整体为单元,并考虑井距、井排与裂缝长度的关系,相继在很多地区形成了油气田整体压裂开发井网。

2.2.4.1 Introduction to Integral Fracturing

2.2.4.1 整体压裂简介

Since the seepage mechanism of fractured formations heavily depends on the cooperation between fracture geometry and orientation, and injection production pattern, different well deployments impact field production. The main features of the integral fracturing are: (1) achieving the maximum economic profit from the entire reservoir, rather than from a single well; (2) integrating well pattern with the fracturing by numerical simulation to optimize field development.

由于油水井压裂后储层内部渗流机理及其变化在很大程度上取决于人工裂缝的参数、方位与注采井网的配合关系,因此不同的井网部署就会产生不同的开发效果。对此,得到了整体压裂的要点:(1)整体压裂的目标是使整个油气藏达到最佳的经济效益,而非某口具体的单井;(2)将井网系统与人工裂缝系统有机结合起来,运用数值模拟方法对区块进行优化设计。

The key of the integral fracturing is to simulate the influence of petroleum reservoir and fracturing parameters on the production performance. To enhance ultimate recovery, an integral fracturing plan is needed to guide the development of oil and gas fields.

整体压裂的核心是模拟油气藏和裂缝不同参数对油气藏生产动态的影响,以最终的开发效益为目标,提出整体压裂的方案,指导油气田的开发。

2.2.4.2 Integral Fracturing Simulation

2.2.4.2 整体压裂模拟研究

There are two methods to simulate the integral fracturing. Before the wide use of computer technology, electric field simulation was applied. According to the similarity principle of water versus electric current, an electric field similar to the reservoir field was established, and the change of the reservoir field was determined by measuring the electric potential to simulate the effect of integral fracturing. At present, numerical simulation is often used . A corresponding model of reservoir and fracture is first built, then the initial conditions and internal and external boundary conditions are applied, and the numerical calculation is performed by computers.

整体压裂模拟研究的方法可以分为两种。在计算机技术普及之前,主要是电模拟研究,依据水电相似原则建立与实际渗流场相似的电场,通过测定电位值,确定渗流场的变化,以此来模拟整体压裂的效果。随着计算机的普及,现多采用数值模拟方法。首先建立油气藏和裂缝相应的渗流模型,再应用初始条件和内外边界条件,最后用计算机进行数值计算。

In term of the numerical simulation of integral fracturing, it is necessary to simulate the production performance and recovery of different fractured well types (fractured injectors or producers) with different fracture lengths and fracture conductivity, and finally simulate the production performance and recovery under the optimal combination of well patterns and hydraulic fractures.

对于整体压裂数值模拟，需要模拟不同压裂井类型(压裂注水井或压裂采油井)、不同裂缝长度及不同裂缝导流能力下的生产动态和采出程度，最后再模拟注采井网系统与水力裂缝系统优化组合下的生产动态和采出程度。

2.3 Hydraulic Fracturing Technologies Related to Fracture Size

Audio 2.3

All things considered, the artificial fractures, created by hydraulic fracturing treatments, are the most important factor that determines the success or failure of one treatment. If the fracture is too large, it may connect with other unfavorable layers, causing sand or water problems; while if the fracture is too small, it may fail to increase productivity. Therefore, reasonable fracture size is something that we should pay close attention to. Furthermore, as previously discussed many times in section 2.1 and 2.2, for low permeability reservoirs, fracture length is the dominant parameter when performing a hydraulic fracturing operation. In contrast, for medium-high permeability reservoirs, by increasing fracture width to improve fracture conductivity is the main way to improve production. Thus, various fracturing technologies related to fracture size have emerged. In this section, firstly introduced is fracture-height control fracturing. Then tip screen out of fracture length and width control is discussed. Finally, some theories about forced closure is introduced.

2.3 控制裂缝尺寸的压裂技术新进展

水力压裂所形成的人工裂缝是压裂施工的一切出发点和落脚点，若人工裂缝尺寸过大，则可能连通其他层段，带来过早出水出砂的问题；若人工裂缝尺寸过小，则可能无法起到增产目的。因此，合理的裂缝尺寸是人们广泛关注的问题。同时，如在2.1节和2.2节中讨论的那样，对于低渗透储层而言，若要提高压裂效果，应以增加缝长为主；对于中高渗透储层，通过增加缝宽来提高裂缝的导流能力则是提高压裂效果的主要途径。因此，各种与裂缝尺寸相关的压裂技术应运而生。本节首先介绍控缝高压裂技术，之后介绍端部脱砂压裂技术，最后简要介绍强制闭合技术。

2.3.1 Height-Control Fracturing

Fracture height remains to be a negligible yet extremely essential parameter compared with fracture length and width since the beginning of hydraulic fracturing technology. Field engineers often calculate fracture length and width iteratively to obtain the optimum parameters but only resort to the height of target layer as fracture height. And this method has its applications when the target layer and the upper, lower layer have major differences. While numerous micro-seismic monitoring found fracture height at other circumstances is not

2.3.1 控缝高压裂技术

自水力压裂技术诞生以来，与缝长和缝宽这两个参数相比，缝高一直是一个容易被忽视但极为重要的裂缝参数。现场工程师经常迭代(重复)计算出裂缝长度和宽度以获得最佳裂缝参数，但仅将目的层的高度作为裂缝高度来计算。当上下隔层与目的层存在明显差异时，这种

contained in target layer. So what stops fracture height growth? And how to control the fracture height to perform a height-control fracturing?

方法是可行的。然而许多微地震监测结果显示,许多情况下的裂缝高并不仅仅局限在目标层中。那么是什么阻止了裂缝在高度方向上的扩展?怎样进行控缝高压裂?

2.3.1.1 Fracture Height Containment

1. Stress Contrasts

The classic formation characteristic of stress contrasts is often used as evidence for fracture height containment (Fig. 2.30). Without question, a significant increase in the minimum principal stress as the fracture propagates up or down into a different formation, can make it very difficult for the fracture to extend. Fracturing fluid does not have sufficient pressure to separate the rock and without a significant increase in net pressure, fractures will extend only laterally, rather than grow into the high-stress region.

2.3.1.1 裂缝高度的控制

1. 应力差

上下地层与目的层存在的应力差是裂缝高度受到限制的重要因素(图 2.30)。毫无疑问,当裂缝向上或向下扩展到不同的地层时,最小主应力的显著增加会使裂缝难以继续延伸。此时压裂液无法使上下地层的岩石破裂,除非净压力显著增加,否则裂缝将仅沿横向扩展,而不能扩展到高应力区域。

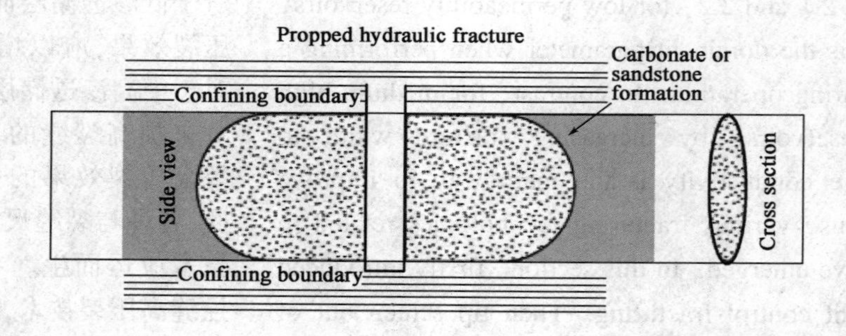

Fig. 2.30 Fracture Height Containment by Boundary Formations

However, how often do these stress contrasts exist? It is not appropriate to assume that an over or underlying stratum will automatically have higher in-situ stress than the reservoir formation. In fact, neighboring layers are just as likely to have reduced stresses as increased stresses. Unless there is some independent evidence to corroborate stress-based fracture height containment, it cannot be relied upon.

然而,这种上下地层的地应力差往往都存在吗?这种假设其实并不合适。实际上,现场作业发现,相邻地层的应力增大或减小的可能性是相同的。除非有独立的数据表明存在增大的应力差,否则不能依赖这个假设。

2. Fracture Toughness and/or Tip Effects

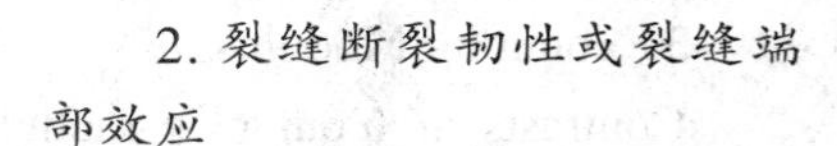
2. 裂缝断裂韧性或裂缝端部效应

In order for the fracture to propagate, the rock must be physically split apart at the fracture tip. In formations where this is easy, the formation is said to be "brittle" and has low apparent fracture toughness (Fig.2.31). The opposite effect, in which relatively large quantities of energy are used to split the rock apart, is referred to as "ductility" with high apparent fracture toughness. Although there is still considerable debate as to what exactly is happening at the fracture tip, there is no doubt that a significant portion of the net pressure is used up at the fracture tip. How much energy is used is controlled by how ductile the formation is.

水力裂缝要继续向前扩展，则处在裂缝端部附近的岩石必须断裂张开。比较容易破裂的地层称为脆性地层，且地层的断裂韧性较低(图 2.31)。相反，如果某地层需要较大的能量才能压开，则称该地层为具有较高断裂韧性的韧性地层。尽管人们对于裂缝端部效应仍讨论不休，但都认为压裂施工压力的很大一部分在裂缝尖端被消耗，具体消耗多少取决于该地层的韧性程度。

Fig. 2. 31 Hydraulic Fracturing of Brittle Formations
(Source: https://abnewswire. com/pressreleases/hydraulic-fracturing-market-global-trends-size-share-growth-and-forecast-2016-to-2022_149109. html)

If the fracture propagates from a formation that is brittle into a formation that is ductile, extra energy will be required to keep the fracture propagating in that direction. If this energy is not available, or if it is easier for the fracture to grow in an alternative direction, then the fracture will not significantly penetrate the ductile formation. Consequently, contrasts in apparent fracture toughness can form the most reliable barriers to height growth, especially for shales. It is possible that fracture height containment often attributed to stress contrasts is in fact due to apparent fracture toughness contrasts.

如果裂缝从脆性地层扩展至韧性地层，则需要额外的能量支持裂缝继续扩展延伸。如果此时地面无法提供额外的能量，或裂缝更容易延另一个方向扩展，那么裂缝就不会进入韧性地层。所以，地层之间的表观断裂韧性差会遏制缝高的增长，尤其是在页岩地层中，这种情况最为常见。对裂缝高度的遏制通常归因于应力差，实际上可能是由相邻地层表观断裂韧性差造成的。

3. Young's Modulus

Contrasts in Young's modulus are not very good at preventing fracture height growth. There is an inverse relationship between fracture toughness and Young's modulus. This means that a rapid increase in E can also coincide with a rapid decrease in apparent fracture toughness, making it easier for the fracture to propagate.

However, fracture width is inversely proportional to E for any given net pressure. Therefore, a rapid increase in E will result in a rapid decrease in fracture width, possibly even to the point where it is too narrow for proppant placement.

4. Fluid Viscosity

In spite of a common industry perception, reduced fluid viscosity cannot be relied upon to reduce fracture height growth. In fact there is considerable evidence from field experiments that formations will produce the same pressure response (indicating similar fracture geometry) regardless of fluid viscosity. There is also very little evidence from micro-seismic analysis that thin fracturing fluids produce less height than highly viscous fluids.

3. 杨氏模量

相邻地层的杨氏排量差并不能有效限制裂缝高度扩展。一般而言,地层杨氏模量和其断裂韧性成反比,即杨氏模量 E 越高,地层的断裂韧性越低,水力裂缝就更容易扩展。

然而,对于给定的净压力,裂缝宽度也与杨氏模量成反比。因此,杨氏模量的增加也会导致裂缝宽度的减小,最终就会导致裂缝过窄支撑剂难以进入。

4. 压裂液黏度

尽管普遍的观点认为控制压裂液黏度在一定范围内可以控制裂缝高度的扩展,实际上,压裂液黏度的降低并不能阻止裂缝在高度方向的扩展。很多现场试验表明,无论压裂液黏度如何,地层都会产生相同的压力响应(即产生相似的裂缝几何形状)。微地震分析也很少有证据表明,低黏度压裂液产生的裂缝高度低于高黏度压裂液。

2.3.1.2 Height-Control Fracturing Methods

When performing a hydraulic fracturing treatment, the reservoir characteristics are already settled. The only way to alter or control fracture height is by adjusting the operation itself.

1. Height-Control Fracturing by Artificial Barrier

The basic principle of this technology is to add diverting agent at the end of the pad. There are two types of diverting agent: buoyant diverter and settling diverter. The former density is lower than the fracturing fluid, and the diverter is concentrated on the top of the artificial fracture; the latter is exactly the opposite. It settles at the bottom of the fracture to form a low-permeability artificial barrier to achieve the goal of controlling fracture height (Fig. 2.32).

2.3.1.2 控缝高压裂技术

在进行水力压裂作业时,地层参数已经确定且无法更改,现场工程师只能通过调整控制施工参数来控缝高。

1. 人工隔层控缝高技术

该技术的基本原理是在前置液末段加入转向剂。转向剂分为上浮式和下沉式两种,前者密度低于压裂液,上浮聚集在人工裂缝的顶部;后者恰恰相反,下沉聚集在裂缝底部,形成低渗透人工隔层,以达到控制缝高的目的(图 2.32)。

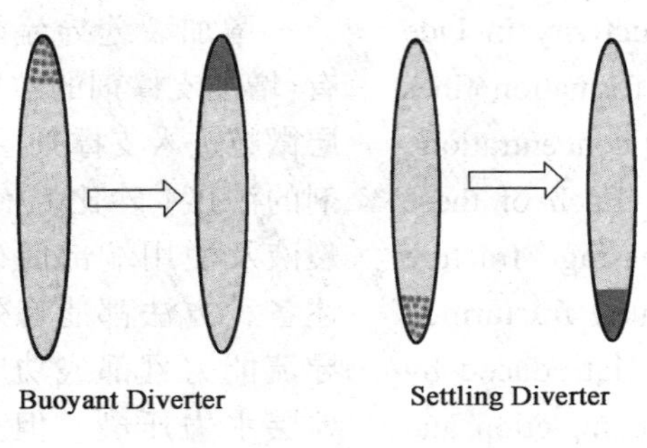
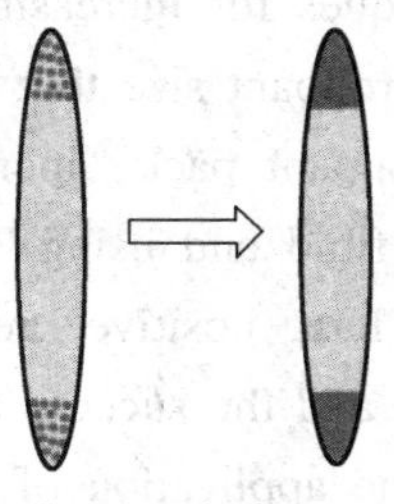

Fig. 2. 32　Artificial Barrier

2. Height-Control Fracturing by Fracturing Fluid Density

By limiting vertical pressure gradient of fracturing fluid, the vertical extension of the fracture can be controlled. To control the upward growth of the fracture, the fracturing fluid with higher density is used to facilitate fracture propagation downward as much as possible (Fig. 2. 33). Similarly, lighter fracturing fluid is used to control the downward propagation of the fracture .

2. 压裂液密度控缝高技术

通过控制压裂液的垂向压力梯度可以控制裂缝的垂向延伸。若要控制裂缝向上延伸,就采用密度较高的压裂液,使裂缝尽量向下扩展(图 2. 33);同理,采用密度较低的压裂液可以控制裂缝向下延伸。

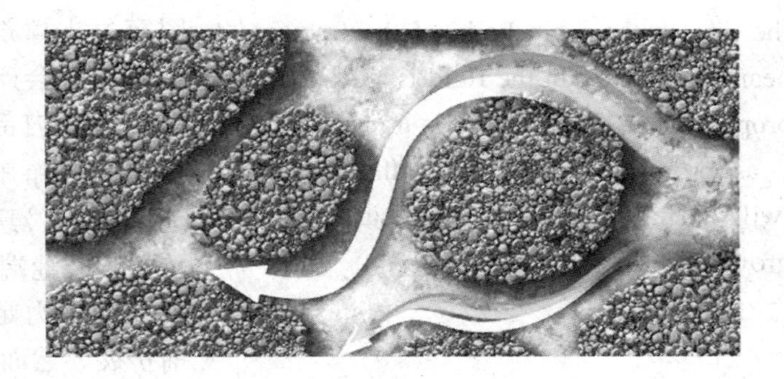

Fig. 2. 33　Fracture Downward Propagation Due to High Density Fracturing Fluid
(Source: https://advances. sciencemag. org/content/fracturing/fracturing fluids)

2. 3. 2　Tip Screenout (TSO)

The capabilities of hydraulic fracturing technology have been extended into softer formations, which often have a medium to high permeability. And hydraulic fracturing in high-permeability reservoirs differs from conventional fracturing in that the objective is more to generate fracture conductivity than length. As permeability rises, it becomes increasingly difficult to produce sufficient width without also generating excessive length.

2. 3. 2　端部脱砂(TSO)

水力压裂技术已经被成功应用于中、高渗透率的疏松地层。对于这些中高渗储层的压裂,最重要的是增加裂缝的导流能力(缝宽)而非长度。但是,要形成足够宽的裂缝而又不使裂缝的长度过长,通常是一个很困难的问题。

Techniques for increasing fracture conductivity include increasing proppant size that do not produce formation fines into the proppant pack, increasing proppant concentration, using clean fluid and using TSO techniques. Each of these techniques has positive results on increasing fracture conductivity and the success of high-permeability fracturing. However, the application of TSO fracturing, introduced by Smith et al. in 1984, has revolutionized the completion and stimulation of wells in higher permeability reservoirs.

增加裂缝导流能力的方法有:增加支撑剂的粒度并阻止地层微粒进入支撑剂层、增加支撑剂的浓度(砂比)、使用清洁压裂液及使用端部脱砂技术。上述各种方法都能有效增加裂缝导流能力且都成功应用于高渗地层水力压裂。但 Smith 等人于 1984 年提出的端部脱砂技术,彻底改变了高渗地层的完井和压裂增产方式。

2.3.2.1 Tip Screenout Concept

2.3.2.1 端部脱砂压裂原理

With reference to Fig. 2. 34, a TSO is achieved by forcing proppant into the fracture tip at a relatively early stage in the treatment. When the first proppant arrives at the fracture tip it forms a "bridge", or proppant plug at the tip. As the proppant collects in the fracture tip, a pressure differential is created by fluid trying to penetrate through the proppant to reach the tip. In the main body of the fracture, $p_{net} > p_{ext}$, which means the energy in the fluid is sufficient to make the fracture propagate, if this pressure is transmitted to the tip. However, as proppant builds up in the tip, the pressure at the tip will fall until it is no longer sufficient to make the fracture grow. This is the TSO.

如图 2. 34 所示,端部脱砂就是强制支撑剂在压裂作业的早期聚集在裂缝的端部。当第一批压裂液抵达裂缝端部时,在端部形成"桥堵"。随着越来越多的支撑剂聚集在裂缝端部,试图穿过支撑剂到达裂缝尖端的压裂液就会产生一个压差。在人工裂缝内部,$p_{net} > p_{ext}$(水力裂缝内部净压力大于裂缝外部地层压力),意味着如果压力能传播到裂缝端部的话,压裂液所具有的压力足以使裂缝继续向前扩展。然而,此时支撑剂聚集在裂缝端部,因此压裂液的压力无法抵达裂缝端部,这就形成了所谓的端部脱砂。

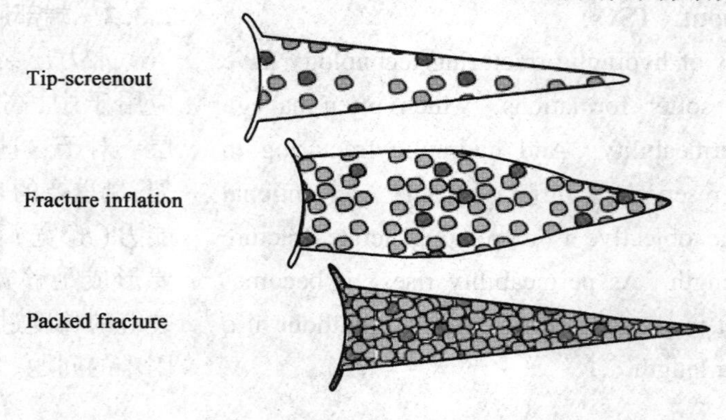

Fig. 2. 34 Tip Screenout

But, fluid is still being pumped into the formation at a constant rate. Given that the TSO will not significantly affect the fluid leakoff rate, the fracture volume has to increase at the same rate, even though the fracture is no longer gaining length and height. This means that the fracture width and conductivity has to increase.

Normally, increasing the slurry concentration after the onset of a TSO is a more efficient way to obtain increased conductivity than by relying on only the increase in width.

但是此时,压裂液仍然被不断地泵入地层,因此裂缝的体积也继续增加。由于裂缝长度方向和高度方向已经被支撑剂封堵,所以裂缝就只能在宽度方向膨胀。

在端部脱砂现象出现后增加支撑剂的浓度能显著提高裂缝的导流能力,因此这两个工艺经常结合使用。

2.3.2.2 Tip Screenout Fracturing

A typical TSO fracturing can be conceptualized in two distinct stages: fracture creation and fracture inflation/packing.

Creation of the fracture and arrest of its growth is accomplished by injecting a relatively small pad fluid (no sand) followed by a slurry containing 1 – 4 lbs of sand per gallon of fluid. Once the fracture growth has been arrested, further injection builds fracture width and allows injection of a high concentration slurry (e. g. 10 ~ 16 ppg). Final proppant concentrations of 20 ft are possible. A usual practice is to retard the injection rate near the end of the treatment to dehydrate and pack the fracture near the well. Rate reductions may also be used to force a tip screenout in cases where no TSO event is observed on the downhole pressure record.

2.3.2.2 端部脱砂压裂技术

典型的端部脱砂压裂可以分成两个阶段:裂缝形成阶段和裂缝膨胀阶段。

通过注入少量的前置液和砂比为 0.12 ~ 0.48g/cm^3 的携砂液,能形成初始水力裂缝并阻止裂缝进一步扩展。当裂缝停止扩展时,进一步注入压裂液会造成裂缝宽度的增加并允许之后注入更高砂比的携砂液(1.2 ~ 1.9g/cm^3)。最终支撑剂的砂比可能会达到 2.4g/cm^3。通常的做法是在接近压裂结束时,降低排量以使井筒附近的压裂液脱水和填充满支撑剂。在井下压力记录中未观察到端部脱砂的情况下,也可以降低泵速来强制裂缝端部脱砂。

And a typical treatment history is schematically illustrated in Fig. 2.35.

端部脱砂作业期间压力变化如图 2.35 所示。

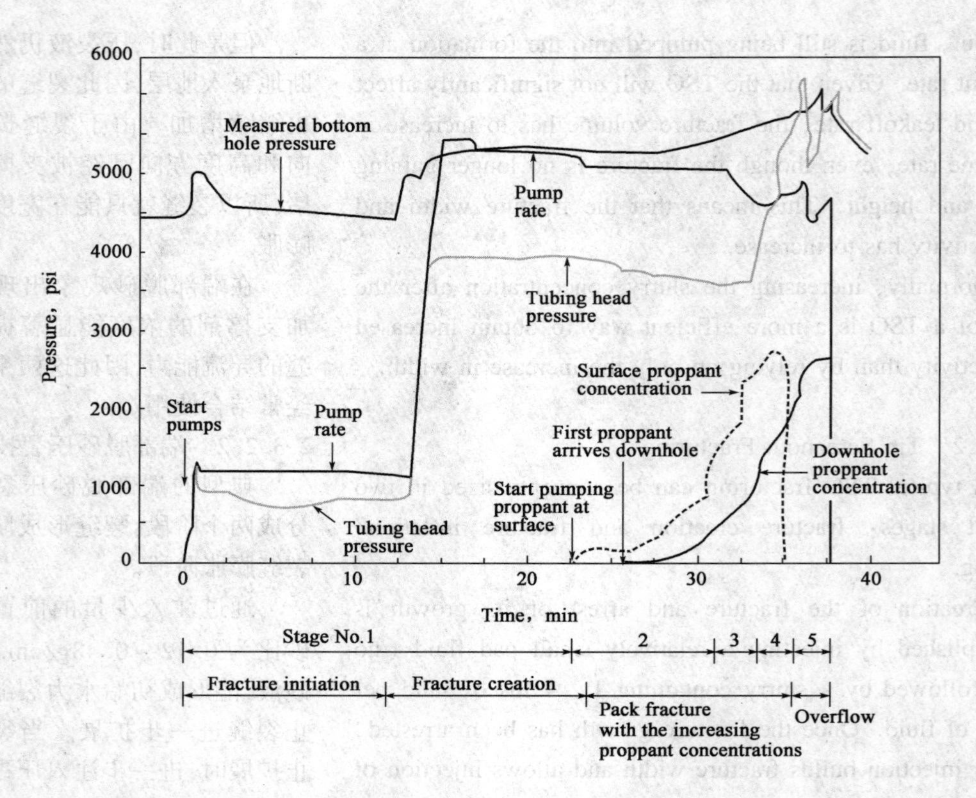

Fig. 2. 35　Treatment Record for Tip Screenout Hydraulic Fracture Stimulation

Table 2. 3 and Fig. 2. 36 illustrate the differences of an equivalent operation between a massive hydraulic fracturing (MHF) in low permeability reservoirs and a high permeability fracturing (HPF) using TSO. The TSO fracture uses only 5% of the amount of proppant and 10% of the fluid volume.

表2.3和图2.36总结了低渗储层水力压裂和高渗储层端部脱砂压裂的区别。与低渗储层水力压裂相比,端部脱砂压裂只需其5%的支撑剂和10%的压裂液。

Table 2. 3　Tip Screenout and Conventional Fracturing Compared

Fracture Type	Conventional	Tip Screen Out
Description	Long and thin (lower conductivity)	Short and fat (higher conductivity)
Width ,in	> 0.25	0.25 ~ 1.5
Length ,ft	500 ~ 1500	50 ~ 500
Proppant Concentration ,lb/ft^2	0.5 ~ 2.0	4 ~ 12

2. 3. 2. 3　Tip Screenout Difficulties

Despite all these advantages TSO have, the technology still has its difficulties when applying.

2. 3. 2. 3　端部脱砂压裂难点

尽管端部脱砂压裂已广泛应用于高渗储层压裂作业,但在具体实施时,该技术也存在难点。

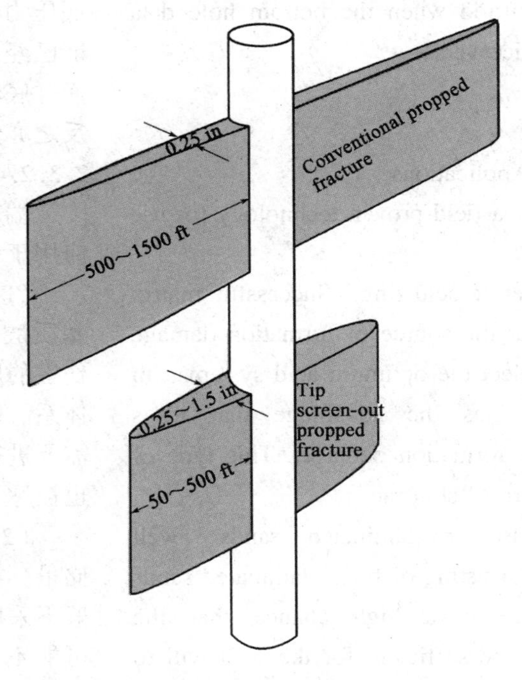

Fig. 2. 36　Tip Screenout and Conventional Fracturing Compared

(1) Reservoir limitation. As the tip screenout occurs, the net pressure starts to rise. The rate at which the net pressure starts to rise is controlled by the formation's Young's modulus: If the rock is too hard, the pressure will rise too quickly and the treatment will soon be over. Therefore, in order for a formation to be a candidate for a TSO treatment, it must have sufficiently high fluid leakoff to allow proppant accumulation in the fracture tip and sufficiently low Young's modulus so that the pressure does not rise too quickly.

(1)地层限制。当缝端开始脱砂时,压裂液净压力逐渐升高。净压力升高的速度由地层杨氏模量控制:如果地层杨氏模量很大,净压力上升很快,端部脱砂很快就会结束。因此,如果某地层需要进行端部脱砂压裂施工,则必须具有高的渗透率,使压裂液滤失进地层从而形成支撑剂在裂缝端部的聚集脱砂;还需要拥有较低的杨氏模量,这样净压力的升高就不会过快。

(2) Detection difficulty. Frequent field experience suggests that the tip screenout can be difficult to model, affect or even detect. Accurate bottom hole measurements are imperative for meaningful treatment evaluation and diagnosis. Calculated bottom hole pressures are unreliable because of the sizable and complex friction pressure effects associated with pumping high proppant slurry concentrations through small diameter tubulars. Surface data may indicate

(2)难以检测。现场实践表明,端部脱砂的过程难以模拟、控制,甚至难以检测。获得准确的井底压力对压后分析和评估至关重要。但在端部脱砂压裂中,由于使用高砂比的压裂液,因此产生的摩擦阻力十分大,这就导致难以准确计算出井

that a TSO event has occurred when the bottom hole data shows no evidence, and vice versa.

底压力。地面压力数据显示可能已经出现了端部脱砂现象,然而井底的压力数据却表明没有;反之亦然。

2.3.2.4 Tip Screenout Applications

2.3.2.4 端部脱砂压裂应用

TSO fracturing is now a field proven technology for use in a number of areas.

端部脱砂压裂技术被广泛应用于多种情况。

(1) Alternative to matrix acidizing. Successful matrix acidizing often requires that the source of formation damage be identified in order to select the optimum acid systems. In contrast, TSO fracturing has the advantage that it is independent of the type of formation damage. This type of treatment is generally called a "skinfrac".

(1)替代基质酸化。进行基质酸化前一般要先分析储层伤害的原因,并配置对应的酸液体系。端部脱砂压裂技术就不需要进行这一步,端部脱砂压裂也被称为解堵型压裂。

(2) Reserve increase in laminated sands. well completion in formations consisting of finely laminated sands is problematic since there is a high chance that the perforation density will not be sufficient for the well will to make contact with all the hydrocarbon bearing zones. Production via a TSO fracture will ensure that recovery is achieved from all the zones (Fig. 2.37).

(2)增加层状砂岩的可采储量。由细砂岩构成的地层一般不采用射孔技术,因为孔眼很可能不足以覆盖所有的含油气区域,通过端部脱砂压裂将确保从所有的含油气层获得产能(图2.37)。

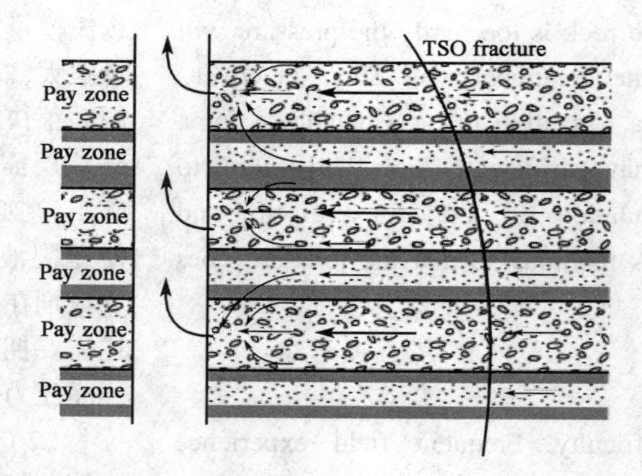

Fig. 2.37 Tip Screenout Connects Extra Reserves. All Productive Sand Bodies Being Produced after Connection via Hydraulic Fracture

(3) Horizontal well completion: this is schematically illustrated in Fig. 2.38 where the large number of closely spaced propped TSO hydraulic fracturing treatments have to be placed along a 1200 m horizontal well. This type of completion has been used on a wide scale for the oil bearing chalk fields.

(3)水平井完井:如图2.38所示,在长达1200m的水平井段紧密排列了大量端部脱砂压裂的裂缝。这种利用端部脱砂压裂作为完井方式的技术已广泛用于含油地层。

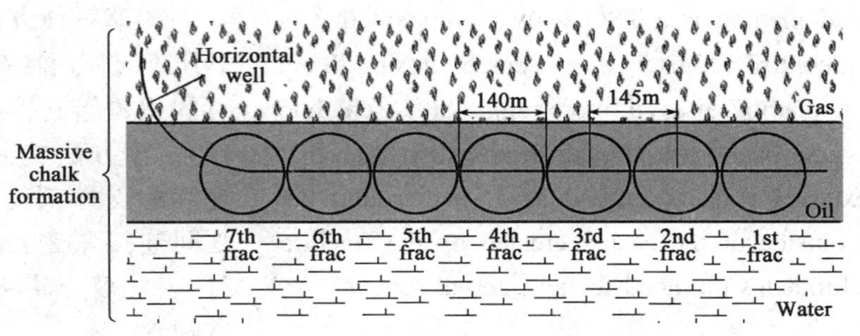

Fig. 2.38　Completion Employing Combined Horizontal Well and Multiple Fracturing Technologies

(4) Sand control: when TSO treatments are in or unconsolidated formations, they incorporate three basic methods to control sand production in the operations (Fig. 2.39).

(4)防砂:当端部脱砂压裂被应用于弱胶结地层时,有三种基本的防砂方式(图2.39)。

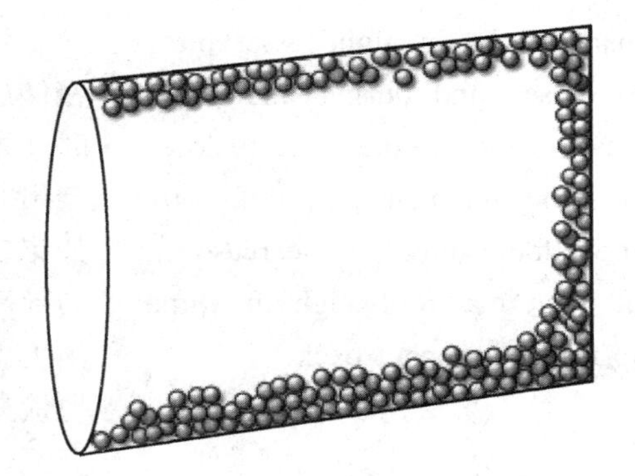

Fig. 2.39　TSO Sand Control

① The first method is a four-step operation. The TSO is performed, the wellbore is cleaned up, the screen assembly is run, and the gravel-pack operation is performed.

② The second method is a one-step operation. The designed fracture treatment is pumped with the screen in place, the crossover tool in the circulating position and the choke on the annulus closed. The annular choke is opened when the gravel-pack volume is pumped at a slower rate to ensure annular packing. After screenout of the casing/screen annulus, the tool is placed in reverse position and the excess slurry is reversed out of the tubing.

①第一种方法包含四个步骤:进行端部脱砂压裂,冲砂清洗井筒,下入筛管组合,进行砾石充填防砂。

②第二种方法只需一步:在筛管就位的情况下泵入压裂液进行压裂。交叉工具此时处于循环位置,环空上的扼流圈关闭。当砾石以较慢的速率泵送以确保环形填充时,扼流圈打开。在套管筛网环空出现端部脱砂后,将交叉工具置于反向位置,并将多余的携砂液从油管中洗出。

③ The third method is a TSO with a proppant that is stabilized with a control additive (e. g. , curable resin, fibers or both). This method uses a more aggressive schedule designed to promote backward packing to an extent such that a continuous external pack is formed as a ring around the annulus on the outside of the casing created by the fracture width, which eliminates the need for an internal screen.

③第三种方法是在端部脱砂压裂时使用特殊的支撑剂(如可固化树脂支撑剂或纤维支撑剂)。该方法在井筒周围形成连续的支撑剂充填层,支撑剂之间连接紧密,地层砂无法通过,因此就不需要使用内部防砂筛管。

2.3.3 Forced Closure

2.3.3 强制闭合技术

Closing wells in for a matter of hours, overnight, or for several days following a hydraulic treatment was the accepted practice for many years. The extended shut-in time was thought to allow the fracture to close, as well as following any viscosified fluids to break completely back to water.

通常在水力压裂后,都要关井几小时、一晚甚至几天,以确保裂缝闭合在支撑剂上,增黏压裂液破胶溶于水中。

However, fractures, particularly in tight reservoirs, may require a long time to close, and during this time, excessive proppant settling may occur. If the fracture loses conductivity near the wellbore, the treatment may fail. Any pinching effect in the near-wellbore area or decrease in conductivity in the proppant pack may outweigh the time delay benefit of fluid cleanup in the proppant pack.

然而,水力裂缝,尤其是致密储层中的裂缝,通常需要较长时间才能闭合。在这段时间内,会出现大量支撑剂沉降现象。如果支撑剂沉降导致井眼附近的导流能力降低,那么此次压裂施工就会失败。近井区的任何扼流效应或支撑剂充填层的导流能力降低都可能超过为净化支撑剂充填层中流体而采用的延时所得效益。

For this reason, today a technique called forced closure is very often applied.

因此,目前经常使用强制闭合技术。

2.3.3.1 Forced Closure

2.3.3.1 强制闭合技术简介

The term forced closure was coined in 1986 to describe a process to speed up closure on the proppant, in order to minimize near wellbore proppant settling and to negate smearing of proppant due to ongoing fracture growth after shutdown. Forced closure consists of flowing fracture fluids back out of the well starting immediately after the end of the pumping at a rate of 10s-of-gallons to several (2 to 3) barrels per minute, depending on the number and size of perforations.

强制闭合一词于1986年被提出,旨在描述加速裂缝在支撑剂上闭合的过程,以减少支撑剂的沉降和压裂液对储层的伤害,并消除停泵后裂缝的继续扩展而造成的支撑剂拖尾效应。具体而言,强制闭合技术是在停泵后,立即以每分钟几十加仑到几桶的速度返排压裂液,以保证获得最大的裂缝导流能力。

Forced closure does not cause rapid fracture closure, but rather involves something akin to reverse gravel packing of proppant at the perforations. Thus, a better term for this process would perhaps be "reverse gravel packing". The dominant mechanism in forced closure is felt to be the creation of a proppant pack opposite the perforations. This would clearly explain the reduced proppant production and improved near well fracture conductivities that have been observed. Forced closure can also promote better grain-to-grain contact for treatments that use resin-coated sand.

强制闭合技术并不会造成人工裂缝的快速闭合,相反,它会在射孔孔眼附近形成支撑剂的反向充填。因此,这一过程用反向砾石充填来描述更为合适。强制闭合技术的机理就是在射孔孔眼附近形成支撑剂充填层。这也就解释了为什么强制闭合技术能阻止支撑剂在生产过程中随油气进入井筒,能提高近井地带裂缝的导流能力。强制闭合技术与树脂覆膜砂共同使用更能提高裂缝的导流能力。

A major benefit of this immediate flowback is that the super charge of fluid pressure assists in fracture cleanup and establishing production. With the conventional shut-in approach, this pressure dissipates before the well is opened to flow. Forced closure also provides some latitude in the fluid breaker design. Overly aggressive breaker schedules can result in premature loss of fracturing gel viscosity and rapid settling of proppant. Ideally, a well should initially produce some amount of unbroken gel following a successful fracture treatment.

同时快速返排压裂液利用了压裂完成后压裂液中残余的能量,并能使压裂井快速投产。如果压裂后关井一段时间的话,压裂液中的能量在开井生产前就会逐步损耗殆尽。强制闭合技术还简化了压裂液中破胶液的设计,因为过早注入破胶液会导致压裂液黏度降低、支撑剂过早沉降现象。在压裂液返排过程中,最理想的状态是先返排一部分未破胶的压裂液。

As a collateral benefit, the artificial pressure built up in the formation by the fracture treatment is often sufficient to clean excess proppant out of the wellbore during the forced closure. This can eliminate the costs that would otherwise be incurred in coiled tubing cleanout sand bailing.

采用强制闭合后裂缝中储存的能量往往足以将井筒中的支撑剂冲离井筒,因此作为附加的优点,强制闭合技术可以节省下入连续油管工具清砂的费用。

Forced closure is extremely beneficial when using energized fluids.

当使用充能流体进行水力压裂时,强制闭合技术十分有效。

2.3.3.2 Should forced closure be used on every well?

2.3.3.2 每口压裂井都该使用强制闭合技术吗?

One of the worst reasons to skip forced closure is to allow a history match of falloff pressures after a treatment. Jeopardizing the success of the treatment to gather data is a misplaced priority.

不进行强制闭合的最主要原因就是为了进行压裂后的衰减压力历史拟合,但是为了收集压力数据而危害整个压裂作业的做法不可取。

There are, however, several situations where forced closure is not applicable or should be avoided. The most obvious one is where the well goes on a vacuum. A second is where the operator is doing multiple stages and cannot afford to spend the time flowing back the well, with large numbers of people and very expensive equipment on location.

强制闭合技术也并不适用于某些特定的情况,如当整个井筒内处于真空状态时,或进行分层压裂时,此时由于还要进行后续层段的压裂施工,不可能停下来进行某一层段的强制闭合。

It is also felt that forced closure is not essential in tight gas water fracs or slickwater treatments, where very rapid settling occurs and the mechanism of fracture conductivity is not from a conventional proppant pack. Many wells performed well without forced closure because of fluids banking in the near-wellbore area.

强制闭合技术在清水压裂和滑溜水压裂中并不是必不可少的。在这些情况下,支撑剂本身会快速沉降并且裂缝导流能力也不像传统地由支撑剂所主导。现场试验表明,许多油气井不进行强制闭合也生产良好。

2.4 Hydraulic Fracturing Technologies Related to Fracturing Fluid and Proppant

2.4 压裂液与支撑剂相关压裂技术新进展

Unquestionably, fracturing fluid and proppant play an essential role in hydraulic fracturing because they determine each step and outcome of the treatment. Needless to say, if there is no fracturing fluid and proppant, hydraulic fracturing ceases to exist.

Audio 2.4

对于水力压裂技术而言,压裂液和支撑剂起着至关重要的作用,这是因为压裂施工的每个环节和结果都与压裂液及支撑剂有关。可以说,没有压裂液和支撑剂,就没有水力压裂技术。

Therefore, the development of fracturing fluid and proppant has a great impact on the development of hydraulic fracturing technology itself. In this section, after an introduction of recent advances in fracturing fluid and proppant, the new hydraulic fracturing technologies related to them are also discussed afterwards.

因此,压裂液和支撑剂的进展对水力压裂技术的发展起着重要的作用。本节首先简要介绍压裂液和支撑剂的新进展,随后介绍与之相关的压裂技术新进展。

2.4.1 Recent Advances in Fracturing Fluid and Proppant

2.4.1 压裂液与支撑剂新进展

During a hydraulic fracturing stimulation, fracturing fluid is mainly composed of three parts: pad fluid, proppant slurry, and displacement fluid (Fig. 2.40). The main

压裂液是一个总称,它由前置液、携砂液和顶替液组成(图2.40)。前置液的作用主要是形

function of pad fluid is to open the fracture and propagate it. And proppant slurry carries proppant into the fracture to hold the fracture open. Finally, displacement fluid is used to displace all the proppant into the fracture.

成裂缝并使其扩展延伸,携砂液的作用是将支撑剂携带进裂缝保持裂缝的张开,顶替液的作用是将所有的支撑剂驱替进入裂缝。

Fig. 2. 40　Fracturing Fluid
(Source: https://ngsfacts. org/fracturing fluid)

Actually, the first hydraulic fracturing operation in 1947 didn't utilize proppant (Fig. 2. 41). And it was then when people realize that the artificial fractures will close sooner after the treatment is over. So proppant is introduced into the industry.

实际上,1947 年第一次正式的水力压裂施工中并没有使用支撑剂(图 2. 41)。从那之后,人们意识到不支撑的水力裂缝在压裂结束后很快就会在地层闭合压力的作用下重新闭合,因此需要某种固体物质来阻止裂缝的闭合。由此,开启了支撑剂的发展历史。

Fig. 2. 41　Fracturing Proppants
Source: https://en. wikipedia. org/wiki/Hydraulic_fracturing_proppants)

2.4.1.1 Conventional Fracturing Fluid and Proppant

Table 2.4 summarized the basic types, functions and requirements of conventional fracturing fluid and proppant.

2.4.1.1 传统压裂液与支撑剂

表2.4总结了传统压裂液与支撑剂的类型、功能、性能及其要求。

Table 2.4 Conventional Fracturing Fluid and Proppant

Fracturing Fluid Types	Proppant Types
(1) Water-based (2) Oil-based (3) Foam-based (4) Acid-based	(1) Sand (natural proppant) (2) Ceramsite (artificial proppant) (3) Coating proppant (resin-coated proppant)
Fracturing Fluid Functions	Proppant Functions
(1) Initiate and propagate the fracture (2) Develop fracture width (3) Transport proppant throughout the length of the fracture (4) Easily produced back to the surface after the fracture treatment is finished, leaving a fracture with the maximum permeability	(1) Support hydraulic fractures (2) Provide flow conductivity for propped fractures
Fracturing Fluid Requirements	Proppant Requirements
(1) Stable, predictable rheology under surface and downhole treating conditions and treatment duration (2) Low friction pressures drop at high pump rates in tubing and flow lines (3) Provide fluid loss control (4) Clean and easily degradable to minimize formation damage to propped fracture (5) Compatible with reservoir formation and fluids (6) Economical / low cost	(1) Uniform size and low density (2) High strength and small breaking rate (3) High roundness and sphericity (4) Less impurity content (5) Economical / low cost

2.4.1.2 Recent Advances in Fracturing Fluid

Major advances in fracturing fluid technology have been made over the past 15 years or so. This has been driven by a growing recognition of the following fluid issues: proppant transport; the damage to the proppant pack from fluid residues; and reservoir compatibility, for high-permeability reservoirs where fracture-face damage is important and for low-permeability reservoirs with respect to the prevention of fluid retention issues.

2.4.1.2 压裂液进展

过去的15年左右压裂液技术飞速发展,这是由于人们对压裂液的作用和存在的问题有了更深层次的认识:支撑剂的运移问题,残余压裂液对支撑剂层渗透率的伤害,压裂液与地层的配伍性(高渗地层要避免对裂缝面的堵塞,低渗地层要避免压裂液在裂缝中的大量残余)等。

1. Viscoelastic Surfactant Fluid Systems

The viscoelastic surfactant (VES) fluid systems utilizes surfactant technology to generate a long, wormlike micellar structure within a base water or brine, which then imparts viscosity to the system (Fig. 2. 42). The systems are extremely shear-thinning, easy to break and non-damaging. They have three additional properties that make them especially suitable for reservoir operations. First, as they break back to water viscosity, they are easy to recover; therefore, they are easy to deploy and apply. Second, the VES systems incorporate surfactant chemistries that act as biocides, low surface-tension modifiers, clay-stabilizers and nonemulsifiers: just add the two surfactants on the fly, and the system is complete. Finally, some of these systems can be recovered, recycled and re-utilized on additional treatments, with obvious cost-saving benefits. The VES systems, however, do have two distinct disadvantages: higher costs and lack of high-temperature stability (under 240 °F).

1. 黏弹性表面活性剂压裂液体系

黏弹性表面活性剂(VES)压裂液体系利用表面活性剂在水或盐水中产生长的蠕虫状胶束结构,以此得到所需黏度(图2.42)。该体系容易剪切变形,利于破胶且对储层无伤害。该体系被广泛应用于实际,主要有三点优势。第一,该体系破胶后黏度与水相当,易于返排出地层。第二,该体系可以与灭菌剂、低表面张力改性剂、黏土稳定剂和非乳化剂等添加剂完美配伍,且只需在施工现场加入这些添加剂即可。第三,该体系的部分可以被回收重复利用,大大降低了施工成本。但是,黏弹性表面活性剂压裂液体系同样具有两大缺点:成本较高且耐高温性能较差(低于240°F)。

Fig. 2. 42　VES Fluid Systems

(Source: https://weldonchem. com/content/ves454. html)

2. CO_2 Fracturing Systems

Carbon dioxide (CO_2) has been used in hydraulic fracturing for many years as a partial-phase component, usually to provide a foamed or energized phase to fluid systems, as it can be pumped in a liquid state by conventional

2. CO_2 压裂液体系

作为多相压裂液组分,CO_2多年来一直被应用于压裂液体系。它通常作为泡沫相或为压裂液系统充能,多通过常规泵送

pumping equipment (Fig. 2.43). However, two relatively recent applications of CO_2 fracturing technology have significantly advanced the application in low-pressure, fluid-sensitive environments.

设备以液态形式泵入(图2.43)。然而近年来CO_2压裂液体系的多被应用于低压、敏感性地层。

The first of these applications is the employment of a CO_2 - methanol system, eliminating the use of water entirely. Proppant is slurried directly into the methanol base and mixed with the CO_2 at very high pressures before being pumped down the wellbore. Within the formation, the CO_2 turns to gas and the methanol should vaporize, leaving little or no fluid residue.

一种CO_2压裂液体系就是CO_2—甲醇压裂液体系。该压裂液体系完全不需要使用水,支撑剂直接被加入甲醇基液中并在非常高的压力下与CO_2混合,然后泵送到井筒中。在该压裂液体系进入地层之后,CO_2由液态变为气态,甲醇也逐渐挥发,地层中几乎没有压裂液残留。

Fig. 2.43 CO_2 Fracturing on Field

(Source: https://sxycpc.com/news.NewsContent)

Another innovative application of CO_2 for fracturing is the use of a 100% CO_2 system. With this method, proppant is mixed directly into a liquid CO_2 base using a special high-pressure refrigerated blending system that looks and operates like a bulk CO_2 tank with a sand auger running along the bottom. Before the treatment, proppant is loaded into the CO_2 blender and liquid CO_2 is then used to cool the proppant down. Once the proppant is at the correct temperature and pressure for liquid CO_2 fracturing, the auger can then be used to add the proppant at controlled rates to a stream of liquid CO_2. Although treatments are small, this technique represents the ultimate in non-damaging, zero residue, easy-to-recover fracturing fluids.

另一种CO_2压裂液体系是只使用CO_2作为压裂液。该体系使用特殊的高压冷冻混合设备将支撑剂直接混合到液态CO_2中。在水力压裂之前,将支撑剂装入CO_2混合设备中,然后利用液态CO_2冷却加压支撑剂。当支撑剂被冷却加压到合适的温度压力后,再使用螺旋推进器将一定比例的支撑剂添加到液态CO_2中。尽管该体系所能进行的压裂施工规模很小,但其完全不伤害储层、零残留、易返排。

3. Nanoparticles-Stabilized Foam Fracturing Fluid

3. 纳米粒子稳定泡沫压裂液

Special attentions are being focused on the capacity of nanoparticles-stabilized foams as carrier fluids to prevent proppant settling, maintain a larger propped area, and improve hydrocarbon productivity from the unconventional reservoirs (Fig. 2. 44).

随着纳米科学的不断发展，人们逐渐开始关注纳米稳定化泡沫作为压裂液的携砂能力，以防止支撑剂沉降，维持更大的支撑面积，并提高非常规油气藏的油气产量(图 2. 44)。

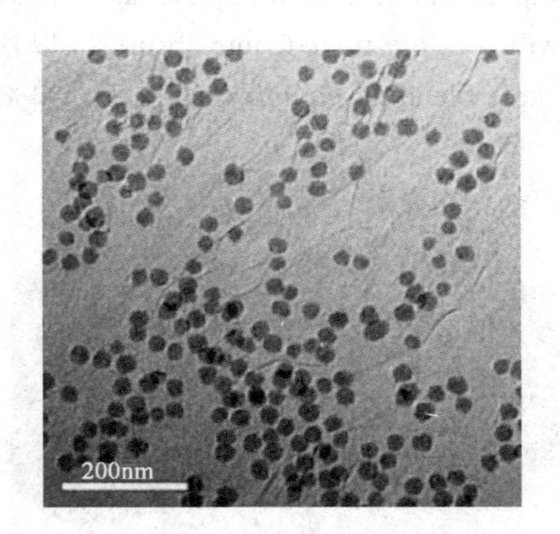

Fig. 2. 44 Nanoparticles-stabilized Foam Fracturing Fluid
(Source: https://foresproppants. com)

Presently, the stability and rheological properties of ultra-low water contents nanoparticles stabilized foam fracturing fluid is not yet explicit due to limited studies and no field trials has been reported so far. Nanoparticles can serve dual purposes of foam stabilizers and nano-proppants during fracturing treatments by nanoparticles-stabilized foams. The nanoparticles could serve as proppants by flowing into the micro-sized fissures and the nano-pore of shale and ultra-tight formations to generate more fractures in the matrix, increasing the conductivity of the fissures and prevents micro-fissures closure after the fracturing. The potential benefits of nanoparticles as foam stabilizers and nano-proppants are yet to be extensively investigated through experimental studies, simulation predictions and oilfield trials. Further experiments are required.

目前，由于研究有限，超低含水纳米粒子稳定泡沫压裂液的稳定性和流变性尚不明确，且未有现场试验报道。在用纳米粒子稳定泡沫压裂液体系进行压裂时，纳米粒子可以起到泡沫稳定剂和纳米支撑剂的双重作用。纳米粒子可以流入微小裂缝和页岩的纳米孔隙作为支撑剂，并能在致密地层基质中产生更多的裂缝，增加裂缝的导流能力且防止压裂后的微裂缝闭合。尽管有上述作用，纳米粒子作为泡沫稳定剂和支撑剂的潜在益处尚未通过实验研究，该压裂液体系还需要进一步研究。

2.4.1.3 Recent Advances in Proppant

1. Low and Neutral Density Proppant

The use of low and neutral density proppants has become increasingly common (Fig. 2.45). Currently, the full potential of these products has yet to be realized in the field. Ultimately, they have the capability to completely revolutionize the way treatments are performed, as it will no longer be necessary to use highly viscous fracturing fluids, nor to use complex blending equipment to add proppant on the fly. Instead, proppant will be pre-mixed into brine carrier fluids with relatively simple blending equipment.

2.4.1.3 支撑剂进展

1. 中低密度支撑剂

中低密度支撑剂在水力压裂中的应用越来越广泛(图2.45)。由于其密度较低,与常规高密度支撑剂相比,中低密度支撑剂并不需要高黏度压裂液携带,更不需要现场复杂的混砂设备进行混砂。相反地,可以用简单的设备将中低密度支撑剂提前加入压裂液中。

Fig. 2.45 Low Density Proppant

(Source: https://proppantprice.com)

2. Ultra-Lightweight Proppant (ULWP)

ULWPs are designed to have neutral density within a fracturing fluid (Fig. 2.46). This means that the proppant particles will remain suspended in the fluid without the requirement for significant fluid viscosity, resulting in two major advantages. The principal benefit is that the proppant can be readily and efficiently placed using a simple brine-based fluid system. Clearly this will significantly reduce the operational complexity and treatment costs, while maintaining the ability of the proppant to stay suspended in the fracturing fluid. The secondary benefit is that the ULWP can be pre-mixed into the brine and held in tanks on the location until it needs to be pumped. This pre-mixing approach eliminates the need for complex blending, proportioning and metering systems on location. This approach will also mean that the concentration of the ULWP can be precisely controlled in a pre-mix situation, thereby avoiding unplanned/ undesirable "events" during pumping.

2. 超低密度支撑剂(ULWP)

顾名思义,该支撑剂(图2.46),密度极低。这样即使压裂液黏度不高,支撑剂仍不会沉降,从而使该种支撑剂有两大优势。最大的优势在于,使用简单的压裂液体系即可有效地悬浮支撑剂。显然,这将显著降低操作复杂度和施工成本,同时保持支撑剂悬浮在压裂液中。另一个优势是能在压裂施工前与压裂液混合。这种提前混合就无需在现场进行复杂的混合配比和计量,也就意味着加入支撑剂的量可以提前、精确地计算,避免了在注压裂液过程中再进行配比混合的不良结果。

Fig. 2.46　Ultra-Lightweight Proppant
(Source: https://proppantprice.com)

The combination of these two benefits dramatically simplifies surface operational complexity, eliminating the blender and proppant-handling equipment in favor of mixing tanks and a booster pump. In addition, ULWPs have been used extensively with slickwater fracturing systems in tight and unconventional gas reservoirs, where the fluid systems are kept as simple as possible in order to reduce the costs, minimize formation damage and maximize fluid recovery.

正因为有上述两点优势,超低密度支撑剂的应用极大地简化了地面操作的复杂性,避免了使用搅拌机和支撑剂处理设备,有利于混合罐和增压泵的工作。超低密度支撑剂已被广泛用于致密和非常规气藏中的滑溜水压裂液体系,因为滑溜水压裂液体系需要保持尽可能简单的配方,以便降低成本、最小化地层伤害并最大化压裂液返排。

3. Fiber Technology

Fiber technology was developed to hold the proppant in the fracture during the production of oil, gas or both and to allow more flexibility in flowback design than possible with curable-resin-coated proppants (Fig. 2.47). These additives work by the physical mechanism of random fiber reinforcement; therefore, chemical curing reactions are not necessary to hold the proppant in place. No combination of temperature, pressure or shut-in time is required. Wells can be flowed back at high rates (dependent on the number of perforations). Also, flowback is possible immediately after the fracturing treatment is completed.

3. 纤维支撑剂

纤维技术在油气的生产过程中将支撑剂固定在裂缝中,并使返排设计具有比使用固化树脂包层支撑剂有更大的灵活性(图2.47)。这些纤维添加剂通过纤维增强的物理机制起作用,因此,不需要化学固化反应将支撑剂保持在适当位置,也不需要合适的温度、压力或关井时间使支撑剂沉降。使用纤维支撑剂时,压裂液可以在压裂完成后立即以高速返排。

Fig. 2. 47　Fiber Technology

(Source: https://www.hydraulicfrackingblog.com/fiber_proppant)

2.4.2　Acid Fracturing

The fracturing technology that employs acid as fracturing fluid without any proppant is called acid fracturing (Fig. 2. 48). Therefore, acid fracturing is actually a technology related to fracturing fluid. However, the first description of hydraulic fracturing applied to petroleum reservoirs is actually acid fracturing (discussed in 2. 1. 1) and the technology is still widely used. So in this section, acid fracturing is detailed introduced.

2.4.2　酸化压裂

用酸液作为压裂液而不加支撑剂的压裂,称为酸化压裂(图2.48),简称为酸压。因此,酸化压裂实际上属于一种压裂液相关压裂技术。但是,由于酸化压裂的出现要早于真正意义上的水力压裂(参见2.1.1节)并且该技术仍被广泛应用,所以在本部分,对酸化压裂技术单独详细进行介绍。

Fig. 2. 48　Acid Fracturing

(Source: https://sloilfield.com/Business-Solutions/Acid-Fracturing-Technical-Service.html)

Hydraulic fracturing with acid (usually hydrochloric acid) is an alternative to propped fractures in acid-soluble formations such as dolomites and limestones. The process relies on acid etching of the fracture face, rather than the placing of proppant, to produce conductivity.

用酸液(多用盐酸)进行水力压裂多应用于酸溶性地层,如白云岩地层和石灰岩地层。酸压主要依靠酸液对裂缝壁面的酸蚀,而非支撑剂的支撑,来提供导流能力。

2.4.2.1 Acid Fracturing Process

To start, a non-reactive pad fluid is pumped into the formation to create a fracture with the desired length. Acid system is then followed (the most commonly used acid fracturing fluid is 28% HCl), so that as much rock volume as possible can be removed. The acid system is also usually viscosified to reduce wellbore friction, to suspend and transport any fines released by the acid, and to retard the acid reaction rate. Crosslinked acid systems are commonly used to achieve these goals. A combination of retarded reaction rate and high pumping rate is intended to get "live acid" as far from the wellbore as possible.

2.4.2.1 酸压步骤

首先,将非反应性的前置液泵入地层产生裂缝并使其扩展到一定长度。之后加入酸液体系(最常用的酸液体系是浓度为28%的盐酸),酸蚀掉尽量多的岩石体积。酸液体系同时需要有一定的黏度以降低井筒摩阻,悬浮、运移酸蚀中产生的地层微粒,并降低酸液与地层的反应速度。一般多使用交联酸液体系实现上述目标。通常,多将减缓酸液与地层反应速率和地面高泵速组合,旨在增大"活性酸"的有效作用距离。

At the end of the treatment, the fracture is allowed to close and the spent acid flowed back as soon as possible. Sometimes a closed fracture acidizing treatment may be performed. This consists of pumping additional acid down the closed fracture, below fracturing pressure, to widen the etched width of the fracture and hence artificially increase conductivity.

压裂完成后,迅速返排酸液使裂缝闭合。有时还要在裂缝闭合后进行酸化处理,通常包括将多余的酸液在低于地层破裂压力的情况下注入闭合的裂缝,以增大酸蚀裂缝的宽度,从而提高裂缝的导流能力。

2.4.2.2 Acid Fracture Conductivity

The most reliable method for estimating acid fracture conductivity was presented by Nierode and Kruk in 1973. Other methods, including laboratory testing, have been developed subsequently, but none have proved to be as reliable. A summary of the method follows.

2.4.2.2 酸蚀裂缝导流能力

计算酸压后裂缝导流能力的最可靠方法是1973年Nierode和Kruk提出的方法,后来,又发展出了各种方法,包括实验测定法等,但这些方法都存在局限性。下面简述N-K法计算裂缝导流能力。

The created fracture conductivity C_f (md · ft), can be calculated using etched fracture width w_{etch} (in), rock embedment strength, S_{RE} (psi), and closure pressure, p_c (psi):

酸蚀裂缝的导流能力 C_f (md · ft)可以通过酸蚀裂缝宽度 w_{etch} (in)、岩石嵌入强度 S_{RE} (psi),和地层闭合压力 p_c (psi)来计算:

$$C_f = C_1 e^{-C_2 p_c} \tag{2.4}$$

$$C_1 = 1.47 \times 10^7 w_{etch}^{2.47} \tag{2.5}$$

The average etched fracture width w_{etch}, can be obtained by calculating the volume of rock dissolved by the acid and dividing by the fracture area. The C_2 term is related to rock embedment strength, S_{RE} as follows:

其中平均酸蚀裂缝宽度 w_{etch} 可以通过计算酸蚀的岩石体积再除以破裂面积来获得。C_2 与岩石嵌入强度 S_{RE} 有关：

$$C_2 = 0.001(13.9 - 1.3\ln S_{RE}), S_{RE} \leqslant 20,000\text{psi} \tag{2.6}$$

$$C_2 = 0.001(3.8 - 0.28\ln S_{RE}), S_{RE} > 20,000\text{psi} \tag{2.7}$$

The rock embedment strength has to be determined experimentally from core or outcrop samples.

岩石嵌入强度必须通过岩心或露头样品进行实验确定。

2.4.2.3 Reaction Stoichiometry

2.4.2.3 酸压化学反应

The main chemical reactions of interest in acid fracturing are those between HCl and calcium carbonate (limestone) or calcium-magnesium carbonate (dolomite) (Fig.2.49). The chemical reaction for limestone is written as

酸压中主要的化学反应是 HCl 和碳酸钙(石灰石)或碳酸钙镁(白云石)之间的化学反应(图 2.49)。HCl 和石灰石的反应方程式为

$$2HCl + CaCO_3 \longrightarrow H_2O + CO_2 + CaCl_2$$

and for dolomite as

HCl 和白云石的反应方程式为

$$4HCl + CaMg(CO_3)_2 \longrightarrow 2H_2O + 2CO_2 + CaCl_2 + MgCl_2$$

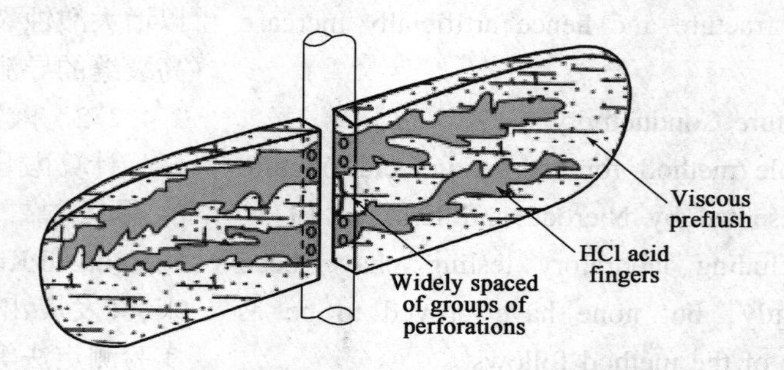

Fig. 2.49 WISPER (Wide Spread Etched Ridges) Process for Acidizing Homogenous Chalks

The first reaction equation indicates that two molecules of HCl react with one molecule of calcium carbonate to form one molecule each of water, carbon dioxide and calcium chloride. The second equation shows that four molecules of HCl react with one molecule of calcium-magnesium carbonate to form two molecules each of water and carbon dioxide and one each of calcium chloride and magnesium chloride.

第一个方程式表明两分子量的 HCl 与一分子量的碳酸钙反应形成一分子量的水、二氧化碳和氯化钙。第二个方程式表明,四分子量的 HCl 与一分子量的碳酸钙镁反应形成两分子量的水和二氧化碳,以及一分子量的氯化钙和氯化镁。

These equations can be used to determine the dissolving power X_C of the acid, which is the volume of rock dissolved per unit volume of acid reacted. The mass dissolving power (i. e. the mass of rock dissolved per unit mass of acid reacted) is first defined as

这些方程式可用于确定酸的溶解能力 X_C，即每单位体积的酸可以溶解的岩石体积。要计算 X_C，先计算质量溶解力 β（即每单位质量的酸可以溶解的岩石质量），β 定义为

$$\beta = \frac{(\text{molecular weight of rock}) \times (\text{rock stoichiometric coefficient})}{(\text{molecular weight of acid}) \times (\text{acid stoichiometric coefficient})} \tag{2.8}$$

For the limestone reaction,

对于白云石而言，

$$\beta = \frac{100.09 \times 1}{36.47 \times 2} = 1.372$$

So that each gram of 100% pure HCl dissolves 1.372g of rock. To obtain the dissolving power, the masses must be converted to volumes:

即每克纯 HCl 可以溶解 1.372 克白云石。为了获得酸的溶解能力，要将质量换算为体积：

$$X_C = \frac{\rho_C \beta C}{\rho_{CaCO_3}} \tag{2.9}$$

where ρ_C and ρ_{CaCO_3} are the densities of the acid solution and calcium carbonate, respectively, and C is the weight-fraction concentration (e. g. 0.28 for 28% acid). For example, the specific gravity of 28% acid is 1.14, whereas that for 15% acid is 1.07. Applying this calculation for the limestone – HCl reaction, X_{15} is 0.082, and X_{28} is 0.161. Similarly, for dolomite the values are 0.071 and 0.141, respectively.

其中 ρ_C、ρ_{CaCO_3} 是酸液和碳酸钙的密度，C 是质量浓度分数（如 28% 的盐酸 C 值为 0.28）。举例来说，28% 酸的密度为 1.14，而 15% 酸的密度为 1.07。将该计算应用于石灰石—HCl 反应，X_{15} 为 0.082，X_{28} 为 0.161。类似地，对于白云石，其值分别为 0.071和 0.141。

The stoichiometric equations for acid reactions provide a relation for the coupling between fracture geometry and acid spending. Because there are many unknowns in acid fracturing, the modeling can be simplified by neglecting the variation in density of the acid and using that of 10% acid, which is a suitable average for most acid fracture treatments. In this case, X_{100} can be approximated as $10X_{10}$ and X_C as

这些方程式还为裂缝几何参数和酸消耗量之间的耦合提供了联系。由于在酸压中存在许多未知因素，因此可以通过忽略酸的密度变化并使用 10% 的酸来简化建模，这对于大多数酸压施工来说是合适的平均值。

CX_{100}. Now, consider the volume of a fracture element of cross-sectional area (width times height) A and length δ_x in which the acid concentration changes by an amount ΔC. The volume of acid spent is $A\delta_x\Delta C$, and the volume $\Delta A\delta_x$ of rock dissolved is

在这种情况下，X_{100} 可以近似为 $10X_{10}$，X_C 可以近似为 CX_{100}。此时，再考虑横截面积（宽度乘以高度）A 乘以长度 δ_x，其中酸浓度改变量为 ΔC。酸消耗量为 $A\delta_x\Delta C$，溶解的岩石体积 $\Delta A\delta_x$ 为

$$\Delta A_{etch} = X_{100} A \Delta\overline{C} \tag{2.10}$$

where A_{etch} is the etched area and $\overline{C}$ is the average acid concentration in the cross section.

其中 A_{etch} 是酸蚀区域的面积，$\overline{C}$ 是通过该面积的平均酸液浓度。

2.4.2.4 Acid Fracturing Limitations

1. Reservoir Limitation

Generally, acid fracturing is limited to hard rock formations since soft formations will plastically deform into the etched width created by the acid, reducing and even eliminating the effects of the treatment. In addition, acid fracture conductivity tends to be significantly less than that can be generated by proppant fracturing. These two facts mean that fracture acidizing is generally performed on lower-permeability, tight carbonate formations, rather than soft, more permeable rocks.

However, it has also been suggested that with modern acid systems, the technique could be applied to sandstone reservoirs.

2. Control Limitation

Acid fracturing may be preferred operationally because the potential for unintended proppant bridging and proppant flowback is avoided. However, designing and controlling the depth of penetration of the live acid into the formation and the etched conductivity are more difficult than controlling proppant placement. Acid penetration is governed by the chemical reaction between the rock and the acid, and conductivity is determined by the etching patterns formed by the reacting acid. In both cases, acid fracturing introduces a dependence on rock properties that is not present in propped fracturing. In addition, the properties that acid fracturing

2.4.2.4 酸压局限性

1. 地层限制

酸压适用于较硬的地层，因为疏松地层酸压后地层会变形导致裂缝闭合，从而消除酸压的效果。此外，由于酸蚀裂缝的导流能力远远低于填砂裂缝，因此酸化压裂多应用于低渗、致密碳酸盐岩地层，而非疏松、高渗地层。

但近年来也有学者提出，通过使用现代酸液体系，酸压技术也可以用于砂岩油气藏中。

2. 控制限制

与水力压裂相比，酸压不存在支撑剂堵塞或回流的现象。但是，设计和控制活性酸的有效作用距离及酸蚀裂缝的导流能力，要比设计和控制填砂裂缝困难得多。活性酸的有效作用距离与酸岩反应速率有关，而酸蚀裂缝的导流能力又由酸岩反应的刻蚀形状控制。无论如何，酸压的计算总与地层岩石的性质有关，并且酸压设计和要控制的

design and control depend on are usually more difficult to determine than other formation properties.

参数都很难确定具体值。

2.4.3 Fracturing Fluid Related Technologies

2.4.3 压裂液相关压裂新技术

Fracturing fluid and proppant related technologies refer to a series of new techniques for fracturing using fracturing fluid or proppant systems that are different from conventional fracturing, such as the use of foam-based fracturing fluids or hydrocarbon-based fracturing fluids, as well as some special proppant systems.

压裂液、支撑剂相关压裂新技术是指使用不同于常规压裂液、支撑剂体系的压裂技术,如使用泡沫为基液的压裂液或烃为基液的压裂液进行压裂,或使用一些特殊的支撑剂体系进行压裂。

2.4.3.1 Slickwater Fracturing

2.4.3.1 滑溜水压裂

Slickwater, also known as friction reduce water, is one of the most commonly used fracturing fluids for hydraulic fracturing in low permeability reservoirs. It is also one of the key technologies for the successful development of shale gas in the United States (Fig. 2.50).

滑溜水又称减阻水,是低渗储层水力压裂最常用的压裂液之一,滑溜水压裂技术也是美国成功开发页岩气的关键技术之一(图2.50)。

Fig. 2.50 Shale Gas Fracturing in United States
(Source: https://ssecon.com/shale-gas/hydraulic-fracturing)

The key to this technology is the composition of the fracturing fluid. In general, water and proppant account for 98% ~ 99% of the slickwater fluid, and only 1% ~ 2% is the additive. However, the role of additives is very important, generally including friction reducers, surfactants and clay stabilizers, which are used to make up for the low viscosity and poor carrying capacity of water. The slickwater fracturing technology employs a large displacement, low sand ratio, and low viscosity method, and often forms complex fracture networks in the formation.

该技术的关键在于压裂液的配置。一般而言,滑溜水压裂液中98% ~99%都是水和支撑剂,只有1% ~2%为添加剂,但添加剂的作用却至关重要。添加剂一般包括降阻剂、表面活性剂和黏土稳定剂等,弥补了水黏度低、携砂能力差等缺陷。滑溜水压裂技术采用大排量、低砂比、低黏度的施工方式,往往能在地层中形成复杂的缝网。

2.4.3.2 Compound Fracturing

Compound fracturing is a technology that different types of fracturing fluid systems are selected specifically for pad fluid, proppant slurry and displacement fluid, respectively (Fig. 2. 51). Generally speaking, low-viscosity slick water or linear guar gum acts as pad fluid to create the fracture, and the cross-linked jel fluid acts as a proppant carrier to carry the proppant into the fracture. Finally, slickwater makes the final displacement. Needless to say, each fracturing fluid system can be flexibly modified to adapt to the fracturing operations of different reservoirs to achieve the best results.

2.4.3.2 混合压裂

混合压裂是前置液、携砂液和顶替液都采用不同类型压裂液体系的一项压裂技术,又称为变组分压裂(图 2.51)。通常而言,低黏度的滑溜水或线性瓜尔胶作为前置液进入地层使裂缝起裂延伸,交联冻胶液作为携砂液携带支撑剂进入裂缝充填,滑溜水作为顶替液做最后的驱替。当然,各段压裂液体系可以灵活变通,以适应不同性质地层的压裂施工,达到最佳的效果。

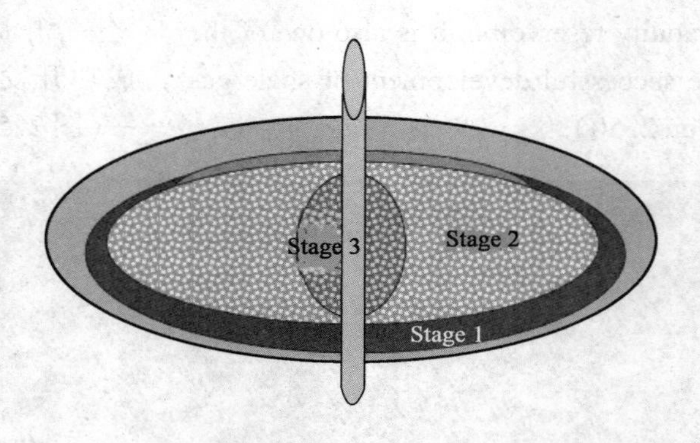

Fig. 2. 51 Compound Fracturing

2.4.3.3 Water-Free Fracturing

Waterfree fracturing is a nomenclature for fracturing operations without the use of water-based fracturing fluids or even the water itself at all.

1. High Energy Gas Fracturing (HEGF)

The fracturing technology of forming a number of radial fractures around the wellbore by high temperature and high pressure gas generated from gunpowder or propellant is called high energy gas fracturing (Fig. 2. 52). It is a new fracturing technology developed after explosive fracturing and hydraulic fracturing.

2.4.3.3 无水压裂

无水压裂,就是不使用水基压裂液甚至完全不使用水进行压裂作业的统称。

1. 高能气体压裂(HEGF)

利用火药或推进剂产生的高温高压气体压裂形成多条径向裂缝的技术称为高能气体压裂(图 2. 52)。它是继爆炸压裂、水力压裂后发展出的一项新型压裂技术。

Fig. 2.52　High Energy Gas Fracturing
(Source: https://bhge.com/upstream/completions/sand-control)

(1) Principles: High energy gas fracturing mainly possesses three stages: ① The in-well pressurization stage, in which the pressure in the wellbore gradually increases, and the energy is stored in the wellbore in the form of compressed liquid or gas; ② Fracture initiation stage, the fractures are opened at this stage, but they are short, and the energy is stored in the wellbore and the compressed fluid in the fractures; ③ Fracture propagation stage, the fractures quickly propagate forward until the energy is consumed. In addition to the mechanical stress caused by high temperature and high pressure gas, high energy gas fracturing also relies on impulsive effect of the explosion which lifts and drops the liquid in the wellbore to reduce the near wellbore damage, and relies on high temperature and chemical reaction to remove wax and easily-blocked particles.

(1)原理:高能气体压裂主要经过三个阶段:①井内增压阶段,在该阶段井筒内压力逐渐增大,能量以压缩液体或气体的形式储存于井筒内;②起裂阶段,此阶段裂缝刚被压开,裂缝较短,能量储存于井筒和缝内的压缩流体中;③延伸扩展阶段,裂缝迅速向前延伸,直至能量消耗完。除高温高压气体的机械应力作用产生裂缝外,高能气体压裂的爆炸使井筒内液体提升回落,产生脉冲冲击作用从而解堵地层,高温作用和化学作用还可以清除蜡质、易堵塞微粒。

(2) Procedures: ① Preparation: including well washing, sand flushing, well drifting, and preparation of killing fluid. ②Projectile assembly: It is recommended to assemble a shell-type or a shell-free gunpowder generator or use liquid gunpowder, and connect the projectiles step by step at the wellhead and install the ignition device. ③Conveying: cable conveyed or tubing conveyed are all feasible whereas the former uses a cable for ignition, and the latter uses a rod. ④After the gunpowder is delivered to the specified position, ignition and fracture. ⑤ After the fracturing operation, the cable is taken out and the whole well shall be ready for production.

(2)施工步骤:①施工准备:包括洗井、冲砂、通井、准备压井液。②弹体组装:可以组装有壳式、无壳式火药发生器或使用液体火药,并在井口将弹体逐级连接,装上点火装置。③输送:选择电缆输送或油管输送,前者用电缆进行点火,后者投棒点火。④将炸药运送到指定位置后,点火压裂。⑤压裂完成后,起出电缆,准备生产。

(3) Application: After years of development, high energy gas fracturing technology is now often used in combination with other technologies. For example, it can be combined with perforating to perform super-positive pressure perforation, which can not only form perforation channels, but can also strengthen the channels to form double-wing fractures, thereby achieving the purpose of protecting reservoirs and increasing productivity. When it is combined with hydraulic fracturing and acid fracturing, the field shows that the composite technology achieves better results than only utilize a single technology.

(3)应用:经过多年的发展,高能气体压裂技术如今多与其他技术联合使用。例如,它与射孔技术复合可以进行超正压射孔,既能形成射孔通道,又能对射孔通道进行强化处理形成双翼裂缝,达到保护储层、提高产能的目的;也可以与水力压裂和酸压复合。现场试验表明,复合技术取得了比单一压裂或酸化更好的效果。

2. Fracturing with Carbon Dioxide

2. CO_2 压裂

Fracturing with carbon dioxide can be divided into CO_2 foam fracturing, CO_2 fracturing, supercritical CO_2 fracturing and CO_2 foam fracturing. There is also a technology uses CO_2 foam fracturing fluid, which is composed of CO_2 and foaming agent added to the water-based fracturing fluid. Therefore, this fracturing technology is basically the same as conventional fracturing technology, and will not be introduced in this book.

CO_2 压裂技术分为 CO_2 干法压裂技术、超临界 CO_2 压裂技术和 CO_2 干法泡沫压裂技术。CO_2 泡沫压裂技术使用 CO_2 泡沫压裂液,它是在水基压裂液中加入 CO_2 和起泡剂构成的,因此它的压裂技术和常规压裂技术基本一致,不做过多介绍。

(1) CO_2 fracturing.

(1) CO_2 干法压裂。

CO_2 fracturing technology employs liquid CO_2 as fracturing fluid and does not involve any water during its whole procedure. The mainly added chemical agent is viscosifier. Since CO_2 is soluble in crude oil, this technology can significantly reduce the viscosity of crude oil and increase the energy of dissolved gas. In addition, CO_2 can also be dissolved in water to form carbonic acid, which inhibits the expansion of clay minerals. Therefore, CO_2 fracturing technology has incomparable advantages. The surface manifold of CO_2 fracturing technology is shown in Fig. 2.53.

CO_2 干法压裂技术使用液态 CO_2 作为压裂液,不含任何水,添加的化学剂主要是增黏剂。由于 CO_2 溶于原油,所以该技术可以大幅降低原油黏度,增加溶解气驱的能量。除此之外,CO_2 还可以溶解在水中形成碳酸,抑制黏土矿物膨胀。因此,CO_2 干法压裂技术具有其他压裂技术不可比拟的优点。CO_2 干法压裂技术地面管汇如图 2.53 所示。

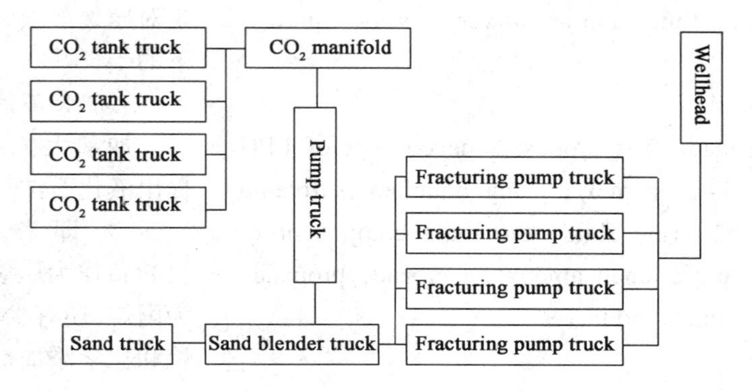

Fig. 2. 53 Surface Manifold of CO_2 Fracturing Technology

(2) Supercritical CO_2 fracturing.

When the temperature and pressure is greater than the CO_2 critical value (T_c =31.1℃, p_c =7.38 MPa), CO_2 will reach the supercritical state and become the supercritical CO_2 fluid. Under the high pressure and high temperature conditions of a fracturing operation, CO_2 can easily reach supercritical state. The density of supercritical CO_2 is close to that of liquid, its viscosity is close to that of gas, its surface tension is close to zero, and its diffusion coefficient is very high, which means it has a strong penetrating ability.

(2)超临界 CO_2 压裂。

当温度、压力大于 CO_2 临界值(T_c = 31.1 ℃、p_c = 7.38 MPa)时,CO_2 就会达到超临界状态,成为超临界 CO_2。在压裂施工的高压高温条件下,CO_2 很容易达到超临界状态。超临界 CO_2 的密度接近液体,黏度接近气体,表面张力接近于0,而且扩散系数较高,具有很强的渗透能力。

Supercritical CO_2 fracturing technology has all the advantages of traditional CO_2 fracturing technology, and has the advantages of lower surface pressure, lower requirements for sand mixing vehicles, and better production. It is the future trend of fracturing with carbon dioxide.

超临界 CO_2 压裂技术具备传统 CO_2 干法压裂技术的全部优点,并且还有施工压力小、对混砂车要求低、增产效果好等优点,是 CO_2 压裂技术的发展趋势。

(3) CO_2 foam fracturing.

This technology uses foaming agent to form nitrogen foam in liquid CO_2, which can increase the viscosity of the liquid CO_2 and protect them from vaporization. Due to the higher viscosity, this fracturing technique provides better control of fluid loss; at the same time, the friction is smaller, allowing the proppant to be pumped into the formation at lower pumping rates. Typical objects treated by CO_2 foam fracturing technology are dry gas wells that are

(3) CO_2 干法泡沫压裂。

该技术利用起泡剂在液态 CO_2 中形成泡沫,既能增加压裂液黏度又能保护液态 CO_2 不蒸发。由于具有更高的黏度,该压裂技术可以更好地控制滤失;同时摩阻较小,使得支撑剂可以在较低的泵送速度下泵入地层。这种压裂技术处理的典型对象

sensitive to fracturing fluids and are at very low pressure.

是对压裂液敏感、处于低压状态下的干气井。

3. LPG Fracturing

3. 液化石油气压裂

GasFrac of Canada first proposed the concept of LPG fracturing (Fig. 2. 54), namely, using liquefied petroleum gas (LPG) as a fracturing fluid, the main component of which is propane, and a small amount of ethane, propane, butane and other chemical additives.

加拿大 GasFrac 公司最先提出液化石油气压裂的概念(图 2. 54),即采用液化石油气(LPG)作为压裂液,其主要成分为丙烷,还有少量乙烷、丙烯、丁烷和化学添加剂。

Fig. 2. 54 LPG Fracturing at the Oil Field
(Source: https://energyindepth.org/just-the-facts-anti-marcellus-activists-distortions-about-natural-gas-jobs-unsupported-by-the-facts/)

A typical LPG fracturing operation is shown in Fig. 2. 55, and the main considerations are safety issues and the problem of keeping the LPG in liquid phase. The proppant is first placed in a container and the nitrogen is circulated throughout the system to prevent air from entering. The LPG is then injected into the proppant container and pressurized with nitrogen to maintain the LPG in a liquid state. The thickened LPG fracturing fluid is then injected into the well. Continue to increase the pressure until the fractures are created and then pump into the proppant slurry. When the injection volume reaches the designed scale, stop the pump, and shut in the well.

典型的 LPG 压裂作业如图 2. 55 所示,施工过程中主要考虑的问题是安全问题和使 LPG 保持液态的问题。首先将支撑剂放入密闭容器中,氮气循环整个系统,密封防止空气进入。再向支撑剂容器注入 LPG,氮气加压使其保持液态。随后向井底注入经过稠化的 LPG 压裂液,持续加压直至裂缝起裂,然后泵入携砂液。待注入量达到设计的规模,停泵,关井。

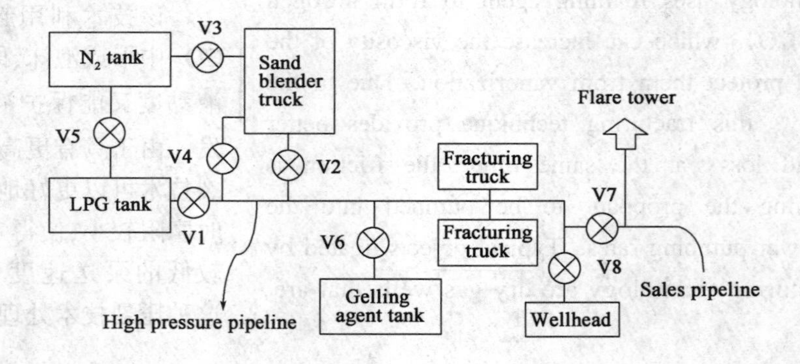

Fig. 2. 55 LPG Fracturing Operation

Compared with conventional fracturing, LPG fracturing technology has the following characteristics:

(1) High fracturing fluid recovery: LPG fracturing fluid has a density of only half of water, and its viscosity and surface tension are also very low. After fracturing, it can be completely recovered.

(2) Long stable production period: LPG fracturing fluid will be in a volatilized form with natural gas, or mixed with crude oil to reduce the density of crude oil and improve recovery. At the same time, the problem of the retention of the hydrous phase does not exist in LPG fracturing, which indicates that LPG fracturing can greatly increase the effective fracture length.

(3) Small reservoir damage: LPG fracturing fluid is highly compatible with reservoir fluids, and it will not cause reservoir damage even if it encounters various types of sensitive clay minerals.

(4) Low cost: Although the cost of LPG fracturing fluid is higher than that of water-based fracturing fluid of the same volume, if the cost of post-treatment of water-based fracturing fluid is calculated, then LPG fracturing fluid has higher cost performance. When the LPG fracturing fluid is flowed back, it can usually be reused or enter directly into the production pipeline.

与常规压裂相比,LPG 压裂具有以下优点:

(1)返排效果好:LPG 压裂液密度只有水的一半,其黏度和表面张力也很低,在压裂后可以进行彻底的返排。

(2)稳产周期长:LPG 压裂液会与天然气以挥发形式互融,或与原油混相降低原油的密度,使采收率提高。同时,LPG 压裂液没有水相滞留的问题,可以极大增加有效缝长。

(3)储层伤害小:LPG 压裂液与储层流体配伍性极高,即使遇到各类敏感性黏土矿物也不会造成储层伤害。

(4)成本低:尽管与相同体积的水基压裂液相比,LPG 压裂液的成本要高,但是如果算上水基压裂液后期处理的成本,那么 LPG 压裂液的性价比更高。LPG 压裂液返排后通常可以重复利用,或直接进入生产管线。

2.4.4 Proppant Related Technologies

2.4.4.1 Hiway Flow-Channel Fracturing

Hiway flow-channel fracturing (Hiway fracturing) is a new technique introduced by Schlumberger in 2010 to maximize the conductivity of sand-filled fractures. This technique maximizes oil and gas production by forming large flow channels within the proppant pack.

2.4.4 支撑剂相关压裂新技术

2.4.4.1 高速通道压裂技术

高速通道压裂技术(Hiway Hydraulic Fracturing)是 2010 年斯伦贝谢公司推出的新技术,旨在最大幅度地提高填砂裂缝的导流能力。该技术通过在支撑剂内部形成大的流动通道,实现油气产能的最大化。

As shown in Fig. 2. 56, hiway fracturing employs pulsed sanding, that is, adding a part of fracturing fluid with proppant and then adding another part of fracturing fluid without proppant until all proppant is pumped into the formation. Compared with conventional fracturing, this technique breaks the notion that the fracture conductivity is provided by the proppant. The fracture is not filled with continuous proppant but with a number of proppant packs. The fracture conductivity is mainly derived from the large channels between the proppant packs. These channels are connected to each other to form a complex network, and with large channels containing small channels, the sand-filled fracture conductivity is greatly improved. At the meantime, the existence of these channels eliminates the influence of fracturing fluid gel residue, proppant debris and formation particles on the conductivity, and can significantly improves the fracturing performance.

如图 2. 56 所示,高速通道压裂工艺采用脉冲式加砂,即加一段含有支撑剂的压裂液后再加一段不含支撑剂的压裂液,直至完成加砂。与常规压裂相比,该技术打破了传统压裂依靠支撑剂提供裂缝导流能力的理念。该技术的人工裂缝不是由连续的支撑剂进行支撑,而是由众多支撑剂团进行支撑,裂缝的导流能力主要来源于支撑剂团之间的大通道。这些大通道相互连通形成立体网络,大通道内包含小通道,极大提升了填砂裂缝的导流能力。同时,这些大通道的存在消除了压裂液破胶残渣、支撑剂碎片和地层微粒对导流能力的影响,显著提高水力压裂增产效果。

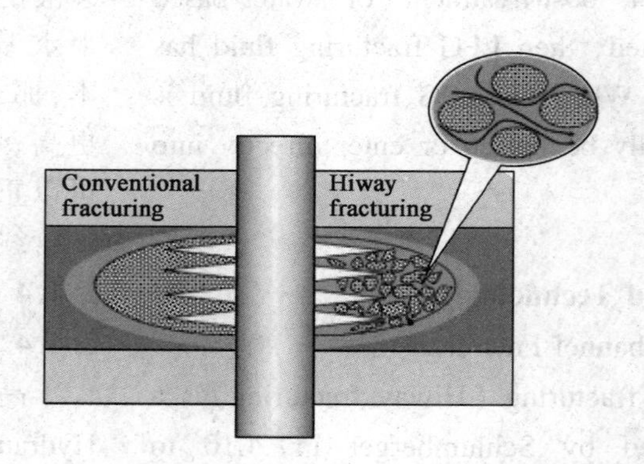

Fig. 2. 56 Conventional Fracturing Compared with Hiway Fracturing

The hydraulic fractures that were fractured by the hiway fracturing operation have the channel size of millimeters, which is more than 10 times the size of the conventional fracture channel. Obviously, the hiway fracturing technology can significantly improve fracture conductivity and fracture anti-pollution ability, reduce the difficulty of sanding, and increase the fracture size.

高速通道压裂形成的填砂裂缝,其通道大小为毫米级,这是传统填砂裂缝通道大小的 10 倍以上。显而易见,高速通道压裂技术能显著提高裂缝导流能力和抗污染能力,降低加砂难度,增大裂缝几何尺寸。

2.4.4.2 Fiber Fracturing

Fiber fracturing is a new fracturing technology that uses fiber sand control proppant (Fig. 2.57) to replace the traditional proppant for fracturing. The fibers and proppants intertwined with each other within the fracture, and are entangled to form a network structure, which effectively blocks the flow of the sand without affecting the flow of the fluid.

2.4.4.2 纤维防砂压裂技术

纤维防砂压裂就是使用纤维防砂支撑剂(图2.57)替代传统支撑剂所进行的水力压裂。如前所述,纤维与支撑剂颗粒在裂缝中相互接触,缠绕形成网状结构,有效阻碍砂粒的通过,但又不影响流体的通过。

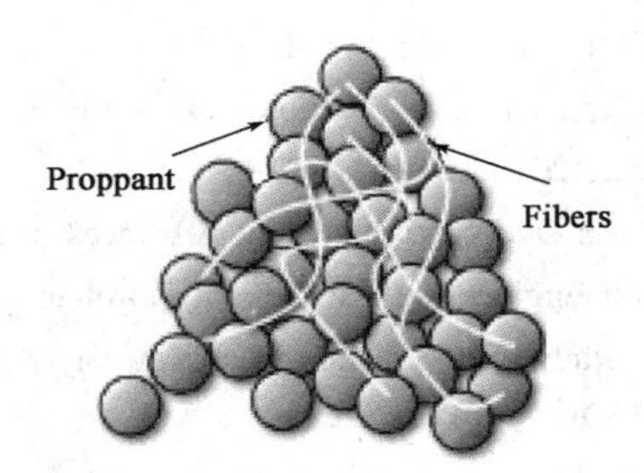

Fig. 2.57 Basic Principles of Fiber Fracturing

In addition to preventing the formation sand from entering the wellbore, the technology can also prevent the proppant from flowing back into the wellbore. Therefore, the technology is not only suitable for fracturing in sand-producing formations, but also recommended for rapid flowback operations after fracturing. Besides, if degradable fiber is used, the fiber will form a channel in the proppant after degradation and improve the conductivity of the fracture.

该技术除了能防止地层砂进入井筒外,还可以防止支撑剂返排回流进井筒。因此该技术不仅适用于易出砂地层的水力压裂,还适用于压裂后快速返排作业。同时,若使用可降解纤维,则压后纤维会在支撑剂中返排形成孔道,提高裂缝的导流能力。

Exercises/练习题

1. Describe the main roles of hydraulic fracturing. /简述水力压裂的主要作用。

2. How can a hydraulic fracturing be evaluated?/如何评价水力压裂的效果?

3. A reservoir has a permeability of 0.1mD. A vertical well of 0.328ft radius draws the reservoir from the center of a radius of 1500ft. A fracture design resulting in length 400ft and conductivity equal to 1000 mD · ft. calculate the fracture equivalent skin effect and the fold of increase in well productivity. /一个油藏的渗透率为0.1mD,半径为0.328ft的直井波及该油藏的半径为1500ft,压裂设计的裂缝长度为400ft,导流能力为1000 mD · ft。计算裂缝等效表皮系数和油井产能增加倍数。

4. What is re-fracturing?/什么是重复压裂?

5. How are the re-fracturing wells selected?/如何选择重复压裂井?

6. Describe the re-fracturing methods. /简述重复压裂方法。

7. How is the fracture height controlled?/如何控制裂缝的高度?

8. How is the tip screenout fracturing implemented?/如何实现端部脱砂压裂？

9. What are the main applications of the tip screenout fracturing?/端部脱砂压裂的主要应用有哪些？

10. Describe briefly the recent advances in fracturing fluid./简述压裂液的进展。

11. Describe briefly the recent advances in proppant./简述支撑剂的进展。

References/参考文献

[1] Cipolla C L, Wright C A. Diagnostic Techniques To Understand Hydraulic Fracturing: What? Why? and How? [J]. SPE Production & Facilities, 2002, 17(1):23-35.

[2] Wood D A. Hydraulic Fracturing Impacts and Technologies: a Multidisciplinary Perspective[J]. Journal of Natural Gas Science & Engineering, 2016, 28:A4.

[3] Veatch R W, Moschovidis Z A. An Overview of Recent Advances in Hydraulic Fracturing Technology[C]. International Meeting on Petroleum Engineering. Society of Petroleum Engineers, 1986:40-50.

[4] Wang X W, Ren S, Song Y G. Study on the distance between pay zones in multi-layer and zonal fracturing [J]. Natural Gas Industry, 2009, 29(2):92-94.

[5] Rodgerson J L, Lopez H, Phil S. Unique Multistage Process Allows Pinpoint Treatment of Hard-to-Reach Pay[J]. Spe Production & Operations, 2006, 21(4):441-447.

[6] Wilson A. New Methods and Workflows Boost Effectiveness of Multizone Stimulation[J]. Journal of Petroleum Technology, 2013, 65(6):89-94.

[7] Shaefer M. Awakening an Old Field:A Case Study of Refracturing Program in the Greater Green River Basin [J]. Spe Production & Operations, 2008, 23(2):135-146.

[8] Carpenter C. Enhanced Gas Recovery by CO_2 Sequestration vs. Refracturing Treatment[J]. Journal of Petroleum Technology, 2015, 67(7):125-127.

[9] Wang T, Xu Y, Jiang J, et al. The technology of annulus hydraulic-jet fracturing with coiled tubing[J]. Natural Gas Industry, 2010, 30(1):65-67.

[10] Castaneda J, Castro L, Craig S, et al. Coiled Tubing Fracturing: An Operational Review of a 43-Stage Barnett Shale Stimulation[C]. SPE Coiled Tubing and Well Intervention Conference and Exhibition. 2010.

[11] Vrabec M, Jordanova G. Analysis of systematic fracturing in Eocene flsch of the Slovenian coastal region [J]. Geologija, 2017, 60(2):199-210.

[12] Duan W, Cong L, Zhang L, et al. Optimization Design of Integral Fracturing Parameters Based on Fine 3-D Geology Model[J]. Journal of Southwest Petroleum University, 2013.

[13] Salah M, Gabry M A, Elsebaee M, et al. Control of Hydraulic Fracture Height Growth Above Water Zone by Inducing Artificial Barrier in Western Desert, Egypt[C]. Abu Dhabi International Petroleum Exhibition & Conference,2016.

[14] Wenbin G U, Pei Y, Zhao A, et al. Application of artificial barrier technology to fracture height control in fracturing wells[J]. Oil Drilling & Production Technology, 2017.

[15] Chekhonin E, Levonyan K. Hydraulic fracture propagation in highly permeable formations, with applications to tip screenout[J]. International Journal of Rock Mechanics & Mining Sciences, 2012, 50(2):19-28.

[16] Dontsov E V, Peirce A P. A Lagrangian Approach to Modelling Proppant Transport with Tip Screen-Out in KGD Hydraulic Fractures[J]. Rock Mechanics & Rock Engineering, 2015, 48(6):2541-2550.

[17] Bochkarev A, Budennyy S, Nikitin R, et al. Pseudo-3D Hydraulic Fracture Model with Complex Mechanism of Proppant Transport and Tip Screen Out[C]. European Conference on the Mathematics of Oil Recovery, 2016.

[18] Hui J J, Nie L, Song P J. Application Analysis of Hydraulic Fracturing Fracture Forced Closure Technology [J]. Liaoning Chemical Industry, 2013.

[19] Varela R A, Ypf S A, Maniere J L. Successful Dynamic Closure Test Using Controlled Flow Back in the Vaca Muerta Formation [C]. SPE Argentina Exploration and Production of Unconventional Resources Symposium, 2016.

[20] Zazovsky A, Tetenov E V, Zaki K S, et al. Pressure Pulse Generated by Valve Closure: Can It Cause Damage? [C]. SPE International Symposium and Exhibition on Formation Damage Control. Society of Petroleum Engineers, 2014.

[21] Al-Muntasheri G A, Liang F, Hull K L. Nanoparticle-Enhanced Hydraulic-Fracturing Fluids: A Review [J]. Spe Production & Operations, 2017, 32(2).

[22] Cipolla C L, Lolon E, Mayerhofer M J. Resolving Created, Propped and Effective Hydraulic Fracture Length[J]. Spe Production & Operations, 2008, 24(4):619 – 628.

[23] Vincent M C. The Next Opportunity To Improve Hydraulic-Fracture Stimulation[J]. Journal of Petroleum Technology, 2012, 64(3):118 – 127.

[24] Barati R, Liang J T. A review of fracturing fluid systems used for hydraulic fracturing of oil and gas wells [J]. Journal of Applied Polymer Science, 2014, 131(16):318 – 323.

[25] Karadkar P, Bataweel M, Bulekbay A, et al. Recent Advances in Foamed Acid Fracturing[C]. SPE 192392, 2018.

[26] Ravikumar A, Marongiu Porcu M, Morales A. Optimization of Acid Fracturing with Unified Fracture Design[C]. Abu Dhabi International Petroleum Exhibition and Conference, 2014.

[27] Aboud R, Melo R C D. Past Technologies Emerge Due to Lightweight Proppant Technology: Case Histories Applied on Mature Fields[C]. Latin American & Caribbean Petroleum Engineering Conference. Society of Petroleum Engineers, 2007.

[28] Liang F, Sayed M, Al-Muntasheri G A, et al. A Comprehensive Review on Proppant Technologies[J]. Petroleum, 2016, 2(1):26 – 39.

[29] Zoveidavianpoor M, Gharibi A. Application of polymers for coating of proppant in hydraulic fracturing of subterraneous formations: A comprehensive review[J]. Journal of Natural Gas Science & Engineering, 2015, 24:197 – 209.

[30] Alotaibi M A, Miskimins J L. Slickwater Proppant Transport in Complex Fractures: New Experimental Findings & Scalable Correlation[C]. Spe Technical Conference and Exhibition. 2017.

[31] Arnold D L. Liquid CO_2-sand fracturing: 'the dry frac'[J]. Fuel & Energy Abstracts, 1998(3):185.

[32] Ishida T, Chen Y, Bennour Z, et al. Features of CO_2 fracturing deduced from acoustic emission and microscopy in laboratory experiments[J]. Journal of Geophysical Research Solid Earth, 2016, 121(11).

[33] Li Q H, Chen M, Jin Y, et al. Application of New Fracturing Technologies in Shale Gas Development[J]. Special Oil & Gas Reservoirs, 2012, 19(6):1 – 7.

[34] 周德胜. 非常规油气储层体积改造裂缝扩展与织网机理研究[M]. 北京:科学出版社,2018.

[35] 张琪. 采油工程原理与设计[M]. 东营:石油大学出版社, 2000.

Horizontal Well Fracturing & Volume Fracturing

水平井压裂和体积压裂

Audio 3.1

Starting in the 1980s and, eventually, widely introduced in the early 1990s, horizontal wells have proliferated and have become essential in oil and gas production. Horizontal wells have the advantages of large oil drainage area, high single well production, high penetration, high reserves, avoiding obstacles and harsh environments. They have become the focus of attention in the research and practice of the petroleum industry. However, people are still facing new challenges in the development of horizontal wells in low and ultra-low-permeability reservoirs. The horizontal well is not hydraulically fractured, resulting in a single well output that is lower than the fractured vertical well when a large amount of money is invested. Horizontal well fracturing has been a problem for the past few decades. After the 1980s, a large amount of research and practice at home and abroad to achieve multi-stage fracturing of horizontal wells to improve single well production.

从 20 世纪 80 年代水平井的引入到如今的广泛应用,水平井已成为石油和天然气生产中必不可少的开采方式。水平井具有泄油面积大、单井产量高、穿透度大、储量动用程度高、避开障碍物和适合环境恶劣地带等优点,在石油行业的科研和实践中成为人们关注的焦点。然而,水平井开发低渗透、特低渗透油气层时,仍面临着新的挑战。研究结果表明,在投入大量资金的同时,对水平井进行水力压裂后的单井产量高于压裂过的直井的产量。在开始的几年中,水平井压裂一直是一个难题,但 20 世纪 80 年代以后,国内外进行了大量的研究与实践,实现了水平井多段压裂,达到了提高单井产量的目的。

About horizontal well production formula, for steady state

关于水平井产能公式,对于稳态情况有

$$q = \frac{K_H h(p_e^2 - p_{wf}^2)}{1424\bar{\mu}ZT\left[A_a + \frac{I_{ani}h}{L}\ln\frac{I_{ani}h}{r_w(I_{ani}+1)} + D_q\right]} \tag{3.1}$$

$$A_a = \ln\frac{a + \sqrt{a^2 - (L/2)^2}}{L/2} \tag{3.2}$$

where K_h is the horizontal permeability, h is the reservoir thickness, μ and Z are the viscosity and gas deviation factor, respectively, T is the reservoir temperature and I_{ani} is a measurement of vertical-to-horizontal permeability anisotropy given by:

其中 K_h 是水平渗透率，h 是储层厚度，μ 和 Z 分别是黏度和气体偏差因子，T 是储层温度，I_{ani} 是垂直—水平渗透率各向异性的比值，由下式给出：

$$I_{ani} = \sqrt{\frac{K_H}{K_V}} \tag{3.3}$$

There are many horizontal well fracturing technologies, but since all these fracturing technologies are implemented within the horizontal parts of the well, the well's completion method determines how it should be fractured. Thus, this book catalogues all horizontal well fracturing technologies into two types based on the completion method of a horizontal well. Namely, horizontal well open-hole completions fracturing technologies and horizontal well cemented completions fracturing technologies. Be advised, the main purpose of horizontal well fracturing is to isolate different sections to enables multi-stage and multi-cluster fracturing.

水平井压裂技术繁多，由于这些技术都是在水平井段实施，因此压裂方式是由完井方式决定的。本书根据水平井的完井方式将所有水平井压裂技术分为两类，即水平井裸眼完井压裂技术和水平井套管完井压裂技术。需要注意的是，水平井压裂的主要目的是隔离不同部位以实现多级和多簇压裂。

3.1 Open-Hole Completion and Fracturing Technologies

3.1 裸眼井完井与压裂技术

The open-hole multi-stage fracturing technology is the current domestic and international advanced technology. This section introduces open-hole completion and open-hole fracturing.

裸眼井分段压裂技术是当前国内外的先进技术，本节介绍裸眼完井和裸眼井压裂。

3.1.1 Open-hole Horizontal Well Completions

3.1.1 裸眼水平井完井

Open-hole completions are utilized in competent, stable formations. Because horizontal wells often span considerable lengths, rock mechanical properties of the reservoir must be fully understood for efficient and effective open-hole completions. The primary collapsing stress in a horizontal well is the vertical overburden component. Horizontal stresses

裸眼完井用于较完善稳定的地层。由于水平井通常跨越的长度较长，因此必须充分了解储层的岩石力学性质，这样才能有效完井。水平井中的主要破坏应力是垂直方向上的上覆岩

reduce the potential risk of vertical collapse. Wellbore instability problems are not normally encountered in low-permeability reservoirs because the rocks are generally quite competent and consolidated; these reservoirs are candidates for open-hole completions.

层压力,但水平井中的水平应力降低了垂直坍塌的风险。低渗透油藏通常不存在井筒不稳定的问题,由于岩石固结能力非常强,这些低渗透储层都可以用裸眼完井。

So, open-hole or "barefoot" completions are effective in formations with low risk of wellbore collapse or sand production, for example in dolomites, hard sandstones or limestones and shale-free siltstones. An open-hole completion involves running casing to the producing horizontal interval and leaving the end of the casing open. Barefoot completions are generally applied only in formations with unconfined compressive strength larger than 10,000 psi. For such completions, it should be expected that for long-term performance as the reservoir depletes, the wellbore must remain open and solids-free. Fig. 3. 1 shows an open-hole completion.

裸眼完井适合用在井壁易坍塌和产砂风险低的地层中,如白云岩、坚固砂岩、石灰岩和没有页岩的粉砂岩。裸眼完井过程包括将套管下到生产层并使套管末端保持开口。裸眼完井一般适用于岩石无围压、抗压强度大于 10000 psi 的地层。对于这样的完井工艺,随着油藏生产,井筒应保持长期性能,保持裸眼并无固体颗粒。图 3. 1 为裸眼完井示意图。

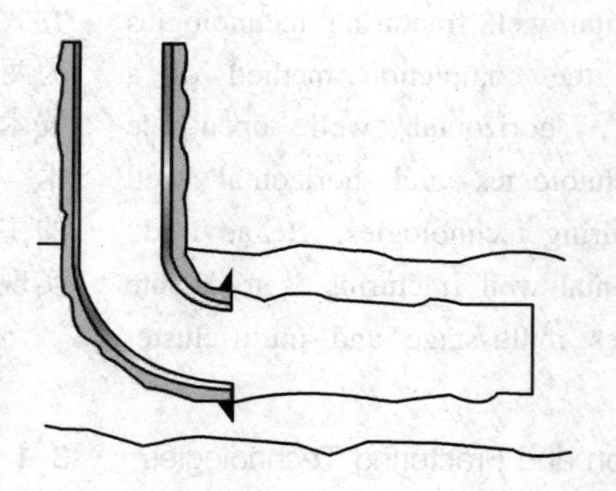

Fig. 3. 1 Open-Hole Completion

For less stable formations, an open-hole completion with a slotted liner run into the horizontal section helps prevent formation collapse near the wellbore (Fig. 3. 2), especially as the reservoir pressure depletes during production. A disadvantage of the slotted liner completion is the difficulty to effectively isolate and stimulate zones or sections in the wellbore.

对于较不稳定地层,通常在水平段采用割缝衬管裸眼完井工艺预防井筒附近的地层坍塌(图 3.2),这种方式对生产后期地层压力衰减作用较大。割缝衬管裸眼完井的不足是难以进行有效层段隔离和进行增产作业。

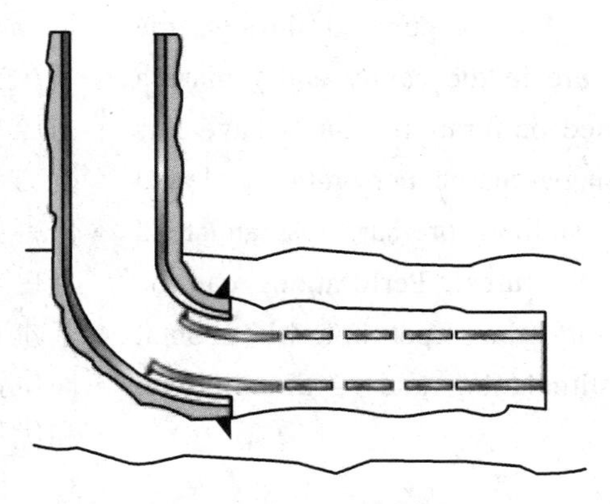

Fig. 3. 2　Open-Hole Completion with Pre-Perforated or Slotted Liner

An open-hole completion with perforated pipe provides wellbore stability (Fig. 3. 3). Limited-entry design is required to design the number and size of perforations to effectively initiate and place multiple hydraulic fractures in a single treatment.

在裸眼井中下入射孔套管能实现较好的井眼稳定性(图3.3),但需要在设计中限制射孔数及尺寸,以便在一次施工中有效地进行多次水力压裂。

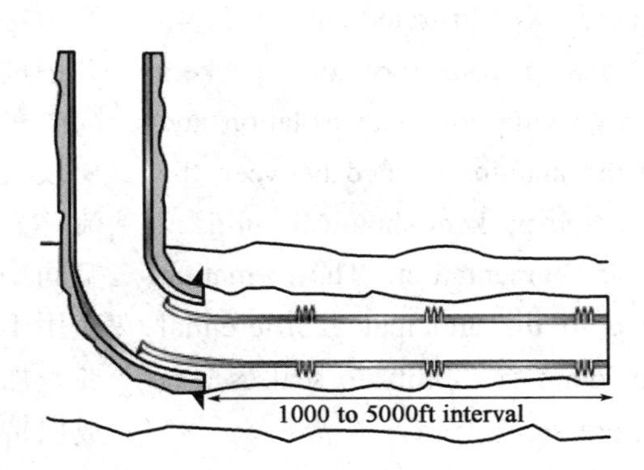

Fig. 3. 3　Open-Hole Completion with Perforated Pipe

3. 1. 1. 1　Open-hole Horizontal Well Perforating

Although it took some time in the early days of horizontal wells to accept the practice, the creation of perforations with a jetting tool results in very effective flow paths from the open hole into the formation. The entire concept of creating perforations in an open hole was at first difficult to accept but it became a relatively common and effective practice. The perforations essentially bypass the damaged zone around the wellbore. A small jetting tool is placed on the end of the tubing or coiled tubing and it is used to create small cavities or tunnels in the formation. After each

3. 1. 1. 1　裸眼井射孔技术

早期人们很难接受裸眼水平井这种用喷射工具给地层开孔形成有效流动路径的开发工艺。裸眼井射孔这个概念起初难以被接受,但随着时间推移它逐渐成为一种有效开发油田的常见方法。射孔的本质是绕过裸井眼周围受损区域,油流通道或连续油管末端的小型喷射工具在地层中形成小型孔洞或通

cavity is created, additional fluid is pumped through the annulus to increase the pressure in the cavity and initiate a small hydraulic fracture. Jetted perforation tunnels have less damage than conventional shape-charge perforations. Jetted perforations have small near-wellbore pressure loss and tend to initiate simple hydraulic fractures. Perforations can be located almost anywhere in the open-hole horizontal completion for effective stimulation and optimum well performance.

道,通过环空泵送的流体来增加其中的压力并实现局部小型水力压裂。射孔形成的通道与传统方法形成的通道相比对地层伤害更小、近井压力损失更小而且更容易进行简单水力压裂。基本可以在裸眼水平完井的任何位置进行射孔,以此达到增产和优化井性能的目的。

3.1.1.2 Zonal Isolation

Another issue that took some time to be recognized is that horizontal wells frequently need zonal isolation because, in spite of the old conventional wisdom, reservoirs are not necessarily homogeneous along the horizontal well path. Zonal isolation in open hole completions, although at times difficult, can be achieved with the installation of formation packers at strategic locations along the slotted liner or perforated pipe in the wellbore, as illustrated in Fig. 3.4. Horizontal well completion testing with formation packers often shows that they do not provide complete isolation and they do not prevent flow in the annulus formed between the pipe and the reservoir. Formation packers should be utilized for zonal isolation or reservoir segmentation. The formation packer needs to be energized to the anticipated differential pressure or constructed to maintain the ability to seal as it is energized by the differential pressure.

3.1.1.2 层间封隔

对水平井进行层间封隔这个观念过了很久才被人们认可。因为水平段的储层不一定是均匀的,所以通常水平井需要进行层间封隔。如图3.4所示,尽管实现较困难,但通过在割缝衬管或井筒中射孔管的关键位置安装地层封隔器,可以达到裸眼完井中的层间隔离。使用地层封隔器的水平井完井测试表明各层没有完全隔离,而且没能防止流体在管道和储层之间的环形空间中流动。地层封隔器一般应用于地层隔离或储层分隔,需要外加能量至预期压差,并设计成保持密封结构。

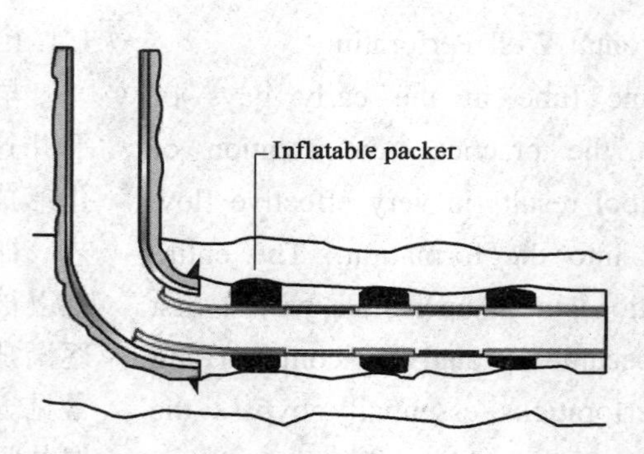

Fig. 3.4 Open-Hole Completion with External Casing Packers

Zonal isolation in open-hole completions can also be achieved with a series of mechanical packers deployed on a production liner with fracturing ports located between the packers to permit stimulation of each interval discretely (Fig. 3.5). The liner does not need to be cemented in place and eliminates the requirement to perforate, fracture, isolate, etc. in a cemented horizontal wellbore. The liner located in the open-hole section allows access to the entire horizontal wellbore rather than encountering problems associated with a barefoot completion. The mechanical packers provide mechanical diversion and isolation at high differential pressures. All of the stimulation treatments along the horizontal wellbore can be pumped in a single continuous operation, which minimizes associated risks and allows optimum efficiency for personnel and equipment. For such completions to work, the rock must be competent and wellbore ovality must be minimal.

通过在生产衬管上配置一系列的机械封隔器也可以达到裸眼完井中的层间封隔,生产管各封隔器间的压裂端口可用来分别压裂各个区间(图 3.5)。生产管无需水泥胶结从而免除了水平井固井中的射孔、压裂、封隔等要求。裸眼井段的尾管能进入整个水平井段,没有裸眼完井的相关缺陷。机械封隔器在高压差下能提供机械分流和封隔。水平井筒的所有增产工艺均能在一次连续操作中进行,最大限度地降低风险,使操作人员和设备的效率达到最大化。对于这样的完井工艺除了对岩石性质有要求外,还需井眼椭圆度极小。

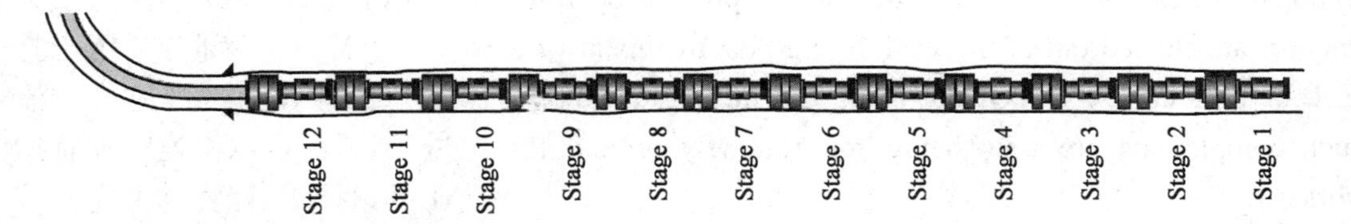

Fig. 3.5 Open-Hole Completion with Mechanical Diversion

3.1.1.3 Novel Open-Hole Completion

The novel open-hole completion is designed specifically for open-hole fracturing of both sandstone and carbonate reservoirs. This innovative, field-proven system greatly increases the effectiveness of fracturing operations by segmenting the lateral and producing mechanical isolation points in the wellbore using high-performance open-hole packers. The system allows precision placement of fracturing (stimulation) fluids to maximize post-fracture productivity of the well.

3.1.1.3 新型裸眼完井技术

新型裸眼完井技术是专门为砂岩和碳酸盐岩储层的裸眼压裂而设计的。这种经过现场验证的创新体系使用高性能裸眼封隔器,通过在水平井筒上产生机械封隔点实现水平井分段作业,大大提高了压裂作业的效率。该技术可以精确设置压裂(增产)流体注入,从而大大提高该井压后产量。

The completion is run as part of an uncemented liner and components are spaced out based on the required number of stages. Once in place, the open-hole packers are hydraulically set. Ball activated fracports are placed between a set of 2 open-hole packers with spacing between the packers varying from 100 ft to 1,000 ft. More than 10 fracports can be run in the open-hole in this type of liner. Initially, all fracports are closed. Starting from the toe section of the well, each fracport is opened by dropping a ball from the surface and the ball isolates the horizontal section below each fracport thus ensuring that at any given time only one zone is being treated. The fracports are designed to have 120% flow area compared to the production liner. Fracturing/stimulation treatments are pumped in separate stages but as a single continuous operation. By eliminating cementing requirements, natural fractures are undamaged and easily stimulated during pumping operations. After the stimulation treatment, the treatment string is designed to stay in the well as part of the permanent well completion (casing). The balls that are dropped to open the fracport are flowed out of the well during post treatment flow back or they can be drilled out using CT milling assembly. Such completions are now being run routinely around the world.

这种完井作为非固井尾管的一部分,其部件基于所需压裂级数布置。裸眼封隔器到位后通过液压坐封。用球来打开的压裂端口布置在2个一组的裸眼封隔器之间,封隔器间距为100~1000ft。这种套管完井在裸眼井中可布置10余个压裂端口。起初关闭所有压裂端口,从水平井趾端开始,通过井口投球依次打开每个端口,并在端口下方用球隔离水平段,确保在任何的给定时间内仅压裂一个区域。与生产衬管相比,压裂端口设计具有120%的流动面积。虽然压裂是对各段实施,但仍能连续作业。该裸眼完井通过减少固井的需求,减轻天然裂缝伤害,使地层在泵送期间易于开启。作业后管柱留在井中作为永久完井(套管)的一部分。用于打开端口的球在压裂液返排时流出井筒,或者可以用连续油管铣削工具钻掉。目前,这类完井技术正在世界各地应用。

3.1.2 Open-Hole Fracturing

Transverse fractures will be initiated and created when the horizontal wellbore is oriented perpendicular to the fracture plane (within 15-degree tolerance). Longitudinal fractures will be initiated and created when the horizontal wellbore is oriented parallel to the fracture plane (within less than 15 degrees). Transverse hydraulic fractures offer the greatest potential to contact and drain a large section of the reservoir. Longitudinal hydraulic fractures may allow the pumping of higher proppant concentrations with a lower risk of screen out.

3.1.2 裸眼井压裂

当水平井筒垂直于裂缝平面(在15°范围内)时,将产生横向裂缝。当水平井筒平行于裂缝平面(小于15°)时,将产生纵向裂缝。横向的水力裂缝极大增加了储层接触泄流区域。纵向水力裂缝可以在出砂风险较低的情况下泵送较高浓度的支撑剂。

Fig. 3. 6 shows properly spaced transverse hydraulic fractures; Fig. 3. 7 shows improperly spaced transverse fractures. Similarly, Fig. 3. 8 shows appropriately placed longitudinal fractures, and Fig. 3. 9 shows haphazardly fractured the horizontal well in the longitudinal direction.

图 3. 6 为均匀铺置的水力压裂横向裂缝;图 3. 7 为不均匀的水力压裂横向裂缝;图 3. 8 为较好铺置的水力压裂纵向裂缝;图 3. 9 为杂乱铺置的水力压裂纵向裂缝。

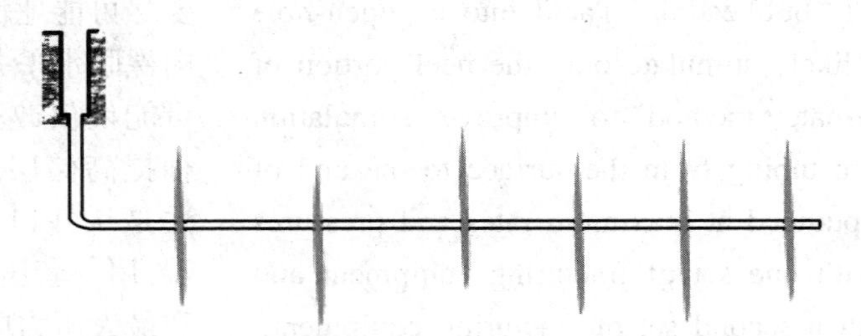

Fig. 3. 6 Ideal Transverse Hydraulic Fractures

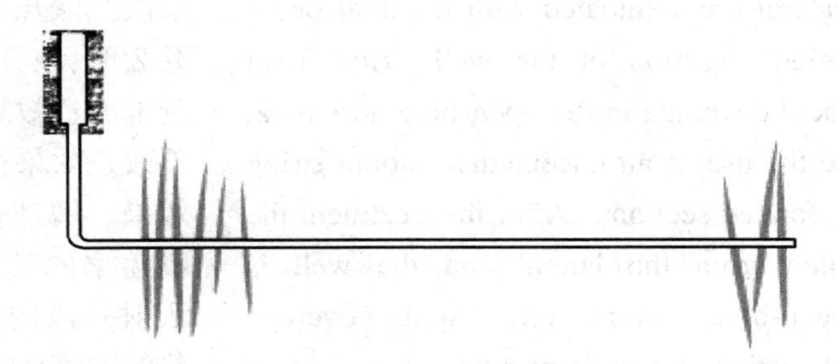

Fig. 3. 7 Non-Ideal Transverse Hydraulic Fractures

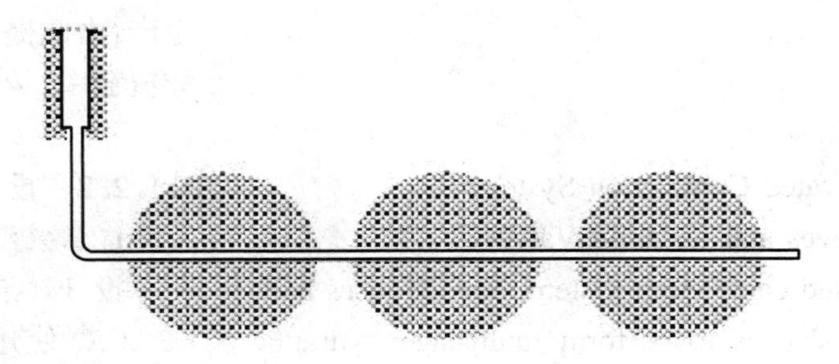

Fig. 3. 8 Ideal Longitudinal Hydraulic Fractures

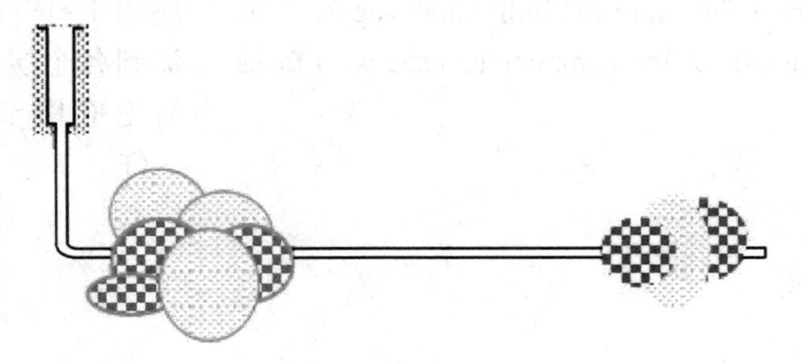

Fig. 3. 9 Non-Ideal Longitudinal Hydraulic Fractures

3.1.2.1 Acid Fracturing Execution

Carbonate reservoirs are often acid-fractured. The fracture geometry created by acid fracturing depends on the acid volume and the injection rate. But live acid must be transported as far away from the wellbore as possible to dissolve and etch the carbonate in order to create conductive fractures. Pumping ("bullheading") acid into an open-hole horizontal well will likely stimulate only the heel portion of the well. An alternate method to improve stimulation effectiveness is to run tubing from the surface to the end of the lateral. Acid is pumped at maximum rates and pressures down the annulus with one set of fracturing equipment and down the tubing with a second set of fracturing equipment. Post-treatment temperature and tracer logs indicate that the heel and toe of the lateral are stimulated with minimal or no stimulation in the middle section of the well. Additional attempts to increase acid coverage in the open-hole horizontal wellbore may involve the use of an uncemented tubing string with limited-entry perforated sections. After the treatment the tubing string is pulled from the lateral and the well is produced as an open-hole completion. Acid coverage throughout the lateral section is thus improved.

3.1.2.1 酸压

碳酸盐岩储层通常要酸压，而酸压产生裂缝的几何形状通常取决于注入酸的体积和注入速率。为在碳酸盐岩储层中溶解或蚀刻出导流裂缝，活性酸需要尽可能地输送到远离井筒处，向裸眼水平井泵入（"挤入"）酸可能仅仅改造了井跟部分。一种提高储层改造效果的替代方法是从井口下油管到水平井趾端，用一套压裂设备以最大速率和最大压力向环空中泵酸，同时用另一组压裂设备沿油管向下泵酸。酸压后的温度和示踪测井表明，该方法对水平井跟部与趾部改造效果较好，但水平井中部改造效果很小，甚至几乎没有效果。增加裸眼水平井筒中酸覆盖率的其他方法还有在限流法射孔段的水平井中使用不固井的油管柱，酸压完成后从水平段中取出油管柱，采用裸眼完井生产，由此提高整个水平井段的酸覆盖率。

3.1.2.2 Ball-Activated Completion Systems

1. Fracture Sleeves and Isolation Valves

The ball-activated completion system (BACS) uses ball-activated fracturing sleeves to perform multistage hydraulic fracturing. Balls and ball seats are used to open the sleeves and divert the fracturing fluid into the individual stages. This section will discuss details of the components used with these types of systems.

3.1.2.2 投球滑套压裂系统

1. 压裂滑套与隔离阀

投球滑套压裂系统（BACS）使用投球打开压裂滑套实现多级水力压裂。球和球座用于打开滑套并将压裂液分流到各个层位。本部分将具体讨论应用于这套系统的各个组件。

1) Pressure-Activated Sleeves

The pressure-activated fracturing sleeve is run at the toe of the well and can be opened by applying pressure into the completion string. When it is time to begin the fracturing job, the appropriate amount of pressure is applied, opening the sleeve and the first-stage fracture stimulation is performed through this sleeve. This sleeve allows access to the formation without through-tubing intervention, which is important because there are ball seats with diameter restrictions above it that may not allow intervention tools to pass.

2) Wellbore Isolation Valve

The wellbore isolation valve (WIV) is run at the toe of the completion when a pressure-activated sleeve is used as the first stage at the toe. The WIV will allow circulation from the completion string to the annulus during run-in so that fluids can be displaced and circulated should the system have trouble getting to the intended depth. When the system reaches the intended depth the ball that corresponds to the WIV is dropped in the well and pumped to the WIV. Applying pressure closes the WIV and it acts as a plug at the toe of the completion. All circulation is shut off between the casing and the annulus, providing well protection, a point to apply pressure against, and to set any hydraulic equipment.

3) Ball-Activated Sleeves

The ball-activated fracturing sleeves are the key component to this completion type. These sleeves contain a ball seat that corresponds to a fracturing ball. Each of these sleeves has a different size ball seat so that each sleeve can be selectively opened. Because each sleeve has an individual ball and ball seat combination for selective opening, the sleeves must be run in order of the smallest ball seat at the toe of the well and the largest at the heel of the well. This is the only way that all of the balls can pass through the ball seats of the sleeves above it.

1)压力打开滑套

压力打开滑套置于水平井趾部,可通过向完井管柱施加压力来打开。当开始压裂时,施加适当的压力把滑套打开,并通过该滑套进行第一阶段的压裂。此类滑套可以在没有过油管干预的情况下实现压裂液进入地层,这一特点很重要,因为在套管内布置有球座,施工工具可能无法通过球座内径。

2)井眼隔离阀

在水平井趾部采用压力打开滑套作为压裂的第一段,井眼隔离阀(WIV)随完井工具下入趾端。井眼隔离阀可以使下套管期间完井管柱与环空形成循环,以便在难以达到预期深度时实现流体置换和循环。当达到预期深度时,投入对应井眼隔离阀的球并泵送到井眼隔离阀,施加压力关闭井眼隔离阀,它像封隔器堵住了完井管柱趾部,切断了套管和环空之间所有循环,起到保护井眼的作用,还可作为一个施加压力的作用位置和设置液压设备的地方。

3)投球滑套

投球滑套是这种压裂系统的关键组成,包含对应于压裂球的球座。这种滑套中球座的尺寸不同,因此每个滑套可以选择性地打开。因为每个滑套具有各自的球和球座组合以便选择性地打开,所以滑套必须按照井趾部处的球座最小和井跟部处的球座最大的方式下入。这是所有球可以穿过其上方球座的唯一方式。

When the ball is pumped down and it lands on the corresponding ball seat, pressuring up will open the fracturing sleeve and expose the fracturing ports on the sleeve. The ball remains on seat during the fracturing job to divert the fluid out of the fracturing ports and to isolate from the previously fractured stage. The fracturing sleeve in the opened and operating position is shown in Fig. 3. 10.

当球被泵入并落在相应的球座上时,加压将打开压裂滑套并暴露出套管上的压裂端口。在压裂期间球保持在球座上并将压裂液分流到压裂端口,并隔离已压裂段。处于打开和操作位置的压裂滑套如图 3. 10 所示。

Fig. 3. 10 Ball-Activated Fracturing Sleeve in the Open Position

The fracturing sleeve remains in the open position after the fracturing job so that the well can produce through the sleeve. To prevent the sleeve from closing and blocking off production, the sleeve locks in the open position as soon as it is opened. The balls will flow back to the surface if there is enough velocity in the production. The ball seats can be produced through, but they are also designed to be milled out if the application calls for it. If the fracturing balls remain in the well, they can be milled out as well.

在压裂作业之后,压裂滑套保持在打开位置使油气可以通过滑套上的压裂端口进行生产。为了防止滑套关闭阻断生产,滑套一打开就锁定在打开位置。如果生产中流速足够大,球将流回地面。油气可以通过球座生产,但如果需要,它们也设计成可以被磨碎。如果压裂球留在井中,也可以将它们研磨掉。

4) Reclosable Sleeves

4)可反复开关的滑套

There is also a reclosable version of the ball-activated fracturing sleeve. This sleeve has the same functionality to divert the fracturing job and isolate it from the stage below with a ball and ball seat in the fracturing sleeve. However, the reclosable sleeve version can be shifted with coiled tubing after the fracturing job. These sleeves have a shifting profile that a shifting tool locks into to shift the sleeve open and closed. It may be preferred or required to mill out the ball

投球压裂滑套包括一种可反复开关滑套,该滑套在利用球和球座隔离其前一层段及实现压裂作业。但是,在压裂作业之后,可反复开关的滑套可以用连续油管关闭。这种滑套带有一种转换器,转换工具与转换器锁定后可转换滑套的打开和闭合

seats before the shifting tool can be used. After the ball seats are milled out, a coiled tubing or work-over string is run in the well with a hydraulic-activated shifting tool that can selectively latch into each sleeve. The sleeves can be selectively closed and opened with this shifting tool. If it is known the seats will need to be milled out, it is best to mill them out before the well is put on production. Producing the well helps to remove the ball seat debris.

状态。转换工具使用前,最好将球座研磨掉。在球座被研磨后,随连续油管或者修井管柱下入井内,被液压激活后可选择性地锁定滑套。这些滑套可以由转换工具来选择性地打开或闭合。如果已知球座需要被研磨掉,最好在井投入生产之前将它们研磨掉,生产过程有助于移除研磨球座碎片。

These sleeves can be used in applications where water breakthrough is possible or in a refracturing application. If water starts producing through a stage, the sleeve can be shifted closed to isolate that stage and save the rest of the well. The fact that all of the sleeves can be reclosed allows for some refracturing options. If it is desired to have new injection points in the well, all sleeves can be closed and an openhole plug-and-perforation job can be performed between the packers. If the refracturing is done through the same injection points, all of the sleeves can be reclosed and the following procedure can be used:

这种滑套可应用于可能发生水窜或需要重复压裂的情况。如果某一段见水,可以关闭该段滑套以隔离该段保持其余部分的生产。对某些重复压裂,所有滑套都可以重新闭合。如果想要在井中产生新的压裂点,则可以关闭所有滑套,在封隔器之间进行裸眼桥塞射孔联作压裂。如果在同一压裂点进行重复压裂,则可以闭合所有滑套,并且按照以下程序操作:

(1) Selectively open a sleeve.
(2) Refracture through the open sleeve.
(3) Reclose the sleeve.
(4) Repeat until all desired stages are refractured.
(5) Open all refractured stages for production.

(1)选择性地打开一个滑套;
(2)在打开的滑套重复压裂;
(3)重新闭合滑套;
(4)重复以上步骤直到所有目的段被重复压裂;
(5)打开所有重复压裂段进行生产。

Only having one sleeve open at a time will force the fracture into the desired stage.

一次只打开一个滑套可以有效地压裂目的段。

2. Openhole Packers

The role of the openhole packers is to isolate the annulus between the completion string and the openhole formation. There are a variety of different packers that have different setting and isolation methods, but their role is the same. All of the packers are designed to conform to the shape of the openhole and irregularities within it. There is no way to test each packer individually, but production results have shown us that the openhole packers do effectively isolate multiple stages in the wellbore.

There are a couple of reasons that these packers may be able to isolate multiple stages even in inconsistent wellbores and with difficult applications. Imagine the packer as an O-ring in the wellbore, which is a fairly accurate analogy. Just like an O-ring, when pressure is applied, the rubber-packing element is energized and will cover an even larger surface area. Also, most of these applications use proppants, so even if the packer does not completely seal due to an irregularity in the formation, the proppant could still plug this path and complete the isolation. It is also important to remember that an airtight seal is not required in this application. As long as the packer creates enough of a blockage to contain the path of least resistance in that stage, the packer will provide effective isolation.

1) Hydraulic-activated packers

One type of openhole packer is the hydraulic-activated packer. This packer relies on applied pressure to set the packer. When the packer sets, the components shift parallel to the packer, forcing the rubber-packing element outward toward the formation. The rubber also has a metal backup ring around it that extrudes with the rubber and conforms to the openhole, enabling a more reliable and higher-rated packer. When the components of the packer shift, the internal mandrel does not shift. This means that the length of the completion string will not change when setting the packers, so an unlimited number of these packers can be set at the same time.

2. 裸眼封隔器

裸眼封隔器的作用是隔离完井管柱和裸眼地层之间的环空。目前有多种不同坐封和分段方式的封隔器,但其作用是相同的。这类封隔器都设计成满足裸眼的形状及其不规则性需求。虽然没有办法单独测试每个封隔器,但是生产结果表明裸眼封隔器确实有效隔离了井筒中多个层段。

这些封隔器即使在不一致的井筒中或复杂情况下也能够隔离多个层段的原因如下。假设封隔器是井筒中的O形圈,当施加压力时,橡胶元件被挤压后可以覆盖更大的表面积。此外,大多数使用封隔器的情况都使用了支撑剂,因此即使封隔器由于地层的不规则性而没有完全起到密封作用,支撑剂仍然可以堵塞该路径并完成封隔。同样重要的是,在这种情况下不需要气密密封。只要封隔器在该阶段产生足够的阻塞,就能够提供有效的隔离。

1)液压封隔器

液压封隔器是裸眼封隔器的一种。此类封隔器依靠施加的压力来坐封封隔器。当封隔器坐封后,部件平行于封隔器移动并迫使橡胶元件向外朝地层扩展。橡胶周围还有一个金属支撑环,可与橡胶一起被挤出环绕裸眼井,因此成为更可靠、更高级的封隔器。当封隔器的部件移动时,内部心轴不会移位,这意味着在坐封封隔器后,完井管柱的长度不会改变,因此可以同时坐封无限个封隔器。

2) Fluid-activated packers

Another packer commonly used for openhole isolation is the fluid-activated or reactive-element packer. This packer relies on fluid to activate and swell the rubber out to contact the formation. When these packers reach the intended depth, an activation fluid is pumped across the packers if it is needed. The packers will swell and contact the formation to provide isolation. These packers are designed to leave enough space between the openhole and the completion tools, so it is easier to get the system to the intended depths.

2)膨胀式封隔器

常常应用于裸眼封隔的另一种封隔器是膨胀式(或反应式)封隔器。这类封隔器依靠流体反应来激活并使橡胶膨胀以接触地层。当封隔器达到预定深度时,反应流体被泵送到封隔器上。封隔器将膨胀并与地层接触以达到隔离效果。设计这类封隔器时应在裸眼井和完井工具之间留出足够的空间,以便这些封隔器更容易达到预期深度。

3. Ball-Activated Completion Systems Fracturing Operations

When the fracturing crew arrives on location, they rig up and apply pressure to open the pressure-activated sleeve. If there is not a WIV and pressure-activated sleeve, fluid will be pumped out of the float equipment and into the formation. The ball corresponding to the first ball-activated sleeve is dropped in the well and pumped onto the ball seat. Applying pressure against the seated ball opens the sleeve. Either way, when the first sleeve is opened, the firststage fracturing begins. After the first stage is fractured, a small amount of additional fluid, called a flush, is pumped through the completion string to clean out any proppant that remains in the liner. While this flush is being pumped, the ball corresponding to the second stage is dropped into the fluid flow and pumped to its ball seat. The ball is dropped into the fluid flow using a manifold or a balldropping head, so the fracturing job does not have to shut down between stages. When the fracturing ball lands on the ball seat, applying pressure opens the sleeve, and the ball isolates from the stage below the fluid through the ports on the sleeve. The second stage is then fractured. The process is then repeated until all stages are fractured (Fig. 3. 11).

3. 投球滑套压裂作业

当压裂人员到达井场时,他们安装设备并施加压力以打开压力打开滑套。如果没有井眼隔离阀和压力打开滑套,压裂液将被泵出管柱并进入地层。对应于第一个球激活滑套的球投入井中并泵送到球座上,对座球施加压力就可以打开滑套。无论哪种方式,当第一个滑套打开时,开始压裂第一段。当第一段被压开后,将少量额外的流体(称为冲洗液)通过完井管柱泵送下来以清除残留在衬管中的支撑剂。当泵送冲洗液时,对应于第二段的球投入流体中并泵送到球座上。球通过管汇或投球器投入流体中,因此压裂作业不必在不同井段之间停止。当压裂球落在球座上时,施加压力打开滑套,球封闭球座上的端口隔离流体下方的层段。接着压裂第二段,然后重复该过程,直到所有段都被压裂(图 3. 11)。

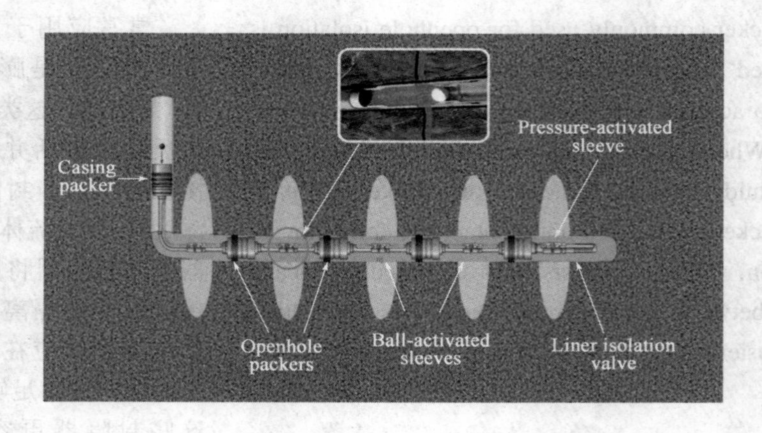

Fig. 3. 11 Fracturing Operations of the BACS

Ideally, the pressure signatures can be seen when the fracturing sleeves open. The ball landing on the seat causes a pressure increase, and opening the sleeve causes a pressure decrease. Unfortunately, this looks very similar to a formation breakdown, and there is no guarantee that this pressure signature can be seen during the fracturing. To assist with being able to see this sleeve shift open, it is recommended to slow the pump rate down as the ball is approaching the ball seat, so that there is a better chance of observing the pressure signatures.

理想条件下,当压裂滑套打开时可以看到压力变化。球落到球座上会导致压力增加,打开滑套则导致压力降低。但不幸的是,这些变化非常类似于地层破裂,并且不能保证在压裂过程中可以看到这些变化。为了能够看到滑套转换至打开,建议在球接近球座时减慢泵速以便更好地观察压力变化。

4. Ball-Activated Completion Systems Post Fracturing Operations

4. 投球滑套压后作业

After the fracturing job, pumping units are moved off location and the well can be put in production. The fracturing balls will flow off the ball seats, and the ball seats can be produced through. Most ball seats have been designed to be milled out easily if it is decided to do so. If reclosable fracturing sleeves are used, they can be functioned with CT at this point.

压裂作业后,压裂装备从现场移走,压裂井可开始投入生产。此时压裂球将从球座落下,流体可以从球座通过。大多数球座都可以设计成很容易被研磨掉。如果使用可反复开关的压裂滑套,可以用连续油管一起作业。

There are instances where the fracturing balls cannot be produced to surface. If there is a differential in pressure between stages, it is possible that the balls could be held on the seat until the pressures balance out. Also, there may not be enough velocity in the production to bring the balls back to surface. The balls would remain in the well and settle into the low spots of the well, potentially creating a debris barrier. If there are concerns about the balls remaining in the well after the fracturing, disintegrating fracturing balls can be used. These fracturing balls will provide the isolation needed during the fracturing job, but will disintegrate in the production fluid afterwards, ensuring that the balls will not hinder production without performing a through-tubing intervention.

有些情况下压裂球不能被带到地面。如果各段之间存在压力差,则在压力平衡被打破前,球可能保持在球座上。此外,生产过程中可能没有足够的流速将球带回地面,球将保留在井中并沉入井的低点,可能会阻碍生产。如果担心在压裂之后球还保留在井中,则可以使用可分解压裂球。这些压裂球同样将在压裂作业期间起到隔离的作用,但压裂之后将在生产过程中分解,确保不会阻碍生产。

5. Alternative Systems and New Technologies

5. 其他替代系统和新技术

Recent advances in the BACS technology allow this completion to use cement to isolate the annulus. The fracturing sleeves have the same functionality as the openhole sleeves, but with minor changes to adapt them to cementing operations. The modifications to these sleeves are primarily to prevent cement from entering the internal components of the sleeve and to prevent it from opening. This system also uses the cemented pressure-activated sleeve that is mentioned under the PNP technologies at the toe of the well.

投球滑套技术(BACS)的最新进展是可使用水泥来隔离环空。压裂滑套具有与裸眼井滑套相同的功能,只需稍作修改即可使其适应固井作业。对这些滑套的改进主要是为了防止水泥进入滑套的内部阻止其打开。该技术也使用了固井的压力打开滑套,具体内容在桥塞射孔联作(PNP)技术水平井在趾端压裂时介绍。

The fracturing process is exactly the same with this system: pumping a fracturing ball to the corresponding seat, opening the sleeve, fracturing that stage, and repeating the process until all stages are fractured. The only difference is in the installation procedures. The openhole packers will not be used in this system, but the sleeves are still installed and spaced out using the casing string. When the system reaches the intended depth, it is cemented in the wellbore.

该系统的压裂过程与其他系统完全相同:将压裂球泵送到相应的球座,打开滑套,压裂该段,并重复该过程直到所有段都被压裂。唯一的区别在于安装过程:该系统中不用裸眼封隔器,但滑套仍需安装并且用套管隔开,当系统达到预期深度时用水泥固井。

3.1.2.3 Cleanup

The first step after proppant fracture stimulation of an open-hole horizontal well is to clean out all the proppant left in the wellbore with coiled tubing and nitrogen, as the fluid velocities generated by natural flow back will not be sufficient to entrain and carry any proppant left in the wellbore. After the proppant is cleaned out, the well should flow back the fracturing fluid. If the well will not flow enough to unload the broken fracturing fluid, then nitrogen can be used to artificially lift the well as long as required. Cleaning the fracture after each perforated interval is stimulated is important; leaving fracturing fluid in the proppant pack may have a detrimental effect on well production. However, it is common practice to stimulate all intervals before flowing back. Therefore, it is important to be able to complete fracturing operations in as short a time as possible.

3.1.2.3 洗井

裸眼水平井支撑剂压裂后的第一步是用连续油管和氮气清除井筒中留下的支撑剂，因为自然回流产生的流体速度不足以完全携带走井筒中剩余的支撑剂。清除支撑剂后，压裂液返排。若井内压力不足以回流压裂液，则可根据实际需要采取注氮气人工举升。不可忽视每个射孔段的压后清洗，因为将压裂液留在支撑剂充填层中会对井产生不利影响。通常的做法是在返排前改造所有层段，因此需尽可能在短时间内完成压裂作业。

3.2 Cased-Hole Fracturing Technologies

Audio 3.2

Compared with an open-hole completion, a horizontal well cased and cemented through the horizontal producing section of the well is generally more desirable and less prone to failures for effective hydraulic fracturing. The horizontal well casing sectional fracturing technology has the advantages of excellent process. This technology satisfies the sectional completion requirements of shale gas horizontal well perforation fracturing. This section will be used for casing completion and casing well pressure. The relevant content of the fracture is introduced.

3.2 套管井压裂技术

与裸眼完井相比，水平井的水平生产井段通过套管和水泥固井通常更加理想且易于进行有效的水力压裂。水平井套管分段压裂技术工艺优良，能够满足页岩气水平井射孔压裂的分段压裂要求，本节将对套管完井及其压裂的相关内容进行介绍。

3.2.1 Cased-Hole Completions

3.2.1.1 Cementing Horizontal Wells

Cementing a well is a reliable method for controlling fracture placement in horizontal wells. Conventional cement slurry formulations can have a negative impact on completion. Such slurry systems have a low solubility in acid, which can cause difficulty in perforation breakdown, inhibit fracture initiation and cause excessive near-wellbore friction effects during stimulation and production.

3.2.1 套管完井

3.2.1.1 水平井固井

固井是控制水平井中裂缝位置的可靠方法。传统的水泥固井可能对完井产生不利影响，这种水泥体系在酸中溶解度低，不易被射孔破坏，妨碍裂缝起裂，在压裂或生产过程中产生过大的近井摩阻效应。

Compared with an open-hole completion, a horizontal well cased and cemented through the horizontal producing section of the well is generally more desirable and less prone to failures for effective hydraulic fracturing. Of course, such a completion is more expensive, and it poses certain challenges on its own. A cased and cemented horizontal section allows fracture initiation points to be controlled in order to place multiple hydraulic fractures. The use of designated fracture initiation points along the horizontal section allows the appropriate spacing of either acid or proppant fracturing treatments. Fig. 3. 12 shows a cemented liner completion; Fig. 3. 13 shows a cemented casing completion.

与裸眼完井相比,水平井水平段的套管固井通常更加理想且易于进行有效的水力压裂。当然,这样完井花费会更高,并且其施工本身也是一种挑战。套管固井水平段需要控制裂缝的起始点以便进行多次水力压裂。沿水平部分在指定裂缝起始点进行压裂能较好地布置酸化或支撑剂压裂。图 3. 12 为衬管固井完井;图 3. 13 为套管固井完井。

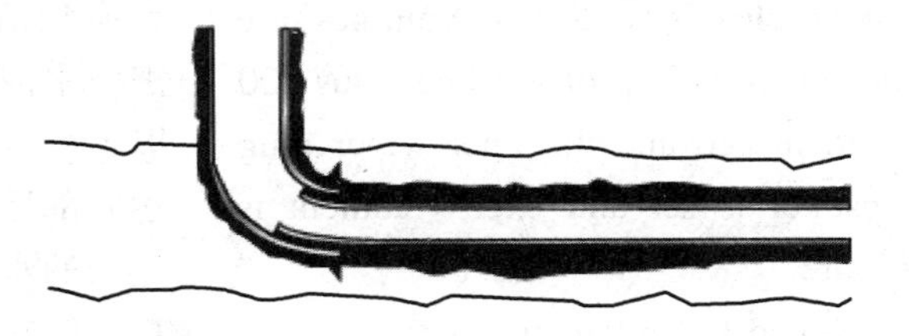

Fig. 3. 12 Cemented Liner Completion

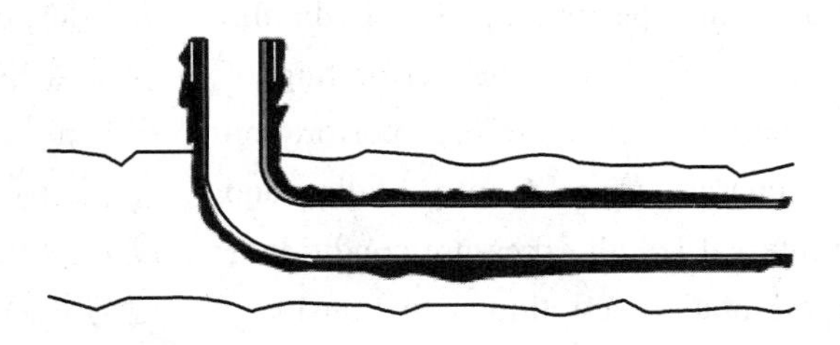

Fig. 3. 13 Cemented Casing Completion

An acid-soluble cement system may be used to cement the liner or casing. Such a system rapidly dissolves when contacted by acid. The enlarged annular space created allows acid to contact the formation adjacent to the perforations with minimal loss of energy and effective zonal isolation of each perforated interval. The acid-soluble cement system can be formulated to achieve the proper density, fluid loss control, compressive strength and free-water requirements for horizontal cementing. Gas migration can be controlled with unique cement additives or by foaming the acid-soluble cement.

酸溶性水泥体系可用于固定衬管或套管。当酸与水泥接触时,水泥迅速溶解,产生更大的环形空间使酸与射孔相邻地层接触,把能量损失减到最小的同时有效隔离各射孔段。酸溶性水泥体系可满足适当密度、控制流体损失、抗压和水平段无自由水固井的要求。可以使用特殊的水泥添加剂,或使酸溶性水泥发泡来控制气窜。

Acid-soluble cement slurry densities can range from 13 to 15.8 lb/gal and can be formulated at lower densities with foam or extenders. The removal of the acid-soluble cement adjacent to the perforations creates an enlarged area in the annulus while providing zonal isolation along the wellbore. The dissolved acid-soluble cement around the casing at the perforations minimizes tortuosity and fracture entry pressure effects. Skin effects, reduced near-wellbore conductivity, and perforation problems encountered with conventional cements are eliminated.

酸溶性水泥密度为13～15.8lb/gal,可以用较低密度的泡沫或增量剂配制。射孔邻近部分的酸溶性水泥溶解后形成扩大的孔眼,其余部分水泥沿井筒提供有效隔离。在射孔处套管周围溶解的酸溶性水泥降低了近井弯曲度和破裂压力,从而避免常规固井中遇到的近井导流能力较小和射孔问题。

Cement slurries are usually batch mixed and the densities are measured with pressurized mud scales before pumping. The slurry follows a weighted spacer which aids mud displacement and hole cleaning. Slurry volumes are calculated to circulate cement to the top of the liner plus 200 feet. After the cement is displaced and the liner wiper plug lands, the liner hanger packer is set and excess cement is circulated out of the wellbore.

水泥通常是分批混合的,并在泵送之前用加压泥浆秤测量密度。加重隔离液有助于水泥置换和清洁孔。计算水泥体积需使水泥循环至衬管顶部200ft以上。水泥顶替和刮衬管塞到达底部后,衬管悬挂封隔器并将多余的水泥从井筒循环出来。

3.2.1.2 Perforating Cemented Completions

3.2.1.2 固井完井射孔

Perforating for hydraulic fracturing is different from perforating for production, and perforating for hydraulic fracturing in horizontal wells varies from the perforating of the vertical well for fracturing. Gun size, perforation phasing, shot density, charge type (entry hole and penetration), perforated interval length, pressure conditions, and fracture-to-wellbore orientation are things to consider. The perforated interval length can affect the success of hydraulic fracturing in transverse and longitudinal laterals.

水力压裂的射孔与生产射孔不同,水平井水力压裂的射孔与垂直井水力压裂的射孔不同。射孔枪尺寸、射孔相位、射孔密度、装药类型(入孔和射孔深度)、射孔段长度、压力条件和裂缝与井筒的方向均是考虑因素。射孔段长度可直接影响横向和纵向水力压裂的成功。

Although for production considerations there are distinct applications for longitudinal vs. transverse configurations, a perforated interval that is too long may lead to the initiation and growth of multiple fractures near the wellbore. Multiple fractures can increase bottom hole treating pressures and near-wellbore friction effects and will decrease fracture width and fluid efficiency. Limiting the perforated interval length for both longitudinal and transverse configurations is always advisable.

出于生产考虑,纵向与横向裂缝布置不同,太长的射孔段可能导致井眼附近产生多条裂缝,多条裂缝会增加井底压力和近井摩擦效应,减少裂缝宽度和压裂效率。因此建议限制纵向和横向裂缝的射孔段长度。

For longitudinal fracturing, re-perforating the well after the fracture treatment is advisable. For transverse fracturing, a limited set of perforations extending up to four times the wellbore diameter or less (usually 1 to 3 feet) reduces or eliminates multiple near-wellbore fractures and especially the tortuosity associated with longitudinal fracture initiation followed by transverse propagation.

对于纵向裂缝压裂,通常在压裂作业确定后再对井进行射孔。对于横向裂缝压裂,射孔段长度需限制在井眼直径的4倍甚至更小(通常为1~3ft)以内,以减少近井的裂缝条数,特别是降低由初始纵向裂缝转向为横向裂缝的弯曲度。

For longitudinal fracturing, oriented perforating guns with 180 – degree phasing, creating perforations that point up and down, are indicated. In certain cases, 0 – degree phasing is very desirable. To avoid penetrating into a water zone, a well drilled in the bottom of the reservoir can be perforated with 0 – degree phasing, with the perforations pointed up. This forces the fracture to migrate upward, reducing the flow of water from the bottom. Abass et al. (1993) even suggested drilling horizontal wells incompetent but nonproducing formations, and then perforating and fracturing, thus producing through the fracture.

对于纵向压裂,有180°定向的定向射孔枪,产生指向上下的孔。某些情况下,需要0°相位射孔;为了避免压裂缝进入水层,在储层底部钻出的井可以进行0°定向射孔,射孔朝上使得裂缝上移,减少底部水侵。阿巴斯等人(1993)曾提出在有潜力但非生产层中钻水平井,进行射孔和压裂从而产生裂缝。

If phased perforations are executed, deep-penetrating perforations are not required because the fracture initiation point is at the cement/formation interface and not along the perforation tunnel. Perforation phasing affects hydraulic fracturing. If the perforations are within 30 degrees of the preferred fracture direction, near-wellbore tortuosity can be minimized.

如果采用相位射孔,则无需深穿透射孔,因为压裂起始点位于水泥(地层)界面处裂缝不是沿着射孔孔道。射孔相位影响水力压裂效果,如果射孔在压裂裂缝方向的30°内,则可以减小近井裂缝弯曲度。

For horizontal well perforating, a typical procedure would involve first a liner cleanout trip where the liner, casing, and tubing are pickled with a solvent and hydrochloric acid. Acetic acid is then spotted in the lateral, and tubing-conveyed perforating guns are picked up and run in the well. Pumping equipment activates the pressure-actuated tubing-conveyed perforating guns. More acetic acid is then pumped into the formation and breaks down each perforation interval.

对于水平井射孔,最典型的做法是先清理衬管,其中衬管、套管和油管用溶剂和盐酸进行酸洗。然后在水平段分散注入乙酸,用油管输送射孔枪。再用泵压设备启动射孔枪,将更多乙酸通过各射孔段泵入地层。

Single-trip, multiple-zone, tubing-conveyed perforating can be used to achieve high-quality perforations. Point-source cluster perforating is designed to achieve a limited-entry fluid distribution for an equal volume of the treatment fluid in each perforated zone. The primary application is acid fracturing. Multiple-trip, multiple-zone, tubing-conveyed perforating may be used in a perforate/stimulate/isolate operation. The fracturing operation may be conducted continuously until complete or scheduled for one or two zones per day due to the operator or well site restrictions.

单次多层的油管输送射孔技术可实现高质量射孔。点源簇射孔用于实现各射孔层中等体积工作液的限流分布,主要应用于酸压。多次多层的油管输送射孔技术可用于射孔、改造或隔离作业。受施工人员和井场的限制,压裂作业可连续进行直到压完所有射孔段,也可按计划每天完成一个或两个层的压裂。

External casing perforating may also be used in horizontal wells. The perforating and zonal isolation systems are attached to the outside of the casing. The external casing perforating system is cemented in place with conventional cementing techniques. The perforating modules are selectively fired to perforate and allow the zone to be stimulated. See Fig. 3. 14. One disadvantage of the system is a limit of cased wellbore diameter, as the external casing perforating gun system maximum casing diameter is 3. 5 inches.

套管外射孔也可用于水平井。射孔和段分隔系统连到套管外,采用传统固井技术完成套管外射孔固井。射孔工具可选择性地点火射孔,并进行相应层段改造,参见图3. 14。该方法的缺点是限制了套管直径,因为套管外射孔枪允许的最大套管直径为3. 5in。

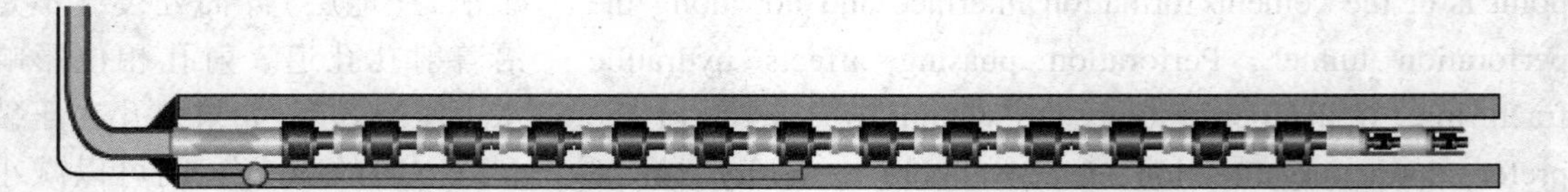

Fig. 3. 14 External Casing Perforating

As with open-hole completions, creating perforations in cemented completions with an abrasive jetting tool leads to large, clean holes within a very short interval. Another tool allows the creation of a radial slot in the casing, which might appear to be an ideal option if not for the potential danger of cutting the casing entirely and shifting it. Both tools perform well in field applications . Coiled tubing provides a method to deploy a bottom hole jetting assembly. The bottom hole assembly is moved to the first zone to be stimulated.

与裸眼完井一样,在套管固井中使用研磨喷射器可在很短层段形成大的、清洁的孔眼。另一种形成眼眼的工具能在套管中形成径向槽,如果没有可能把套管完全切割开及发生套管移动的危险,它将是一种理想的选择。这两种工具在现场应用都较为广泛。连续油管提供了一种下入井底喷射装备的方法,就是将井底装备移至第一层进行压裂改造。

3.2.1.3 Zonal Isolation in Cased Completions

To isolate a zone for treatment, drillable composite bridge plugs are conveyed with coiled tubing and set electrically or hydraulically. A hydraulically set bridge plug is the faster method. As soon as pumping stops, coiled tubing can be rigged up to run the bridge plug to the depth and set it. The advantage of setting the bridge plug electrically is that the use of electric line coiled tubing allows a plug to be set on depth with a collar locator for more accuracy.

In the external casing perforating system, flapper valves provide zonal isolation. The flapper valves are protected from stimulation fluids during fracturing operations by a sliding sleeve. Another method of isolation is used in some applications where coiled tubing deploys a jetting tool to cut perforations.

Limited entry fracturing uses perforation friction pressure to control the stimulation fluid distribution into each perforated interval. Limited entry fracturing has been successfully used in vertical wells. In horizontal wells, limited entry fracturing incorporates fluid friction pressure loss along the long lateral, little or no change in hydrostatic pressure and perforation erosion effects caused by proppant laden fluid. Pipe friction pressure in long horizontal wells may limit the maximum pumping rate. Perforation erosion effects, which may only have a minor effect on fluid distribution in vertical wells, may significantly affect the fluid distribution in a horizontal stimulation treatment.

3.2.2 Fracturing of Cased-Hole Completions

3.2.2.1 Acid Fracture Execution

As in open-hole fracturing, the live acid must be transported as far away from the wellbore as possible to dissolve and etch the carbonate in order to create conductive fractures. But in cased holes, such treatments are a lot easier to accomplish than in open holes.

3.2.1.3 套管完井的层间封隔

为了在压裂中将各层隔离，通过连续油管输送可钻式复合桥塞并利用电或液压坐封。液压桥塞是最快捷的方法。一旦停泵，就可装配连续油管下桥塞至指定深度并坐封。电缆桥塞的优点是电缆连续油管通过接箍定位器可以将桥塞放置到更精确的深处。

套管外射孔系统中挡板阀实现了层间封隔。压裂作业期间，通过滑套保护挡板阀免受压裂液影响。这种隔离方法用于一些连续油管喷射器切割射孔。

限流法压裂用射孔摩阻来控制每个射孔段压裂液的分布，已成功用于垂直井。水平井中，限流压裂整合了沿横向流体摩擦压力的损失、很少甚至可以忽略静水压力，以及携砂液引起的射孔侵蚀效应。长水平井中的管道摩阻可能会限制最大排量。射孔侵蚀效应对垂直井中的压裂液分布影响较小，但对水平井增产作业中的压裂液分布影响较大。

3.2.2 套管完井压裂技术

3.2.2.1 酸压

在裸眼压裂中，为了压出导流裂缝，活性酸必须尽可能远离井筒以防止溶解和腐蚀碳酸盐。但是在套管井中酸压比在裸眼井中更容易实现。

In addition, diversion techniques can be far more easily applied to acid fracturing, than to proppant fracturing. Therefore, it is feasible to pump an acid fracture treatment into a wellbore containing several open perforated intervals. Generally consisting of a combination of ball sealers and viscosity-contrast techniques, allow the fracturing fluid to be efficiently diverted away from the weakest interval to the next weakest interval. If several diversion stages are employed alternately with several acid fracture stages, it is possible to place acid fracture treatments effectively over several perforated intervals.

此外,与支撑剂压裂相比,分流技术更容易应用于酸压。因此,将酸压技术用在包括几个射孔段的井中较为可行。通常结合封堵球和黏度对比技术,能让压裂液从最薄弱段向次薄弱段分流。如果需在多个压裂段中交替分流,则在几个射孔段内进行酸压更为有效。

3.2.2.2 Plug-and-Perforate Completion Systems

3.2.2.2 桥塞射孔联作压裂系统

The plug-and-perforate (or plug-and-perf, also abbreviated PNP) completion system is the multistage completion technique that was originally used to unlock these unconventional plays, and is still the most common type of multistage completion system used today. This system traditionally uses cement to isolate the annulus, perforations to breach the cemented liner to provide a fluid-flow path during fracturing and production, and fracturing plugs to isolate from the previously fractured stages and divert the fluids out of the perforations. When the fracturing job is finished, the plugs are milled out to put the well on production.

桥塞射孔联作(PNP)压裂技术是最初用于开采非常规储层的多级压裂技术,并且仍然是目前最常见的多级压裂技术类型。该系统一般是利用水泥来隔离环空,射孔穿破水泥固结的衬管以在压裂和生产期间提供流体流动通道,并且使用压裂桥塞隔离已经压裂过的层段使流体从射孔中流出。当压裂作业完成后,将桥塞研磨掉,油气井投入生产。

1. Fracturing Plugs

1. 桥塞

Fracturing plugs are designed as temporary isolation devices. The plugs are set in the case to isolate and divert the fluids, and then the plugs are removed from the well. The plugs are available in two types: ball-isolating plugs and self-isolating plugs, each of which has advantages depending on the type of completion in use. Both types of plugs are illustrated in Fig. 3.15.

桥塞是一种临时封隔装置,安装在套管内以隔离和转移流体,随后从井中移除。桥塞有两种类型:球封隔桥塞和自封隔桥塞,每种桥塞都有各自的优势,具体使用哪种取决于完井类型。两种类型的桥塞如图 3.15 所示。

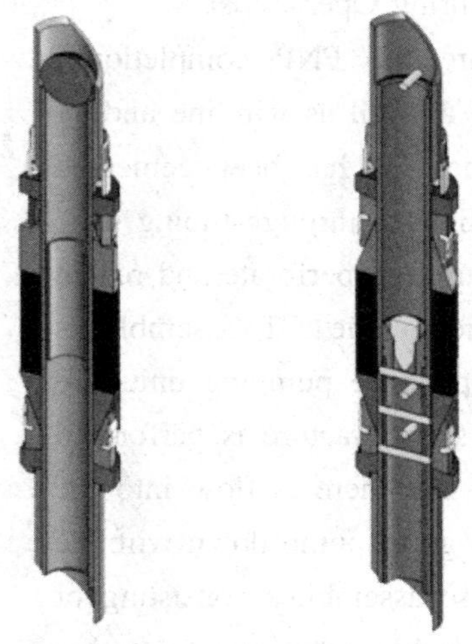

Fig. 3. 15　Ball-Isolating and Self-Isolating Fracturing Plugs

(1) Ball-isolating plugs. The ball-isolating plugs rely on a fracturing ball to provide the isolation. There is still flow to the previously fractured stage through this plug until a ball is dropped from the surface and it lands on the seat for isolation.

(1)投球桥塞。投球桥塞依靠压裂球来提供隔离。压裂液仍然能够通过桥塞流到已压裂段,直到球被投入并且落在球座上以进行隔离。

(2) Self-isolating plugs. The self-isolating plugs have an isolation device installed inside the mandrel. There are a couple of different types of isolation devices, such as a poppet or a caged ball. These plugs allow flow from below, but not from above. As soon as these plugs are set in the case, there is isolation from the previously fractured stage.

(2)自封隔桥塞。自封隔桥塞在心轴内装有隔离装置。一般有几种不同类型的隔离装置,例如提升阀或笼式球。这种桥塞允许压裂液从下方流入,但不能从上方流入。一旦这种桥塞安置到套管内,就会隔离先前压裂过的层段。

2. Perforating the Well

With the PNP method, it is preferred to perforate multiple clusters of perforations per stage to "fracture through" to increase efficiency. This "clustering" is made possible by using select firing perforation systems. These systems have multiple sets of perforation guns that can be fired selectively using electronic signals sent through the wireline. When the first set of perforations is shot, the assembly is moved uphole. Then, when the wireline is in position, another electronic signal is sent to fire the second set of guns. This process is repeated until all the clusters are perforated.

2. 射孔

基于桥塞射孔联作压裂,一般先每段射孔多簇来提高压裂效率。通过选择点火射孔系统使得"簇"成为可能。这些系统具有多组射孔枪,这些射孔枪可以通过电缆发送的电子信号选择性地点火。当完成第一组射孔后,整个组件向上移动。然后,当电缆到合适位置时,发送另一电子信号以点火第二组射孔枪。重复该过程直到完成所有簇射孔。

3. Plug-and-Perforation Fracturing Operations

When it is time to fracture the PNP completions, pressure pumping unit is required as well as wireline and/or coiled tubing (CT). Because the well has been cemented into place and there is no circulation, a through-tubing trip, usually performed on CT, is required to perforate and regain communication with the formation. The CT assembly is pulled out of the hole, and the pressure pumping units are connected to the well. The first stage fracture is performed through these perforations. Now that there is flow into the formation again, it is common to go to pump down wireline assemblies . Fig. 3. 16 shows these assemblies consisting of perforation guns, a setting tool, and a composite fracturing plug.

3. 桥塞射孔联作压裂作业

当桥塞射孔联作进行压裂时,需要压裂泵及电缆和(或)连续油管(CT)。由于已经用水泥固井,没有液体循环的环空,因此通常需要油管来循环液体。一般采用连续油管射孔并重新获得与储层的连通。将连续油管组件从井内提出,并将压力泵单元连接到井口。第一段的压裂通过这些射孔进行作业。当压裂液进入储层后,通常会泵入电缆组件。图 3. 16 显示了这些组件,其中包括射孔枪、安装工具和复合压裂桥塞。

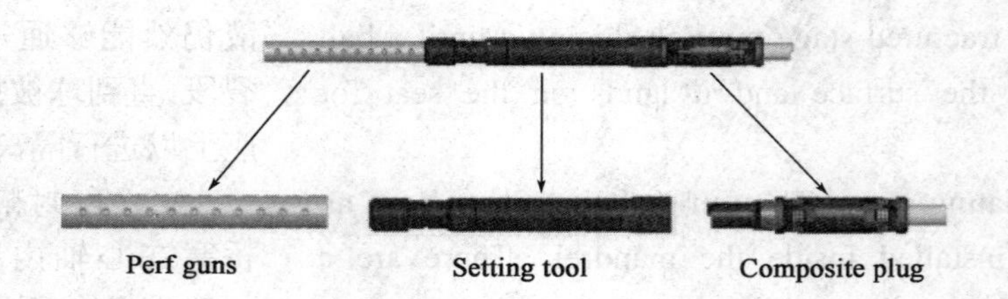

Fig. 3. 16 Wireline Pump-Down Assembly

This assembly is pumped down to the intended depth and the plug is set by sending an electronic signal to the setting tool. The setting tool will set the plug and then release from it. If it is a self-isolating plug, the plug can be pressure tested before the perforations are fired. The perforation guns are pulled uphole to the intended depth, and the first cluster of perforations is fired using another electronic signal through the wireline. The assembly is pulled uphole to the next cluster depth and the second cluster is fired by sending another electronic signal. This process is then repeated until all of the perforation clusters are shot and then the wireline is retrieved. The wireline unit is disconnected and the pressure pumping units are set up. A stage using a self-isolating plug can begin fracturing right away. If a ball isolating plug is used, the ball is dropped into the well and pumped onto the plug. When the ball seats on the plug, there will be an abrupt pressure increase and

将该组件泵至预期深度后,通过向安装工具发送电子信号来安置桥塞,安置好桥塞后安装工具脱离。如果是自封隔桥塞,则可以在射孔之前对桥塞进行压力测试。将射孔枪向上提至预定深度后,通过电缆发射出点火信号点火射孔第一簇。将组件向上提到下一簇深度后,再次发送信号射孔第二簇。重复该过程,直到射出所有射孔簇,然后取回电缆。电缆单元取出后连接压力泵单元。使用自封隔桥塞的段可以直接开始压裂作业。如果使用投球桥塞,则向井中投入球并泵送到桥塞上。当球落在桥塞上时,压力会突然增

that indicates isolation from the stage below. When isolation is achieved, the fracture begins. After the fracturing is performed on this stage, the pressure pumping units are disconnected from the well and the wireline unit is set up again with a new wireline assembly, and this process is repeated until all stages are fractured. Fig. 3. 17 illustrates a well using PNP.

加,这表明已完成与下段的封隔。当实现封隔后开始压裂作业,在该段进行压裂之后,压力泵单元与井断开,并且下入新的电缆组件重复该过程直到所有段都被压裂。图 3. 17 展示了使用桥塞射孔联作的井。

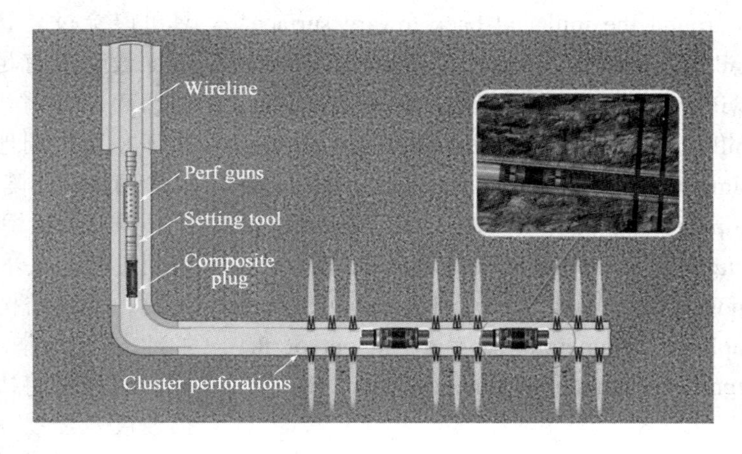

Fig. 3. 17 Wellbore Diagram of a Plug-and-Perforation Job

4. Milling Out the Composite Bridge Plugs

After the fracturing job is completed with PNP, the plugs will have to be milled out and removed before the well is put on production. Most mill-out applications rely on weight-on-bit (WOB) while milling, but it can be difficult to get enough WOB when milling out in horizontal wells with coiled tubing. In this application, the mill-out is designed to try to reduce the dependency on WOB drilling as much as possible. This section discusses the components and the process of milling out the plugs.

4. 铣掉复合桥塞

在使用桥塞射孔联作完成压裂作业之后,在将井投入生产之前必须将桥塞铣掉并移除。这一操作多数都依赖于铣削时的钻压(WOB),但是在使用连续油管铣削时,很难获得足够的钻压。在这个应用中,铣掉设计旨在尽可能地减少对钻压的依赖。下面讨论铣削桥塞的工具组成和过程。

1) Bottom hole Assembly Design and Mill Selection

1)井底钻具组合设计和铣掉选择

The milling BHA consists of a motor and mill. The motor is selected based on the fluids that will be circulated and the amount of torque output that is needed.

井底钻具组合由电动机和铣削工具组成。基于内部循环的流体和所需的扭矩输出量选择电动机。

A variety of mill designs are effective in this application. The mills will vary, but some parameters stay mostly the same: mill size, shape of the mill face, and the cutting structure. The mill size is typically designed to be 95% to 99% of the drift diameter of the casing to ensure the mill can pass through the casing. The profile of the mill is designed so that there is enough flow area to circulate the milled out debris around the mill and back toward surface. The face of the mill needs to be concave. A convex face that points outward will vibrate and kick away from the plug, hindering the mill-out process. In addition, the cutting structure is very important. The cutting structure needs to be aggressive on the plug, and very durable. The durability of the cutting structure will determine how quickly the mill will wear down and how often it will need to be retrieved from the well to be replaced. Numerous trips in and out of the well can add a significant amount of time and cost.

在这个应用中可以选择多种铣削设计,铣削工具可能会有所不同,但有些参数基本保持不变,例如铣削工具尺寸、工具表面形状和切削结构。通常将铣削工具尺寸设计为套管偏移直径的95%~99%,以确保铣削工具可以穿过套管。铣削工具的剖面设计成有足够的流动面积使磨碎的碎屑在铣削工具周围循环并返排到地面。铣削工具的表面需要凹陷,因为朝外的凸面会振动并从桥塞上崩开,阻碍铣削过程。此外,铣削结构非常重要。铣削结构需要应对桥塞的冲击,要求非常耐用。铣削结构的耐用性将决定铣削工具磨损速度及需要从井中取出更换的频率,多次进出井会增加大量的时间和成本。

2) Fluid Selection

The fluids should be designed to help with the milling process as well as to remove the debris. The fluid should be viscous enough to circulate the debris and bring it back to surface. The fluids should also reduce the amount of friction of the coiled tubing and at the mill face to ensure all of the applied weight and power is being transferred to the mill, rather than lost to friction. Reducing friction will also help optimize the pump pressure, and a properly designed fluid will prevent differential sticking.

2)流体选择

流体应设计成有助于铣削过程及去除碎屑;流体应足够黏稠来循环碎片并将其带回到地面;流体还应该减少连续油管和铣削工具面的摩擦,以确保所有施加的重量和动力都传递到铣削工具,而不是损失在摩擦上,减少摩擦的同时也有助于优化泵压;合适的流体设计还可防止压差卡铣削工具。

5. Alternative Systems and New Technologies

There are two recent technological advances developed for PNP completion that are focused on improving the efficiencies during the fracturing job and allowing access to new applications. These technologies are the pressure activated fracturing sleeve and the large-bore fracturing plugs.

5. 其他替代系统和新技术

桥塞射孔联作压裂有两项新技术,主要集中在提高压裂作业效率,这些技术包括压力打开滑套和大口径压裂桥塞。

The pressure-activated fracturing sleeve can be used to replace the trip required to perforate the first stage. It is deployed on the completion string and installed at the toe of the string. When it is time to begin the fracturing job, the pressure pumping units set up and apply pressure to open this sleeve. When the sleeve is opened, the first-stage fracturing begins. This sleeve saves the time and cost of deploying a wire line unit and performing a through-tubing trip to perforate the first stage.

压力打开滑套可用于代替第一级射孔所要经过的下入井内过程。它部署在完井管柱上并安装在管柱的趾部,当需要开始压裂时,安装压力泵并施加压力以打开滑套,开始第一段压裂。这种滑套节省了第一级射孔部署电缆和过油管作业的时间及成本。

The large-bore fracturing plug is a ball-isolating plug that is designed to remain in the wellbore after the fracturing job and to be produced through. It is used in conjunction with a disintegrating fracturing ball that provides isolation on the plug during the fracturing job, but it will disintegrate in the production fluid afterwards. Once the fracturing is complete, the balls disintegrate and the well can be put on production without milling out the plugs.

大口径压裂桥塞是一种投球隔离桥塞,压裂作业后留在井筒中并且可以进行油气生产。它与可分解压裂球结合使用,在压裂作业期间起到封隔作用。一旦压裂完成,球就会分解,不用铣出桥塞就可以直接投入生产。

3.2.2.3 Coiled Tubing-Activated Completion Systems

3.2.2.3 连续油管分段压裂系统

Coiled tubing-activated completions systems (CTACS) use coiled tubing (CT) to achieve multistage isolation. There are two primary methods with CT, using either an abrasive perforator or CT-activated sleeves to access the stages in the well. A coiled-tubing packer or a sand plug can be used to isolate from the previous stage.

连续油管分段压裂系统(CTACS)使用连续油管实现多级封隔。连续油管有两种主要使用方法,即通过研磨射孔器或连续油管激活滑套进入井中的各个段。连续油管封隔器或砂塞可以用来隔离前一段。

1. Components of Coiled Tubing-Activated Systems

1. 连续油管分段压裂系统的组成

This section describes the components commonly used with coiled tubing systems, including components such as fracturing sleeves and abrasive perforators, and how they are used in different applications under different conditions.

下面介绍常用于连续油管系统的组件,包括压裂滑套和研磨射孔器等,以及它们在不同条件下、不同场景中的使用。

(1) Coiled tubing-activated fracturing sleeves.

(1) 连续油管分段压裂压裂滑套。

The CT-activated fracturing sleeves provide the flow path for the fracturing fluids to enter each stage. There is a variety of these types of sleeves in the industry. One type is the mechanically shifted sleeves that rely on mechanical force from the CT to shift the sleeve open. The CT tools lock into the sleeve and applying downward force opens the sleeve.

连续油管分段压裂压裂滑套为压裂液进入每段提供了流动通道。石油工业中有各种各样的滑套。一种是机械移位滑套,依靠来自连续油管的机械力

Another option is the pressure-balanced sleeves. These sleeves have internal pressure ports at the top and the bottom of the sleeve. As long as these ports have the same amount of pressure applied to each one, they remain pressure balanced and in the closed position. These sleeves are opened by using a CT packer to create a pressure imbalance across these ports. The CT packer is run through the liner into the fracturing sleeve, and it is set in between the two pressure ports. The ports are now isolated from each other and pressure is applied to the annulus. Because of the CT packer isolation, the top port will have applied pressure, but the bottom one will not. This creates the imbalance and the sleeve shifts into the open position, exposing the fracturing ports.

打开滑套,连续油管工具锁定滑套,向下施加力打开滑套。另一种是压力平衡滑套,这些套筒在顶部和底部都有内压式端口,只要对每个端口施加相同的压力,它们就保持平衡并处于关闭状态,通过使用连续油管封隔器在这些端口间产生不平衡压力,便可以打开这些滑套。连续油管封隔器穿过衬管后进入压裂滑套,并放置在两个压力端口之间。端口间彼此隔离时,压力则施加到环空中。由于连续油管封隔器的隔离,顶部端口将受到压力,但底部端口不会。这样压力不平衡后滑套移动到打开位置,露出压裂端口。

(2) Abrasive perforator.

An abrasive perforator is an alternative to conventional perforating guns. The abrasive perforator is a CT tool that creates holes in the casing by pumping fluid and sand through the CT and into the casing in an abrasive manner. As shown in Fig. 3. 18, this tool uses a water and sand mixture to cut the casing with an abrasive jet coming through the nozzle. When the casing is cut, these perforations are used to divert the fracturing fluid into the formation at that point. The abrasive perforator can be seen in Fig. 3. 19.

(2)研磨射孔器。

研磨射孔器是传统射孔枪的替代品,是一种连续油管工具,它通过将流体和沙子泵送通过连续油管以磨蚀方式进入套管,在套管中产生孔。如图3.18所示,该工具使用流体和沙子混合物通过喷嘴喷射磨料射流来切割套管。当套管被切割后,这些射孔点可将压裂液转移到地层中。研磨射孔器如图3.19所示。

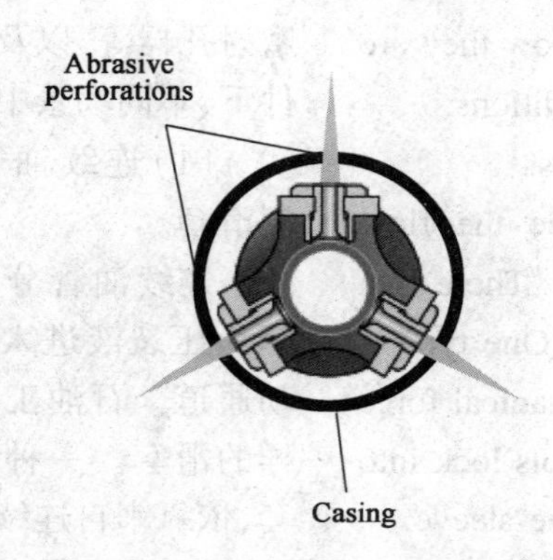

Fig. 3. 18 Abrasive Perforator Cutting through the Casing

Fig. 3. 19 Abrasive Perforator

(3) Coiled tubing packers.

The CT packer is used to create a mechanical barrier to isolate below the CT assembly. This will isolate the previously fractured stage and divert the fracturing fluid through the fracturing sleeve ports or abrasive perforations. The packers are also used to isolate the top and bottom ports on the pressure-balanced sleeves so that the sleeve can be opened. The CT packers for this application are specially designed to rely on weight as little as possible, just like the plug mill-out application. When the CT is extended to certain lengths in the horizontal lateral, it can be very difficult to get weight down to the bottom of the CT string.

(3)连续油管封隔器。

连续油管封隔器用于形成机械屏障以隔离下方的连续油管组件,这样可以隔离已压裂段并使压裂液通过压裂滑套端口或磨蚀穿孔转移。封隔器还可用于隔离压力平衡滑套的顶部和底部端口,以便打开滑套。就像铣出桥塞的应用一样,用于此类应用的连续油管封隔器专门设计成尽可能少地依靠重量。当连续油管在水平横向延伸到一定长度时,可能很难将重量传递到连续油管管柱的底部。

(4) Sand plug.

Another way to isolate the abrasive perforation or the fracturing sleeve is to pump a sand plug at the end of each stage. A sand plug is created by increasing the sand concentration in the fluid to a high enough level that it plugs off and fluid can no longer enter the formation through that injection point. Because fluid cannot pass into the formation at that injection point, it is isolated from the next stage fracturing without the use of the CT packer. The sand plugs do require a clean out trip with the CT when the fracturing is complete.

(4)砂塞。

另一种隔离研磨射孔或压裂滑套的方法是在每段的末端泵入砂塞。砂塞是通过将流体中的砂子浓度增加到足够高以使其堵塞并且流体不再能够通过该注入点进入地层而产生的。由于流体不能在该注入点处进入地层,因此在不使用连续油管封隔器的情况下将其与下一段压裂隔离。当压裂完成后,还需要用连续油管清理砂塞。

(5) Casing collar locator.

The casing collar locator (CCL) is used to determine the location of the CT BHA in the wellbore, so the CT tools are placed for opening the sleeve or the abrasive perforator placed for cutting. Between each connection of casing, there is a casing collar that connects to the next joint. When the connection is made between joints, there is a small gap between the two joints of casing. The CCL is part of the CT BHA and has spring-loaded latches that touch the inside of the casing as it moves up and down the wellbore. When these latches pass over a casing collar, the springs push the latches into the gap and the CCL locks into place. There is enough force applied through the springs that it will require

(5)套管接箍定位器。

套管接箍定位器(CCL)用于确定连续油管井底工具组合在井筒中的位置,以此安置连续油管工具来打开滑套或者研磨射孔器以进行切割。在套管的每个连接之间,有一个连接到下一个接头的套管套环。当接头之间进行连接时,套管的两个接头之间存在小的间隙。套管接箍定位器是连续油管井底工具组合的一部分,并且具有弹簧加压的闩锁,当其在井筒中上下移

pulling tension with the CT unit to release the CCL from the collar. This feature allows each casing connection to be located to know where the CT assembly is in the wellbore. The CCL and the spring-loaded latch are shown in Fig. 3.20.

动时会接触到套管内部。当这些闩锁越过套管接箍时，弹簧将闩锁推入间隙中，定位器锁定到位。弹簧可以施加足够的力，因此需要用连续油管提供拉动力以从接箍中释放定位器。这样便可以定位到每个套管的连接处从而知道连续油管组件在井筒中的位置。套管接箍定位器和弹簧加压锁闩如图 3.20 所示。

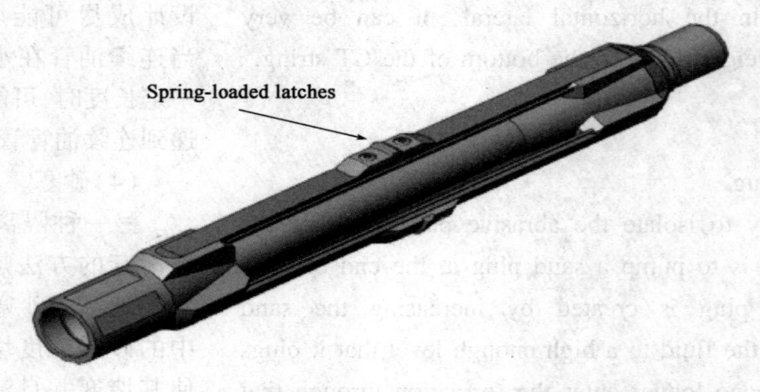

Fig. 3.20 Casing Collar Locator with Spring-Loaded Latches

If a premium casing thread is used, there will not be a gap between each joint of casing. In this application a locating sub will be used to identify the depths in the hole. A locating sub is a small piece of casing that has a groove machined into it that mimics the gap between the joints of casing openhole, and will catch the CCL as it passes over it. This sub would be run in and then spaced out with the completion string, strategically placing it to help identify where the sleeves and other tools are in the wellbore.

如果使用优质套管螺纹接头，则每个套管接头之间不会有间隙。在这种应用中，将使用定位子来识别井的深度。定位子是一小块套管，刻有一个凹槽来模仿裸眼套管接头之间的间隙，并在套管接箍定位器经过它时将其捕获。该定位子下到井中后与完井管柱间隔开，策略地放置它可以帮助识别滑套和其他工具在井筒中的位置。

2. Coiled Tubing-Activated Completions Fracturing Operations

2. 连续油管分段压裂作业

CT-activated systems use annular fracturing behind coiled tubing to achieve multistage hydraulic fracturing. There are a variety of tools available to achieve isolation and diversion and this section will go through the operations for each type of setup.

连续油管分段压裂系统在连续油管后面使用环空压裂来实现多级水力压裂。有多种工具可用于实现封隔和压裂液转向，下面介绍每种类型装置的操作。

1) Fracturing Using Abrasive Perforations as the Injection Point

To fracture this type of completion, pressure pumping and CT are required. When it is time to perform the first fracture, the CT assembly is deployed in the well using the CCL to identify where the BHA is in the wellbore and locate where to abrasive perforate. Then, a sand-and-water mix is pumped through the CT and out of the nozzles on the abrasive perforator. This creates a force that cuts the casing by abrasion. The CT packer is then set and the first stage fracture begins through the annulus of the CT and the casing. When the stage is finished, the fracturing is shut down and tension is applied to the CT to unset the CT packer. The assembly is then pulled uphole to the depth of the next stage and the process is repeated until all stages are fractured. This type of completion is shown in Fig. 3.21.

1)以研磨射孔为注入点进行压裂

压裂这种类型的完井需要压力泵和连续油管。当需要进行第一次压裂时,利用套管接箍定位器将连续油管组件部署在井中以识别井底工具组合在井筒中的位置并定位研磨穿孔的位置。然后将砂和水混合物通过连续油管泵入并从研磨射孔器上的喷嘴中泵出,这样就磨蚀切割套管。再放置连续油管封隔器,通过连续油管和套管间的环空开始第一段压裂。当这段压裂结束时,对连续油管施加张力以解封封隔器,然后将整个组件向上拉到下一段的位置,重复该过程直到所有段都被压裂。这种压裂方式如图3.21所示。

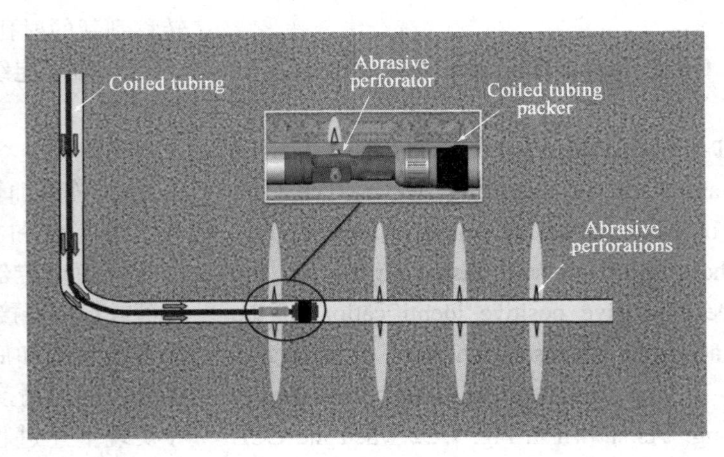

Fig. 3.21 Example CTACS Using an Abrasive Perforator to Create Entry Points into the Casing

After the abrasive perforation is performed, the annular fracturing will need to have a step-rate increase, rather than pumping aggressively right away. During the perforating, all the sand that was used to cut the casing remains in the casing, and this sand will be picked up by the fluid flow and moved into the formation. This additional sand in the fracturing fluid can plug up the injection point and create an accidental sand plug that causes a loss of fluid flow to the

在研磨射孔后,需要梯度增大排量进行环空压裂,而不能立即大量泵入压裂液。在射孔过程中,用于切割套管的所有砂都留在套管中,这些砂随后将被流体带入地层。压裂液中这些额外的砂可能堵塞注入点并产生砂塞,导致流体不能进入地层,

formation, also known as a screen-out. The screen-out can be cleaned out by circulating fluid through the CT annulus, but this creates nonproductive time. It is better to ramp up the fracturing flow rates in increments to disperse the sand in smaller volumes and avoid screen-outs.

也称为脱砂。砂可以通过连续油管的环空被循环压裂液清除，但这会产生非生产时间。最好的方法是以增量的方式逐步提高压裂液流速以便将砂分散避免脱砂。

Another alternative to this method is using sand plugs to isolate each stage rather than the CT packer. The CCL would locate the depth in the wellbore, and the abrasive perforation would be used to cut the injection point through the casing. When the casing is cut the annular fracturing would be performed, and a sand plug would be pumped at the end of the fracturing job to intentionally create a screen-out so that fluid could not enter that injection point any more. The CT assembly is moved uphole and the casing is cut again. Because the previous injection point has a sand plug, the new perforation is the only open injection point in the well, so the fluid diverts into the new perforation.

该方法的另一种替代方案是使用砂塞来隔离每段,而不用连续油管封隔器。套管接箍定位器可以定位井筒中的深度,而研磨射孔将切割穿过套管形成注入点。当套管被切割后,将进行环空压裂,并且在压裂作业结束时泵送砂塞以脱砂,使得压裂液不能再进入该注入点。随后连续油管组件向上移动并再次切割套管。由于先前的注入点有一个砂塞,新的射孔是井中唯一的开放注入点,因此流体会被转移到新的射孔中。

2) Fracturing Using Coiled Tubing-Activated Sleeves

2) 利用连续油管打开滑套压裂

When the CT-activated sleeves are used, the CT BHA is used to locate and open the fracturing sleeve. The CCL locates the sleeve by using the casing collars or locator subs. The sleeves will have short joints of casing on the top and bottom of the sleeve to give positive identification of the sleeves. This is achieved because each joint of casing is approximately 40 – ft long, but the short joints on the sleeves are only 6 – ft. long. As shown in Fig. 3. 22 when the CCL hits another casing collar after pulling up only 6 ft., it is obvious that the assembly is in the sleeve.

当使用连续油管打开滑套压裂时,连续油管井下工具组件将用于定位和打开压裂滑套。套管接箍定位器通过使用套管接箍或定位器接头来定位滑套,滑套的顶部和底部有短的套管接头以便对滑套进行识别。能够实现上述过程是因为滑套的每个接头长约 40ft,但套管上的短接头仅长 6ft。如图 3. 22 所示,当套管接箍定位器在拉起仅 6ft 后接触到另一个套管接箍时,显然工具组件还在滑套中。

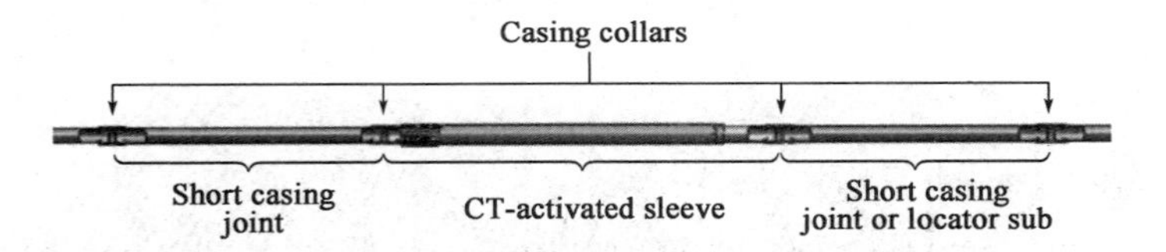

Fig. 3. 22 The CT-Activated Sleeve Is Installed on Short Casing Joints

When the assembly reaches the correct depth, the sleeve will be opened. If it is a mechanical sleeve, the coiled tubing latches into the sleeve and mechanical force will be applied to shift the sleeve open. If it is a pressure-balanced sleeve, the CT packer is set and it isolates the top and bottom ports of the pressure-balanced sleeve. As shown in Fig. 3. 23, pressuring up the annulus causes the top set of ports to have applied pressure that the bottom set does not have. This causes a pressure imbalance in the sleeve, and opens the sleeve. The other sleeves in the completion string are not isolated, so the ports stay in a pressure-balanced and closed position.

当组件到达正确的深度时，滑套将被打开。如果是机械滑套，则连续油管闩锁滑套并施加机械力以打开滑套。如果是压力平衡滑套，则坐标连续油管封隔器并隔离压力平衡滑套的顶部和底部端口。如图 3. 23 所示，向环空加压会导致顶部端口承受压力，而底部端口不承压。这会导致滑套中的压力不平衡，进而打开滑套。完井管柱中的其他滑套没有被隔离，因此端口保持压力平衡和关闭位置。

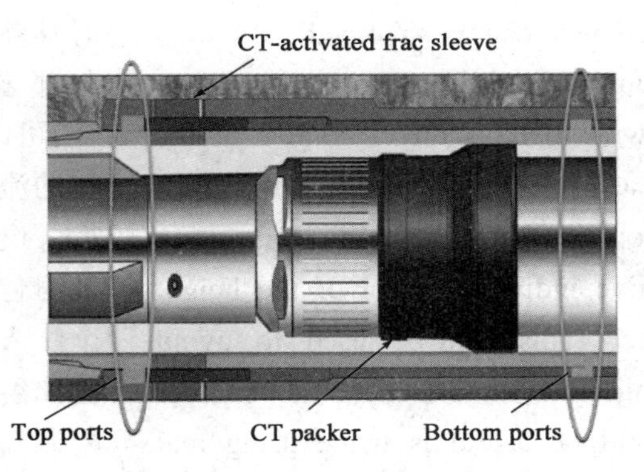

Fig. 3. 23 Setting the CT Packer between the Top and Bottom Ports on the Fracturing Sleeve

When the sleeve is opened, the annular fracturing begins, using the CT packer to isolate from below and divert the fluid out of the ports. When that stage is finished, the fracturing is shut down, tension is applied to the packer to release it, and the assembly is moved to the next sleeve. The process is repeated until all stages are fractured. The completion is seen in Fig. 3. 24.

当滑套打开时，开始环空压裂。放置连续油管封隔器从下方隔离并将压裂液从端口转移到环空。该段作业结束后停止压裂，向封隔器施加张力以解封，并将组件移动到下一个滑套。重复该过程直到所有段都被压裂，如图 3. 24 所示。

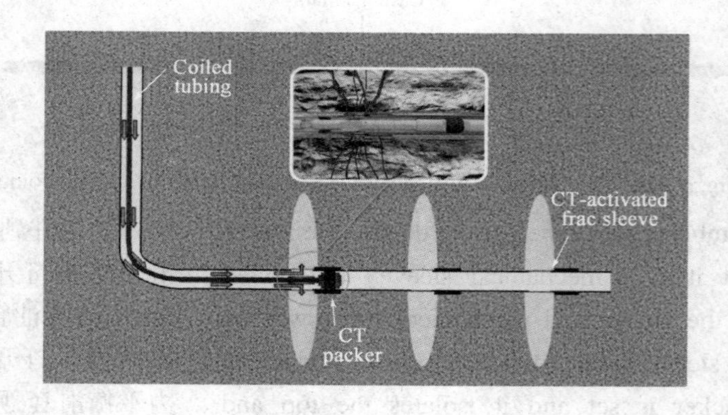

Fig. 3. 24 CTACS Using CT-Activated Fracturing Sleeves

The fracturing does shut down between stages, but it is still an efficient process. The CT packer will need time to equalize and go into a relaxed position. Once the CTpacker is equalized, the BHA is moved uphole to the next stage. Shutting down the fracturing job and being able to move instantaneously has benefits. Like the CT deployed PNP, there is direct control of how much fluid is displaced into each stage so over displacement can be avoided. By only fracturing out of one injection point and having the capability to instantaneously shut down, there is direct control of the volume of fluid placed in each injection point. If it is known how the fracturing will grow in that formation, controlling the volume of fluid will give indirect control of the height and length of the fracture. This can be used to avoid fracturing into water-bearing zones or offset wells. Another scenario in which this would be useful is while using real time microseismic monitoring. If the microseismic monitoring shows the fracturing growing into a nonproductive or waterbearing zone, the fracturing can be shut down and moved to the next stage to save time and money and avoid that zone.

压裂施工在每段作业后都会停止,但它仍然是一个高效率的过程。连续油管封隔器需要一定时间来平衡并进入适当的位置。一旦连续油管封隔器达到平衡,组件组合就会向上移动到下一段。停止压裂作业并及时移动到下一段是有好处的,如与部署有连续油管的 PNP 一样,可以直接控制每段中置换流体的流量,从而可以避免过度置换。仅从一个注入点压裂并能及时停止则可以直接控制到每个注入点中的流体体积。如果已知裂缝将如何在地层中延伸,控制流体的体积将间接控制裂缝的高度和长度,这样可以避免压裂到含水区或分支井中。另一种有用的方案是使用实时微震监测,如果微震监测显示裂缝延伸到非产油区或含水区,则可以停止压裂作业并移至下一段以节省时间和支出并避开该区域。

While annular fracturing with CT, there will be a small amount of fluid being pumped in the CT and out of the abrasive perforator. The primary purpose of doing this is to apply internal pressure so that the annular pressure will not collapse the CT during the fracturing. This column of fluid is at a low enough pump rate to deliver accurate real-time downhole pressure monitoring capabilities, because the static nature of this fluid does not have the friction losses that are seen in the high rates in the annulus. This downhole pressure data can help you detect, and prevent screen-outs.

利用连续油管进行环空压裂时,将有少量流体泵入连续油管并从研磨射孔器中泵出。这样做的主要目的是施加内部压力,保证环空压力在压裂期间不会使连续油管塌陷。这些流体的泵速要足够低,可以提供准确的实时井下压力监测,因为压裂液不会引起高流速所产生的摩擦损失。这些井下压力数据有助于检测并防止脱砂。

Should a screen-out occur, fluid can be circulated through the CT to the annulus to clean out the excess sand in the wellbore. The abrasive perforator is still on the CT assembly even though the sleeves are the primary means into the formation. It is used as a contingency option if there is a screenout that cannot be recovered from or if another stage needs to be added. If either of these scenarios occurs, the abrasive perforation will be used to cut a new injection point into the casing.

如果发生脱砂,流体可以通过连续油管循环到环空中以清除井筒中多余的沙子。虽然滑套是进入地层的主要手段,但研磨射孔器仍然在连续油管组件上。如果出现不能恢复的脱砂或者需要增加压裂段,则它可作为应急选项,出现了这些情况中的任意一种,研磨射孔器可以在套管上切割新的注入点。

3. Alternative Systems and New Technologies

3. 其他替代系统和新技术

Another alternative with the CTACS is using a straddling device on CT to isolate individual perforations. In this scenario, a casing string would be installed into the well and cemented into place. Perforation guns would then be deployed to perforate each of the entry points into the well. A CT assembly would then be run in the well with a BHA containing a straddle tool. The straddle tool is capable of isolating above and below each of the perforated entry points. The straddle tool is placed over a set of perforations and the fracture is pumped through the CT and between the straddling tools. This is illustrated in Fig. 3. 25.

连续油管激活完井系统的另一种替代方案是在连续油管上使用一种横跨装置来隔离各个射孔。在这种情况下,在井中安装套管柱并用水泥固井,然后在井中部署射孔枪射穿每个射孔点。之后在井中下连续油管组件,其中井下工具组合包含横跨工具,横跨工具能够封隔每个射孔入口的上方和下方。将横跨工具放置在一组穿孔上,在连续油管和横跨工具间进行压裂,如图 3.25 所示。

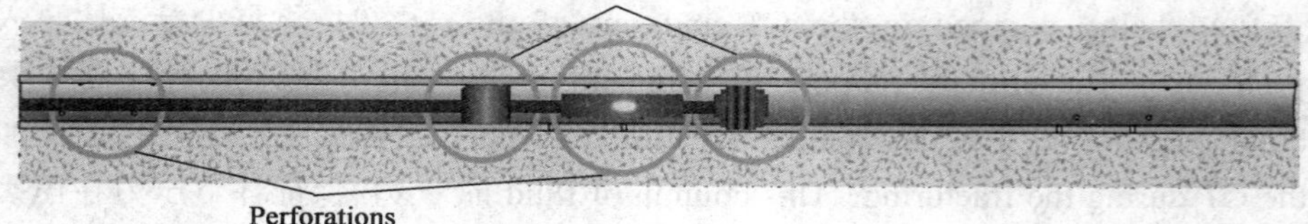

Fig. 3. 25 Coiled Tubing Straddle Tool Isolating Perforations

This technique is also being used in refracturing applications for a previously fractured well that is no longer producing at an economic rate. The straddle tool is positioned over the existing perforations or the previously opened fracturing sleeve, and the refracturing is performed through the CT.

该技术还可应用于不能再经济生产井的重复压裂。将横跨工具放置在已有射孔或先前打开的压裂滑套上,并且通过连续油管进行重复压裂。

3.3 Hydrajet Fracturing

Audio 3.3

Multistage hydrajet-fracturing combines hydrajet perforating and hydraulic fracturing to perform separate, sequential fracture stimulations without mechanical packers. It can reasonably place fractures according to geological condition, and then accurately treat them. Without packer, it uses dynamic isolation to seal flow into the target, saving operating time and lowering operating risk. Therefore, the process not only especially adapts to stimulate open hole, but effectively treats slotted liner completion. From 2005 to 2019, more than 1000 oil and gas wells have been successfully treated using this technology at home and abroad. Fig.3.26 shows the mechanism of Hydrajet Fracturing.

3.3 水力喷射压裂技术

多级水力喷射压裂结合了水力喷射射孔和水力压裂,在没有机械封隔器的情况下进行单独的连续压裂,可以根据地质条件合理进行造缝、压裂。该技术不用封隔器,采用动态隔离将压裂液密封到目标层中,从而节省了操作时间降低了操作风险。该技术适用于裸眼井,能够有效地处理割缝衬管完井。从 2005 到 2019 年,国内外已成功处理了 1000 多口油气井。图 3. 26 为水力喷射压裂机理。

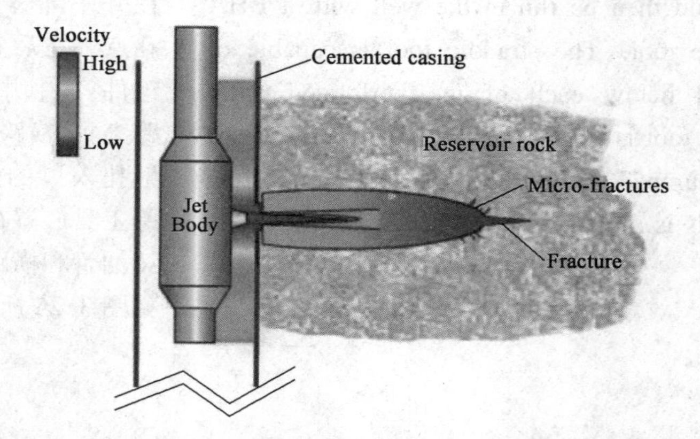

Fig. 3. 26 Mechanism of Hydrajet-Fracturing, Annulus Fluid Is Drawn into the Cavity without Mechanical Parkers

Hydrajet-fracturing technique (HJF), as a relatively new stimulation method, does not require mechanical packers and only one service trip can achieve several stages fracturing treatment. Now HJF has proven to be an economical, effective, and low-risk multistage fracturing process. The HJF method was put forward in 1998 by Jim B. Surjaat-majda et al., Halliburton engineers. They did many laboratory experiments and oilfield tests in Texas and New Mexico of United Stated successfully. In the ten years, HJF application has expanded around the world. There are many successful cases in complex architectural wells, either hydraulic fracturing or acid fracturing treatment. Until 2008, worldwide, over 200 wells have been stimulated using HJF technique by Halliburton Service Corp.

水力喷射压裂技术(HJF)作为一种相对新的增产方法,不需要机械封隔器,只需一次操作即可实现多级压裂。现在HJF已被证明是一种经济有效、低风险的多级压裂工艺。HJF由Hallibuton 工程师 Jim B. Surjaat-majda等人于1998年提出。他们成功地在得克萨斯州和新墨西哥州进行了许多室内实验和油田试验。十年后HJF的应用已遍布全球。无论是水力压裂还是酸压裂工艺,在复杂井也有许多成功案例。2008年,Halliburton Service Corp 使用HJF在全球范围内对200多口井进行了增产改造。

In 2002, the research about hydrajet-fracturing technology was ranked as the High-tech Research and Development Program of China. China University of Petroleum, as a major institute, studied abrasive jetting mechanism, planned fracturing process in details and designed down-hole tools for varied wellbore configurations. On July 2007, Gas Production Engineering Research Institute of Petro China Southwest Oil & Gasfield Company achieved the hydrajet-fracturing process, using 2 – in coiled tubing and special hydra-jet tool, working together with China University of Petroleum. This would be the first hydrajet-fracturing performance with coiled tubing in China.

2002年,HJF研究被列入我国高技术研究发展计划。中国石油大学研究了研磨喷射机理,详细设计了压裂过程,并设计了适用于各种井筒的井下工具。2007年7月,中石油西南油气田公司天然气生产工程研究所与中国石油大学合作,采用2in连续油管和特殊喷射工具实现了水力压裂喷射工艺,这是我国用连续油管进行水力喷射压裂的首次尝试。

3.3.1 Hydrajet-Fracturing Mechanism

In oilfield applications, hydrajetting is used to abrade or penetrate various substances, including steel, cement, and rock formations. Hydrajet-fracturing integrates with abrasive jet perforating, hydra-jet fracturing and hydradynamic sealing as a whole. The brief procedure is following. Firstly, once the base fluid with abrasive (normally quartz sand) is pumped down tubing, abrasive jet begins to create

3.3.1 水力喷射压裂机理

在油田应用中,水力喷射用于磨铣和喷射穿透各种物质,包括钢铁,水泥和储层岩石等。水力喷射压裂工艺是集射孔、压裂、隔离一体化的新型增产改造技术,简要程序如下:首先将带有研磨材料(正常的石英砂)的

perforation tunnel towards the reservoir rock near wellbore. In the end of the tunnel, jet would tend towards stagnation. Setting aside the fact that the jet often carries a substantial amount of sand or other abrasives, the jetting operation appears much like Fig. 3. 27. Fluid is forced from a jetting tool through a small orifice into the annulus. The pressure in the jetting tool C must be higher than the pressure in the annulus A. The fluid ' s high-pressure energy within the tubing is transformed into kinetic energy. According to the Bernoulli Effect, the pressure inside the tunnel will be higher than annulus pressure. The next step is to shut down the casing valve. System pressure increases sharply until exceeding the initiation of fracture pressure (IFP), fractures will be initiated from the tip of tunnel. In fact, this process accomplishes rapidly. Almost at the same time, we begin to inject the basic fluid down the annulus. The above process contains the mechanism of multistage fracturing without mechanical packers.

基础流体泵入油管,研磨喷射器开始在井筒附近岩石上穿孔,喷射流体在孔道末端停滞。实际作业中射流经常带有大量沙子或其他磨料。喷射操作如图3.27所示,流体从喷射工具中喷出并通过一个小孔进入环空。喷射工具C的喷射压力必须高于A中的压力,油管内流体的高压能量将转化为动能。根据伯努利效应,孔道内的压力高于环空压力。其次关闭套管阀,系统压力会急剧增加直到超过起裂压力(IFP)为止,裂缝从孔道端起裂。实际上这个过程非常快,几乎与向环空注入流体同一时间开始。上述过程包含了无机械封隔器的多级喷射压裂机理。

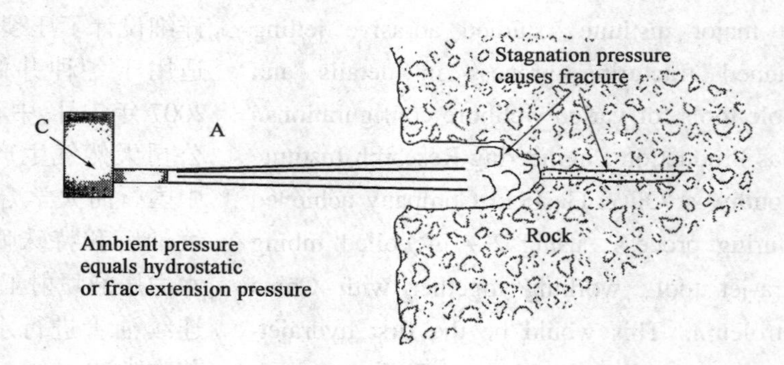

Fig. 3. 27 Hydraulic Jetting Process

3. 3. 2 Hydrajet Fracturing Tool & Operating Process & Field Application

3. 3. 2. 1 Hydrajet Fracturing Tool

There are two types of hydrajet fracturing tools applying to different HJF process respectively. They are the trailing-tool and ball/sleeve-tool. As Fig. 3. 28 showing, the trailing-tool can be divided into five parts: guide shoe, perforated tube, one-way valve, jet body, and centralizers. Some cone nozzles are fixed in the jet body. According to the demand of flow rate, we can optimize diameter, number,

3. 3. 2 水力喷射压裂工具、施工步骤和现场应用

3. 3. 2. 1 水力喷射压裂工具

有两种类型的喷射工具分别适用于不同的水力喷射压裂工艺,分别是拖曳工具和球(滑套)工具。如图3.28所示,拖曳工具可分为五个部分:引鞋、射孔管、单向阀、喷枪和扶正器,一些喷枪内含锥形喷嘴。根据流

and phase of nozzles. Generally, if the flow rate is over 2.3 m^3/min, the nozzle scheme consists of six 6mm/8mm nozzles in two planes, evenly phase 60°. If the flow rate is low relatively, the nozzle scheme would be changed into three 6mm nozzles, evenly phase 120°. The wear resistance of nozzle material determines the tool life directly.

量要求可以优化喷嘴直径、数量和相位。通常如果流速超过2.3m^3/min，喷枪由6个6mm（8mm）喷嘴组成，相位均为60°。如果流速相对较低，喷嘴将改为三个6mm喷嘴，相位均为120°。喷嘴材料的耐磨性直接决定了工具寿命。

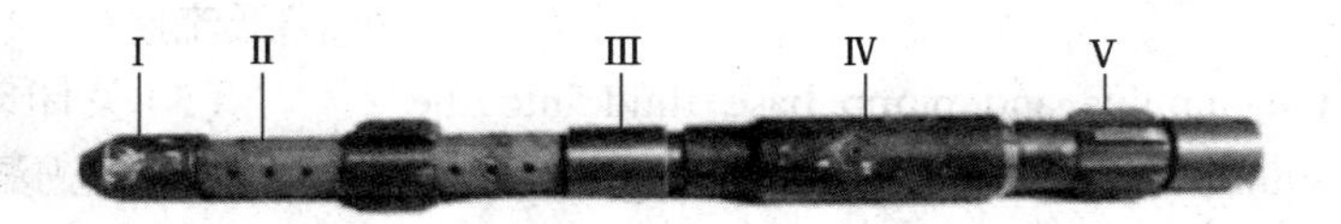

Fig. 3.28 An Assembled Hydrajet Fracturing Tool

Ⅰ—guide shoe; Ⅱ—perforated tube; Ⅲ—one way valve; Ⅳ—jet body; Ⅴ—centralizer

3.3.2.2 Operating Process

3.3.2.2 水力喷射压裂施工步骤

There are two kinds of hydrajet-fracturing operating process, trailing frac-string process and non-tripping frac-string process. The former is suitable for oil well multistage fracturing with low reservoir pressure, few proppant and short span among zones. The most advantage is that only one down-hole tool could achieve several stages fracturing, saving tool cost vastly.

水力喷射压裂工艺有两种，拖曳压裂管柱工艺和不动管柱工艺。前者适用于储层压力小、支撑剂少、层段跨度短的多级油井压裂，最大的优点是只用一个井下工具就可以实现多级压裂，大大节省了成本。

Non-tripping frac-string process demands for ball/sleeve-tools. Several tripping balls with different diameter can control sliding sleeves on/off in different desired fracturing location, realizing multistage fracturing. This process is especially fit for high-pressure gas well and complex architectural well stimulation treatment. In some cases, frac-strings served as completion strings directly during the post-fracture treatment, not only saving production cost but also accelerate production process.

不动管柱工艺需要使用球（滑套）工具，几个不同直径的球可以在不同的压裂位置控制滑套开关实现多级压裂。该工艺特别适用于高压气井和复杂井的增产作业，在一些情况下，压裂后作业时压裂管柱可直接作为完井管柱，不仅节省了生产成本，还能加快生产过程。

Take the non-tripping frac-string fracturing process as an example. The detail procedure is following:

以不动管柱工艺为例，施工步骤如下：

(1) Accurately run frac-string to the treating location, keeping the error less than 0.5 m.

(2) Circulate down the tubing with low flow rate to clean up.

(3) Pump the fluid mixed with quartz sands down tubing then begin to abrasive jet perforating, keeping annulus open. It will take 15 ~ 20 min to acquire a perfect perforating tunnel.

(4) Displace low-crosslinked gel and reduce flow rate slowly.

(5) Shut in the annulus and pump base fluid into the annulus continually until the end of stage fracturing, in order to complement the leakage during fracturing.

(6) Raise flow rate in tubing and start to squeeze gelled pad.

(7) Pump carrying fluid down the tubing and raise proppant concentration in multistep. This process needs adjust to flow rates in tubing and annulus simultaneously to avoid actual pressure exceeding limited pressure.

(8) Inject displacement liquid after finishing proppant input.

(9) Stop pumps and wait for fracture closure until system pressure no longer drops.

(10) Open annulus valve slowly, controlling pressure drop, then drop the tripping ball into tubing and drive it in the low flow rate until open the sliding sleeve correspondingly. Repeat the process (2) ~ (9), beginning to the next stage fracturing.

The difference between trailing process and non-tripping process is only the tenth step. In the trailing process, the system pressure should be lower than allowable pressure of snubbing unit. Then move hydra-jet tool to the next fracturing location.

3.3.2.3 Field Application

Since the first HJF application in China, there are quite a few successful cases in eight oilfields, such as Daqing oilfield, Sichuan gasfield, and Zhongyuan oilfield.

(1)压裂管柱准确下到设计位置,误差小于0.5m。

(2)低流速循环油管进行洗井。

(3)将与石英砂混合的流体泵入油管,保持环空处于开放状态进行喷射射孔,获得理想的射孔孔道需要15 ~20min。

(4)驱替低交联凝胶,缓慢降低流速。

(5)关闭环空并将基液泵入环空直至压裂结束,以此补充压裂过程中的压裂液漏失。

(6)提高油管内的流体流速,向前挤稠化的前置液。

(7)将携砂液多级泵入油管提高支撑剂含量。该过程需要调整油管和环空中的流体流速避免实际压力超过限定压力。

(8)泵入支撑剂后再泵入顶替液。

(9)停泵等待裂缝闭合直到系统压力不再下降为止。

(10)缓慢打开环空阀,控制压降,将球投入油管内并低流速驱动,直至打开相应的滑套为止。重复步骤(2) ~ (9),开始下一阶段压裂。

两种压裂间的区别仅是第(10)步。在拖曳过程中,系统压力应比加压设备的允许压力低,然后将液压喷射工具移动到下一个压裂位置。

3.3.2.3　现场应用

自我国首次采用水力喷射压裂技术以来,大庆油田、四川气田、中原油田等8个油田已取得了许多成功案例。

HJF technology is commonly used in wells with thin oil layers and incapable of fracturing stimulation; and ultra-low permeability tight reservoirs to reduce downhole seepage resistance, wells with conventional perforation difficult to produce or perforation of various casing wells with serious oil layer pollution, and pre-fracturing pretreatment to reduce formation fracture pressure. It is also suitable for wells with a depth of less than 4,000 meters. However, it is generally not suitable to carry out acid-sensing reservoirs with increased acidification.

水力喷射压裂技术通常用于油层较薄、无法进行压裂增产的井,能降低特低渗透致密油藏的井底渗流阻力,也可用于常规射孔难以生产的井、油层污染严重的各种套管井,还能用于压裂前期预处理来降低地层破裂压力等,适用于井深小于4000m的井,一般不宜用于实施酸化增产的酸敏油藏。

3.4 Multistage Fracturing by Swellable Packer

3.4 可膨胀封隔器分段压裂技术

Swellable packers provide the needed isolation for a bare formation design for stimulation. The swellable elastomer will absorb fluid, creating swell pressure and a hydraulic seal; the outside diameter of thepacker can be minimized for maximum clearance during installation. They are typically manufactured on a joint of casing of the same size, weight, and grade as specified for completion. The connections on the packer mandrel are cut as specified. Solid end rings on both sides of the elastomer provide protection to the elastomer as well as provide an anti-extrusion barrier when the differential pressure is applied.

Audio 3.4

可膨胀封隔器为裸露储层提供了所需的隔离环境从而能进行增产作业。可膨胀弹性材料可以吸收流体,产生膨胀压力,实现液体密封。为了获得最大间隙将封隔器的外径最小化,通常在完井指定的相同尺寸、重量和等级的接头上进行安装。封隔器芯轴上的连接按规定切割,弹性材料两侧的实心端环对弹性体有保护作用,还能作为施加压差时的抗挤压挡板。

The swelling of these elastomers can be achieved in the presence of either hydrocarbon or aqueous based fluid depending on the design selection. Fig. 3. 29 shows the multistage fracturing by swellable packer for horizontal wells.

根据设计需要,这些弹性体遇油或遇水膨胀。图3.29为利用可膨胀封隔器实现水平井分段压裂的多级压裂工艺。

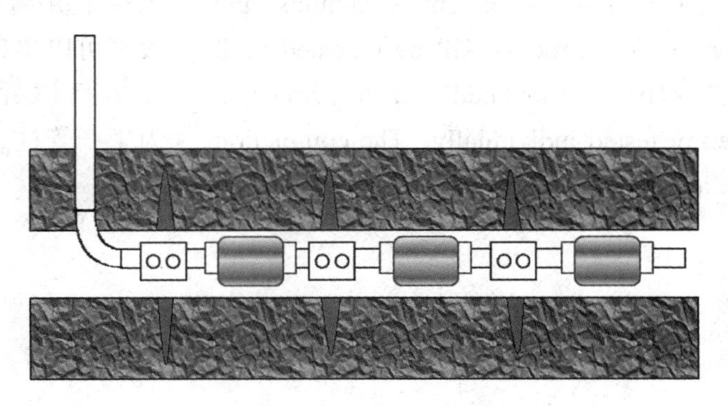

Fig. 3. 29 Multistage Fracturing by Swellable Packer

Swellable packer is a packer that swells when it is in contact with well bore fluids. The swellable part is a special rubber that swells when immersed in well fluids (water, oil, or mixture of both) as the liquid diffuses into the rubber. The rubber can expand 2 to 3 times its original volume in oil. Swell time and volume can be controlled. Up to date, oil swellable parker is the most popular one which swells faster than water or mixture swellable parkers. Once the completion string is fully set in place, diesel (may water or water-diesel mixture) is pumped through the completion string into the annular to soak the packer rubber. After full circulation is achieved, the rubber packers are expanded. The expanded rubber will conform to well hole irregularities like ovalities and washouts.

可膨胀封隔器是封隔器的一种,当其与井筒流体接触时会膨胀。可膨胀部分是一种特殊的橡胶,浸入井液(水、油或两者的混合物)中时,液体扩散到橡胶中后使其膨胀。橡胶体积可以膨胀至原来体积的2~3倍,膨胀时间和体积可以控制。迄今为止,遇油可膨胀封隔器是最受欢迎的,因为它比遇水可膨胀封隔器或混合式可膨胀封隔器膨胀得更快。一旦完井管柱就位,柴油(也可以是水或水基—柴油混合物)就可通过完井管柱泵入环形空间中浸泡封隔器橡胶。在实现完全循环后,橡胶开始膨胀。膨胀橡胶用于冲蚀过的井和井眼不规则的井。

A small ball is then inserted into the fracturing fluid and is pumped down inside the completion string. The ball will go through all previous internal holes of slide sleeves until it seats on the specific hole seat. The pumping pressure acts against the seated ball and pushes the sleeve sliding. The port aperture to the string-wellbore annulus appears to the fracturing fluid once the sleeve slides away. Through the opened to annular port, the fracturing fluid flows into the annulus and fracture the parker isolated interval of the well. After the bottom interval of a horizontal well has been fractured, a slightly larger ball is dropped into the well. The ball is then pumped to seat and open the next sliding sleeve, and fracturing fluid goes into the neighbor annulus and fracture the next interval. The process will be repeated until the entire wellbore has been stimulated. Each stimulated stage can be produced or tested individually. The completion string is left in place and works as a production string during production.

将小球放入压裂液并泵入完井管柱,球通过前面的滑套内孔眼直到落在相应球座上。泵压作用在球座上使滑套滑动。一旦打开滑套,连通管柱与环空的端口就暴露在压裂液中。压裂液通过这个端口流入环空并压裂被封隔器隔离的层段。水平井底部压裂后,将略大的球投入井中,球落入球座并打开下一个滑套,压裂液进入邻近环空并压裂下一段。重复这个过程,直到整个井筒被压裂。每个压裂段都可以单独进行生产或测试。完井管柱保留在原位,生产时作为生产管柱。

The oil swellable packer is generally connected in series to the fracturing string like a "sugar gourd" and goes down the well with the column. When the packer reaches the specified position, the crude oil soaked rubber tube is pumped in the annulus. The oil-swellable packer slowly expands after inhaling the crude oil, gradually adhering to the wall of the well (the wall of the open hole or the inner wall of the casing) and sealing the annulus between the fracturing column and the wall of the well to achieve the purpose of isolating the section.

遇油可膨胀封隔器一般如"糖葫芦"一般串联在压裂管柱上,随管柱一同下入井内。当封隔器达到指定位置后,在环空中泵入原油浸泡橡胶管,橡胶管吸入原油后缓慢膨胀,逐渐贴住井壁(裸眼井壁或套管内壁)并将压裂管柱与井壁之间的环空封住,从而达到隔离井段的目的。

3.5 Just-in-Time Fracturing

Just-in-time fracturing is a multistage fracturing method that was developed by ExxonMobil in the late 1990s that enabled the rapid delivery of multiple stimulation treatments required for the thin lenticular sands of the Piceance Basin, Colorado. The method offered several advantages over existing technologies: (1) it provided effective single-zone stimulation that ensured that each zone was treated to its operational limit, (2) it allowed for treatments to be pumped at high flow rates yet using significantly less horsepower, and (3) it enabled the stimulation of multiple target zones via a single deployment of down hole equipment. The method has also been patented and licensed to a number of service companies. Leveraging the extensive experience with JITP in vertical and S-shaped tight-gas wells, horizontal well applications were recently implemented in the Fayetteville shale, Arkansas. Such a step change in JITP operations; namely, the ability to successfully execute treatments in both horizontal and vertical wells, has created an opportunity space for the further optimization of completion practices in other unconventional resources such as tight oil and coal bed methane.

Audio 3.5

3.5 即时压裂技术

即时压裂是一种多级压裂方法,由埃克森美孚公司在20世纪90年代末开发,能够实现科罗拉多Piceance盆地薄透镜状砂岩所需的多种增产改造。该方法相比现有技术的几大优点是:(1)它能进行有效的单层增产改造,确保每层都被改造到开发极限;(2)它能在高流速泵送下进行作业,但使用的马力显著降低;(3)通过布置单一井下设备可以实现对多个目标层增产。这个方法已获得专利并且许多专业服务公司已持证使用。开始即时压裂用于垂直和S形致密气井,在阿肯色州的费耶特维尔页岩的水平井中获得成功。即时压裂在水平井和垂直井中的成功作业,为进一步优化其他非常规资源(如致密油和煤层气)的完井创造了空间。

JITP relies on ball-sealers to isolate each perf cluster and divert fracturing fluids to the subsequent interval. Close coordination is required between wireline and frac crews to individually perforate, divert, and fracture intervals along the lateral. The wireline gun assembly is positioned in the first set of perforations. Guns are fired and the first treatment is pumped. At the end of treatment, ball sealers are pumped down to seal off the open perforations. When ball sealers seat on the perforations, a sharp rise in wellbore pressure is observed ("ball-out") and guns are fired on the second perforation cluster. Without shutting down the pumps, the gun assembly is moved to the next perforation cluster while fracturing the newly-created set of perforations. The process is repeated throughout the lateral. When all guns are spent or ball action is no longer effective, a frac plug is set above the stimulated zones to provide isolation from the next set of treatments. Continuous pumping allows for uninterrupted operations as well as positive pressure on the ball sealers to facilitate effective fluid diversion. Fig. 3. 30 illustrates the mechanism of JITP.

即时压裂用封堵球来隔离每个射孔群并将压裂液分流至后面井段。压裂人员需要密切协调电缆的操作，从而可以单独沿着横向进行射孔、分流和压裂。电缆射孔枪组件在第一组射孔位置点火。预处理后，将封堵球泵送至畅通的射孔孔眼封堵。当封堵球位于炮眼上，能观察到井筒压力急剧上升("疏通")时，在第二射孔段上发射电缆射孔枪。在不关泵的情况下，压裂产生新射孔孔眼后将枪组件移到下一射孔段，重复该过程直至整个井筒全部压裂完为止。当用完所有电缆射孔枪或封堵球不起作用时，在增产区域上方设置桥塞将下一段隔离。连续泵液可以保证不间断运行，同时保持封堵球上的正压来促进有效的分流。图 3.30 为即时压裂的压裂机理。

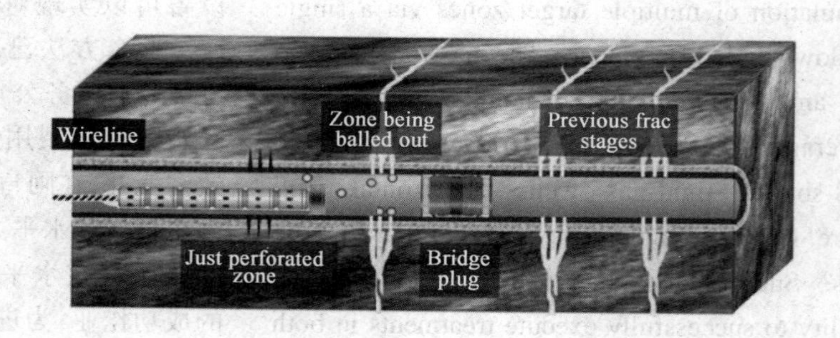

Fig. 3. 30 How JITP Works

To date, JITP has been applied in numerous fields in a variety of wellbores and fluid systems (Table 3. 1).

迄今为止，即时压裂已应用于各种井筒和压裂液类型(表3.1)。

Table 3.1　Range of Successful JITP Applications

Uses	Hydraulic fracturing, acid stimulation
Well Trajectories	Vertical (up to 14,500 ft TVD), S-turn (up to 2,000 ft lateral), horizontal (up to 6,000 ft lateral), and deviated (~45-degrees)
Casing Sizes	4.5in to 7in, sometimes perforating through 2 strings (4.5-inch and 7-inch)
Annular Isolation	Cemented, uncemented with isolation packers
Fluid Systems	Slickwater, linear gel, crosslinked gel, foam, acid
Proppants	White sand, resin coated (pre-and post-cured), ceramic, 100 mesh to 20/40
Pumped Volumes	Up to 5,000 bbls fluid per stage (140,000 bbls per well), 80,000 lbs sand per stage (4 Mlbs per well)
Pump Rates	Up to 50 bpm (Can be higher based on casing size and number of perforations)
Wellhead Pressure	Up to 10,000 psi (limited only by wellhead pressure limits)
Diversion Methods	Ball sealers, bioballs, bridge plugs, frac plugs, sand plugs, bio-degradable plugging material
Ball Sealer Types	Various specific gravities, syntactic foam core, phenolic, polyurethane, bio balls
Perforating Guns	2in to 3in OD, +/-7, 60, 90, 120, and 180-degree phasing
Number of Stages (Perf Clusters)	Up to 14 per event (guns per string), up to 65 per well
Wireline	5/16in wireline, using up to 2 wireline units for multiwell simultaneous frac

Some of the benefits of implementing JITP as a multistage fracturing method are:

(1) This allows for the individual placement of fractures where needed at controlled rates, pressures, and volumes customized for each perforation cluster.

(2) Better control also means that, if desired, it is possible to reduce the risk of fracture growth into offset wellbores to minimize well-to-well fracture interference.

(3) Controlled injection rates in each individual set of perforations increase the potential for more effective fracturing.

(4) Conventional treatments will require approximately 100 bpm pumping rates that are split into 4 ~ 7 perforation sets (depending on operator practices and type of play). Therefore, the effective rate that goes into each perforation set ranges from 15 ~ 25 bpm. Because JITP treats individual perforation sets, it can use much lower surface rates while allowing each cluster to be stimulated at a controlled and known rate.

即时压裂作为一种多级压裂方法,其优点有:

(1)在水平井压裂时对各缝通过每个射孔簇可控的速率、压力和体积进行单独设计。

(2)控制更有效,降低了裂缝增长或偏移到邻井的风险,从而最大限度地减少井间干扰。

(3)控制各射孔簇注入速率提高了压裂效率。

(4)常规的压裂作业大约需要 100 bbl/min 的泵送速率,分成 4 ~ 7 个射孔簇(取决于操作员的经验和油层类型)。泵入每个射孔簇的有效速率范围为 15 ~ 25 bbl/min。因为即时压裂分别对每个射孔组作业,可以在较低的地面速率下保证每个射孔簇以可控并以已知的速率进行压裂作业。

(5) Fracture initiation is performed at high overbalance pressure during perforation, which could reduce the need for acid stages.

(5) 可以在超压下射孔启裂，减少对酸化的需求。

(6) Significant reduction in horsepower requirements on location due to lower surface rates at same treatment pressures.

(6) 在相同作业压力下表面速率较低，因此对井场的功率要求也显著降低。

(7) Horsepower has direct impact on associated completion cost; offset by increased pumping times due to lower rates employed with JITP.

(7) 功率对相关的完井成本有直接影响，补偿了由于使用即时压裂技术时排量较低而增加泵送时间带来的费用。

(8) Advantageous in places where pumping equipment is not readily available for high-rate treatments.

(8) 在泵无法进行高流速作业的地方具有优势。

(9) Effective fracture placement by pumping proppant in each individual set of perforations.

(9) 通过在单独的射孔簇中泵送支撑剂，有效设置裂缝位置。

There is another benefit associated with JITP from a performance forecasting and optimization perspective with regards to reducing uncertainty in the number of created fractures. To date, the optimization of shale gas resource development, which depends on fundamental understanding of drainage area of a well and stimulated rock extent, has been severely limited due to the uncertainty in parameters such as permeability (K), number of created fractures (n_f) and fracture half-length (x_f). The unique aspect of JITP is that fracturing operations occur with wireline continuously present in the wellbore, utilizing ball sealers for fluid diversion. After perforating and treating the first stage, ball sealers are dropped from surface while the gun is positioned at the next stage. When the balls seat on the perforations, a sharp rise in pressure is observed, indicating that it is time to perforate the next stage. Pumping continues without interruption until all guns are spent. At that time a frac plug is generally set to isolate the previously treated stages while a new gun is run. Up to 14 stages have been successfully treated in one gun deployment which is constrained by the height of the available crane and lubricator. Wells with over 60 stages have been routinely completed using JITP, and up to 35 stages have been individually treated in a 24 - hour period with the use of multiwell simultaneous frac.

从性能预测和优化角度看，即时压裂还有另一个好处，即降低造缝数量的不确定性。目前为止，优化页岩气资源开发要求人们对井的泄油面积和压裂区域有基本了解，但渗透率(K)、产生裂缝的数量(n_f)、单侧裂缝长度(x_f)等参数的不确定性严重限制了优化开发。即时压裂的独特之处在于，通过井筒中的电缆能连续进行压裂作业，并利用封堵球分流。在第一阶段进行射孔和增产改造后，将射孔枪移置下一段，从井口投入封堵球，当球座在射孔上时，压力急剧上升表明可以进行下一段射孔。在射孔枪全部射孔结束之前，可持续泵送。在下入新射孔枪前需要坐封桥塞对压裂过的井段进行隔离。受起重机和润滑器可用高度的限制，一般一趟射孔枪可以成功处理多达14个井段。使用即时压裂技术，对水平井普遍可完成60段以上作业。在多井同步压裂中，24h内可对多达35井段完成单独压裂。

3.6 The Casing-Conveyed Perforating Fracturing Technology

Typically, the casing string contains integral isolation devices, perforating guns external to the casing, and actuate the isolation devices remotely. The perforating gun is required to be cemented together with the casing before being drilled into the well. After cementing, start perforating, fracturing, and ball-sealing, repeat the above process to achieve the next stage of fracturing. This section will detail the casing-conveyed perforating frac technology.

Audio 3.6

The Casing-Conveyed Perforating Technology is a new completion method for multiple intervals requiring individual stimulation. This system, a Casing-Conveyed Perforating System, was designed to improve effective stimulation of productive intervals by allowing individual zone stimulation in a rapid, cost effective manner.

The perforating guns are designed to shoot into and through the casing as well as into the formation. The guns are usually fired via external hydraulic control lines. The isolation devices (i. e. flapper valves) are completely compatible with conventional primary cementing operations, as well as subsequent fracture stimulation operations. The flapper valves are actuated when an interval is perforated, and serve to isolate lower intervals during fracture stimulation operations. The frangible isolation devices can be removed with slickline or coiled tubing. Use of the system enables other technologies, such as continuous bottom hole pressure measurement and downhole chemical injection, to be easily and economically incorporated. In this section, the term "module" or "perforating module" refers to a unit assembly, which includes the isolation valve, perforating gun, and related hardware which are picked up and run as a single unit. In the basic casing-conveyed perforating system, the perforating guns are hydraulically fired, using incrementally

3.6 套管外射孔压裂技术

套管外射孔压裂技术的特点是套管柱包含整体隔离装置、套管外射孔枪远程驱动隔离装置。套管外射孔压裂技术把射孔枪与套管一起下入井中一并固井,然后启动射孔,实施压裂,投球封段;重复上述射孔、压裂、封段过程,实现下一段压裂。这项技术的优点是省略了下射孔枪的过程,缺点是射孔位置的固定。本节将对套管外射孔压裂技术进行详细介绍。

套管外射孔技术是一种用于单井多级增产改造的新完井方法,旨在通过快速、经济有效的方式提高储层增产改造效果。

射孔枪用于射入并穿过套管后进入地层。射孔枪通常由外部液压线控制,封隔器(即挡板阀)与传统的固井和随后压裂改造是完全兼容的。一段射孔完成后,开启挡板阀来隔离已压裂段。可用钢丝绳或连续油管移除易碎的隔离装置。使用该系统时可以使用其他技术,如井底压力连续测量、井下化学剂注入。在这一节,术语"模块"或"射孔模块"是指组件单元,包括封隔器、射孔枪和相关硬件设备,它们作为一个单元被取出和下入。在基本的套管外射孔系统中,射孔枪是液压驱动发射的,用逐渐增加的压力发射上一井眼的射孔枪。例如,2000psi higher pressures to fire the uphole guns. For example, 2,000 psi (14 MPa) surface pressure would be utilized to fire the lowermost gun, and the next perforating gun would be actuated when pressure reaches 3,000 psi.

(14 MPa)的井口压力可以驱动最下部的射孔枪,当压力达到3000psi时,发射下一个射孔枪。

The casing-conveyed perforating system provides a means to stimulate multiple low quality pay intervals cost effectively, other benefits included:

套管外射孔压裂技术提供了一种多个低产油层有效增产的方法,其主要优点有:

(1) Significant reduction in total completion time and acceleration of first production.

(1)极大地减少了完井时间并且加快了初期生产。

(2) Less bypassed pay and improved stimulation quality in a stacked-pay environment.

(2)几乎没有绕过含油层,提高了含油层的增产效率。

(3) Monobore well designs, which help prevent liquid loading and facilitate rigless well repairs.

(3)单井设计有助于防止积液,并且便于无钻具修井。

(4) Lower fracturing fluid volume requirements due to smaller tubulars, and the displacement fluid for one stimulation stage becoming the pad fluid for the next stimulation stage.

(4)由于更小的管件减少了压裂液体积,前一个增产阶段的顶替液成为下一增产阶段的前置液。

(5) Lower frac horsepower requirements by only stimulating a single interval at a time.

(5)一次单层增产降低了压裂设备功率要求。

(6) Improved safety, well control, and environmental operations because the equipment is remotely actuated without having to convey equipment inside the casing.

(6)因为设备都可远程控制且套管内部不需要输送装置,提高了安全性和井控效果,而且更环保。

(7) Lower total development costs due to the reduction in tubular requirements, rig time, and associated services.

(7)管件设备简单、钻进时间短和相关检修少使得开发费用低。

3.7 Volume Fracturing

Audio 3.7

Volume fracturing is a series of large-scale fracturing treatments in horizontal wells. The fracturing concept is that the shale contains a large amount of siliceous layers with brittleness, and the layers can form a certain volume of complex seams different from conventional fracturing when encountering damage, which is complicated by horizontal wells and segmental fracturing. The fracture network increases contact with reservoirs and facilitates the release of natural gas from the shale.

3.7 体积压裂技术

体积压裂就是在水平井中进行一系列的大规模的压裂。其压裂理念是由于页岩内硅质含量高的层段具有脆性,遭遇压裂破坏时,与常规压裂不同,能够形成复杂的裂缝网络。裂缝网络的形成可以增大与储层的接触,有利于页岩气的释放。

Fisher et al. studied the correlations among the total fracturing fluid volume, seam size and gas well productivity in Barnett shale development. When Mayerhofer et al. investigated the microseismic monitoring and fracture propagation in the Barnett shale reservoir fracturing reconstruction in 2006, the concept of "stimulated reservoir volume" was firstly proposed. And the relationship between different SRV and cumulative gas production and cluster spacing were studied. Parameters such like fracture conductivity are also involved. According to a large number of research results, the larger the stimulated reservoir volume, the more obvious the increase of the yield. The stimulated reservoir volume has a significant positive correlation with the increase in yield.

Fisher 等人研究了 Barnett 页岩开发中压裂液总量、缝网尺寸及裂缝几何形态与气井产能间的相关关系。Mayerhofer 等人于 2006 年研究了 Barnett 页岩储层压裂改造中微地震监测资料和裂缝扩展，首次提出了"储层改造体积(SRV)"的概念，并对不同 SRV 与累积产气量的关系、簇间距及裂缝导流能力等参数进行了研究。根据大量研究结果可知，SRV 与增产效果具有显著的正相关关系，SRV 越大，增产效果越明显。

3.7.1 Introduction and Development of Volume Fracturing

3.7.1 体积压裂简介及发展

In recent years, the rapid development of unconventional resources has made the technique one step forward, especially the appearance of volume fracturing, which has produced a big influence on the increase of oil/gas production. In 2007, North America began to use the fracturing technology of muti-stage and muti-cluster in horizontal wells, and has made it become the main means of unconventional oil/gas development since then, the exploitation of Barnett shale regarded as the most representative. In China, volume fracturing has been widely applied in Sichuan' shale area and has brought larger gas production than that by traditional fracturing technology in past several years. Recently, this technique also has played a great role in developing the tight oil/gas in some oilfield of Shanxi, China. Volume fracturing technique has effectively reconstructed the unconventional oil/gas reservoir while new problems of frequent casing deformation failure were triggered. In Canada, twenty-eight wells showed different casing damage after large-scale volume fracturing in Quebec, the same phenomena found in American Marcellus shale gas field as well. The casing problem had also serious in China that over ten wells presented the same issues during fracturing job, which hindered the progress of stimulated reservoir volume (SRV).

近年来，非常规资源的快速发展使得相关技术向前迈了一大步，特别是体积压裂的出现，对油气产量的增加产生了很大的影响。2007 年，北美地区开始用多级和多簇的水平井压裂技术，自此这种技术便成了非常规石油和天然气开发的主要手段，其中 Barnett 页岩的开发最具有代表性。在中国，体积压裂已在四川页岩区广泛应用，并且在过去几年带来了比传统压裂技术更高的天然气产量；这项技术在陕西某油田开发致密油气方面也发挥了重要作用。体积压裂技术虽然能有效地开发非常规油气藏，但同时也出现了常见套管变形破坏的新问题。在加拿大，在魁北克进行了大规模的体积压裂后，28 口井显示出了不同的套管损坏问题；美国马塞勒斯的页岩气田也出现了相同的现象；在中国套管变形破坏的问题也很严重，十多口井在压裂作业期间出现了相同的问题，影响了储层体积改造的进程。

The purpose of volume fracturing technology is to achieve the stimulated reservoir volume (Fig. 3. 31). Volume fracturing has the characteristics of large reconstruction range, more stimulation stages, complicated and effective fracture produced and so on. These features make it both similarities and differences with other operations related to the problem of casing deformation. Volume fracturing is to avoid fractures in the near wellbore and widen the far.

体积压裂技术的目的是实现储层体积改造(图 3. 31)。体积压裂具有压裂增产范围大、增产阶段多、产生的裂缝复杂有效等特点。这些特征与和套管变形问题相关的其他操作问题具有相似性及差异性。体积压裂能避免近井多裂缝,开启远井多裂缝。

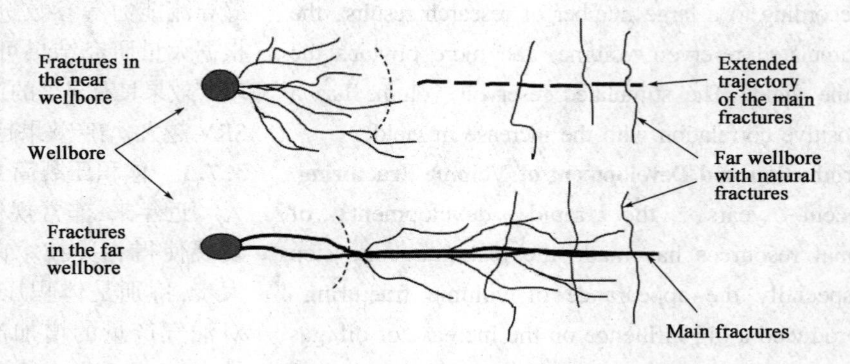

Fig. 3. 31 Volume Fracturing

Wu et al. proposed the difference between the definition of volume fracture in broad and narrow sense. In narrow sense, volume fracturing technology focuses on the purpose of creating fracture network through fracturing methods. While generating one or more main fractures, first secondary fractures are created simultaneously via shear force. Second secondary fractures are formed based on the first secondary fractures and so on. Therefore, fracture network can be developed by intertwining main fractures and multiple orders of secondary fractures. As a result, the reservoir is stimulated in three dimensions. In broad sense, volume fracturing technology also involves the multi-layer fracturing of vertical wells to improve the vertical profile of utilization and multi-stage fracturing of horizontal wells to enhance the flow conductivity and the drainage area of the reservoir.

吴奇等人提出了体积压裂技术的狭义和广义区分,狭义的体积压裂技术是针对通过压裂手段产生网络裂缝为目的的改造技术而言的,通过压裂对储层实施改造,在形成一条或者多条主裂缝的同时剪切产生第一级次生裂缝,在第一级次生裂缝的基础上剪切形成第二级次生裂缝,依此类推,让主裂缝与多级次生裂缝交织联通形成裂缝网络系统,实现对储层在长、宽、高方向的三维改造。广义的体积压裂技术还包含提高纵向剖面动用程度的直井分层压裂技术和提高储层渗流能力及增大储层泄油面积的水平井分段压裂技术。

3.7.2 Volume Fracturing Design SRV & FCI

Complex fracture network models are more suitable for shale volume fracturing design. For the complexity of fractures networks, there are usually two parameters that are used to describe, namely FCI and SRV.

3.7.2.1 SRV

1. Variable Stimulated Reservoir Volume Concept

Wells in Eagle Ford Shale are completed by massive hydraulic fracturing stimulation during which 4,000,000 ~ 5,000,000 pounds of proppant and 10,000 ~ 15,000 bbls of fracture fluids are pumped. Knowing the existence of complex natural fractures network in the reservoir, the large proppant and fluid volumes are believed to form a complicated and heterogeneous SRV geometry around the wellbore. The hydraulic fracturing related microseismic data set is used to define the geometry and the quality of the SRV created by the fracturing stimulation. To accomplish the SRV quantification, we employed microseismic density in terms of points per cubic feet. The model area is discretized into 50 ft. × 50 ft. × 50 ft. grid blocks then the number of points in each grid block is expressed as the 3D microseismic density for that grid block. Areas that have higher microseismic density are believed to have better stimulation than the areas that have lesser density. Based on the microseismic density and the connectivity of the grid blocks the SRV can be divided into three regions:

(1) Hydraulic SRV: It is the volume which is covered by all of the fracturing related microseismic events (Fig. 3.32). Shape of this SRV is highly irregular and is dictated by the pattern and the shape of the pre-existing structures and rock mechanical properties. This SRV is believed to be created by the fluids of the hydraulic fracturing but some portions of it may not contribute to the well's production.

3.7.2 体积压裂设计参数 SRV 和 FCI

复杂裂缝网络模型更适用于页岩体积压裂设计,对于裂缝网络的复杂性,通常采用两种参数来进行描述,即裂缝储层改造体积 SRV 和复杂指数 FCI。

3.7.2.1 SRV

1. SRV 的概念

Eagle Ford 的页岩井是通过大规模水力压裂实现增产的,作业中泵送了 4000000 ~ 5000000lb 支撑剂和 10000 ~ 15000bbl 压裂液。由于了解到储层本身存在复杂的天然裂缝网络,因此认为泵入的支撑剂和流体在井筒周围会形成复杂的非均质结构。与水力压裂相关的微震数据集用于定义由压裂增产措施形成的 SRV 的几何形状和质量。为实现 SRV 量化。采用微震密度,模型区域离散化为 50ft × 50ft × 50ft 的网格块,每个网格块中的点数表示该网格块的3D 微震密度。通常认为具有较高微震密度的区域比具有较小微震密度区域的增产效果好。基于微震密度和网格块的连通性, SRV 可以分为三个区域:

(1) 水力 SRV:即所有与压裂相关的微震事件覆盖的体积(图 3.32)。SRV 的形状是高度不规则的,由预先布置的井网、形状及岩石力学性质决定。这种 SRV 被认为是由水力压裂液产生的,它的某些部分可能井的生产没有帮助。

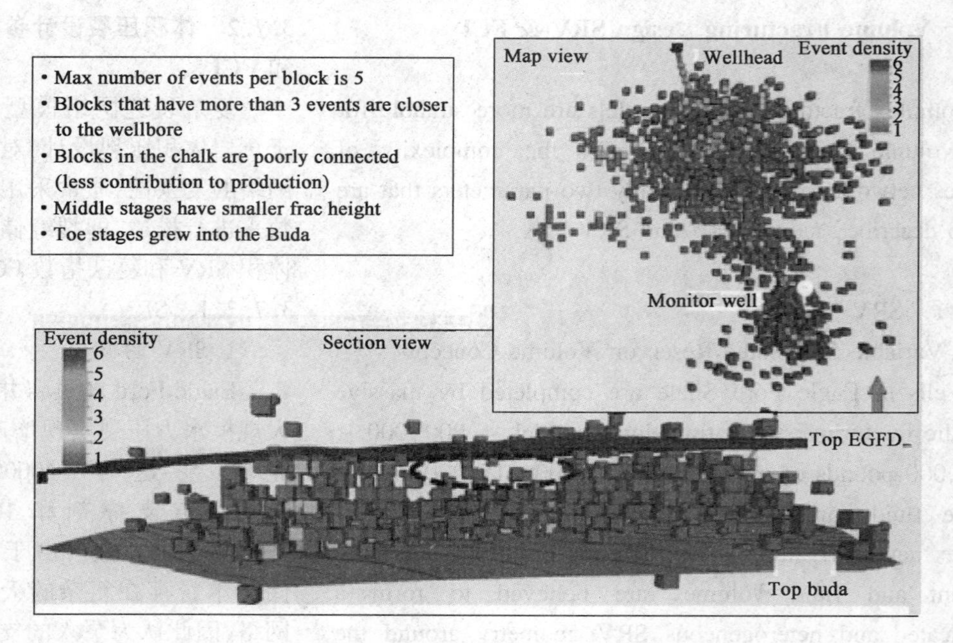

Fig 3.32 Geometry of the Hydraulic SRV

(2) Conductive SRV: This is the volume that is formed by the blocks which have more than one microseismic event (Fig. 3.33). This volume is believed to have effective permeability greater than the background matrix. EUR expected from a well is dependent on the size of the conductive SRV.

(2)导流 SRV:由具有一个以上微震事件点的小区域构成的体积(图 3.33),具有大于基质的有效渗透率,生产井的最终可采储量取决于导流 SRV 的规模。

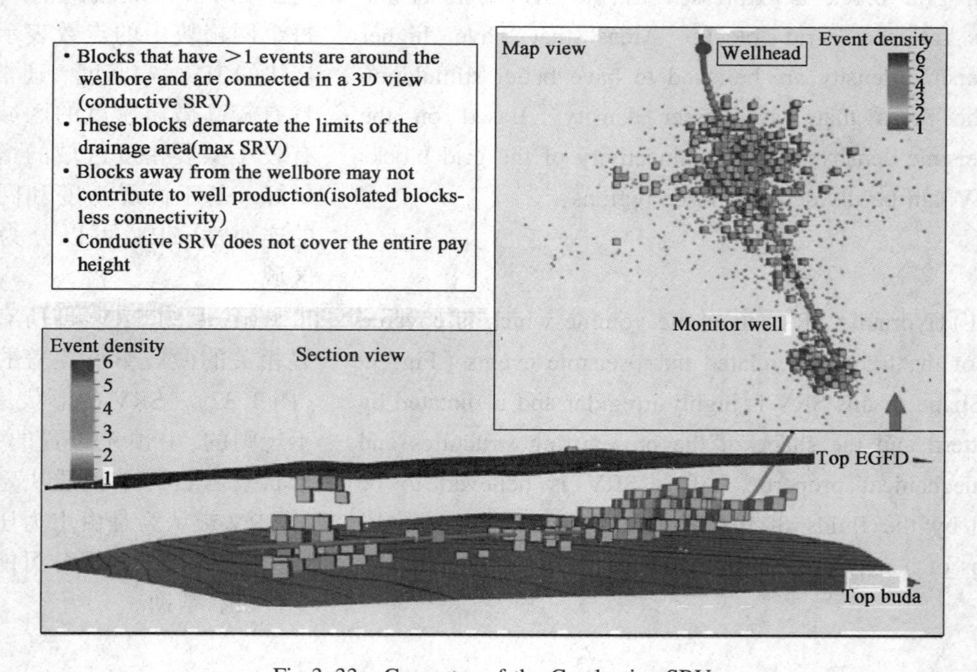

Fig 3.33 Geometry of the Conductive SRV

(3) Flush SRV: The flush SRV is demarcated by the blocks that contain more than two microseismic events (Fig. 3.34). This SRV is located very close to the wellbore and it is interpreted to have the highest effective permeability. The size of this SRV dictates the initial flush production during the first 30-60 days of well production. The high quality of this region is responsible for the high initial productivity often exhibited by Eagle Ford wells.

(3)涌入 SRV:由具有两个以上微震事件点的小区域构成的体积(图 3.34)。这种 SRV 靠近井筒,并具有极高的有效渗透率,其规模决定了在初始30~60 天内的生产产量。这种 SRV 的高质量是通常 Eagle Ford 井表现出高初始生产量的原因。

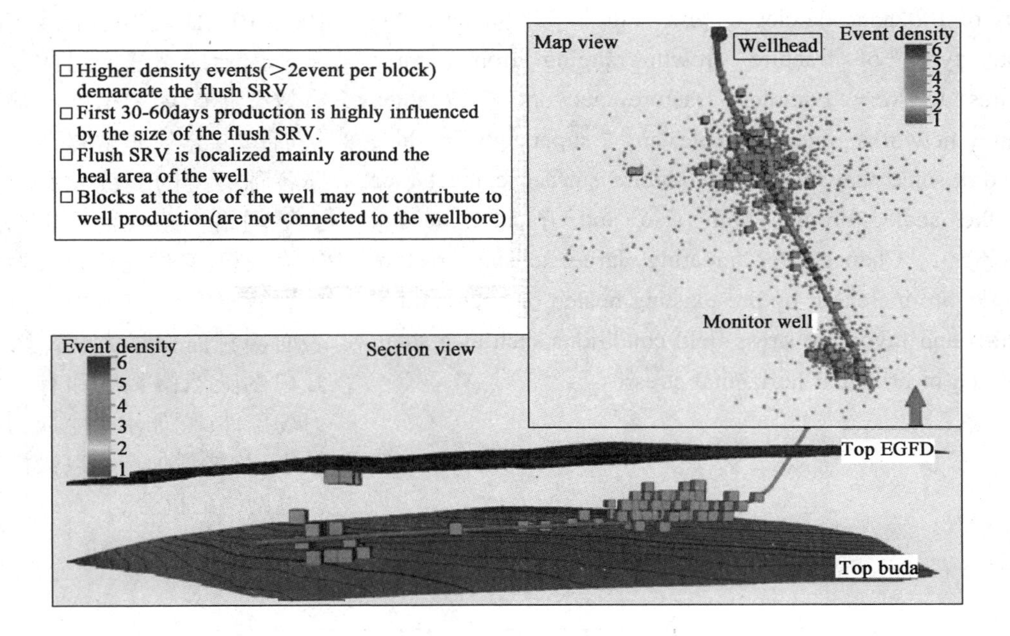

Fig 3.34 Geometry of the Flush SRV

Estimation of the size, geometry and pore volume of each SRV will help to define the hydrocarbons-in-place available for the well during the flush production and during the long term EUR. Hydrocarbons-in-place obtained for each SRV can be verified by the values of the total gas in place derived from the flowing material balance once the well has significant production history. The plot of normalized pressure against the square root of time in conjunction with the flowing material balance plot verified that well performance data can be correlated with the microseismic derived SRV volumes.

估计每个 SRV 的规模、几何形状和孔隙体积有助于确定涌入阶段的可采储量,以及长期最终可采储量(EUR)。一旦一口井有较长的生产历史,可通过动态物质平衡方程计算每个 SRV 的地质储量。归一化压力时间平方根关系图与动态物质平衡图证实了井性能与微震导出的 SRV 相关。

2. SRV Calculation

Horizontal well geometry provides other optimization opportunities. Longer laterals and more stimulation stages can also be used to increase fracture network size and stimulated reservoir volume. Mayerhofer et al. (2006) performed numerical reservoir simulations to understand the impact of fracture network properties such as SRV on well performance. Well performance can be related to very long effective fractures forming a network inside a very tight shale matrix of 100 nano-darcies or less. Fig. 3.35 illustrates the various types of fracture growth ranging from simple fractures to very complex fracture networks. Complex fracture networks are desirable in "super-tight" shale reservoirs since they maximize fracture surface contact area with the shale through both size and fracture density (spacing). Chances for creating large tensile fracture networks are increased by pre-existing healed or open natural fractures and favorable stress-field conditions such as a small difference in principal horizontal stresses.

2. SRV 计算

水平井的几何形态在压裂效果上有优势,较长的水平段和较多的储层改造段可用于增加裂缝网络的规模和储层改造体积。Mayerhofer 等人在 2006 年通过油藏数值模拟研究裂缝网络特性(如 SRV 等)对气井性能的影响,发现气井性能也与在 100×10^{-9}D 或更小的致密页岩基质内形成裂缝网络的较长的有效裂缝有关。图 3.35 显示了从简单裂缝到非常复杂裂缝网络各种类型的裂缝发育。复杂裂缝网络在超致密页岩储层中较为理想,它们能通过裂缝尺寸和裂缝密度(分布)把与页岩的表面接触面积最大化。闭合或开口的天然微裂缝和有利的应力场条件(例如水平主应力差异微小)均能增加产生巨大的拉伸裂缝网络的机会。

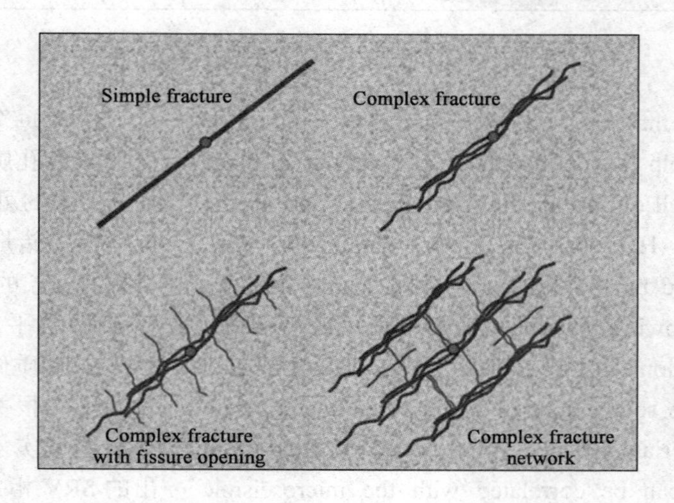

Fig. 3.35 Types of Fracture Growth

The calculation of SRV (a 3 - dimensional structure) also requires an estimate of the stimulated fracture network height in each discrete bin within the contacted shale section. This calculation is also performed within the selected bins and is performed by calculating the network height as the difference between the shallowest and deepest event within the specific bin and top and bottom of the shale section (Fig. 3. 36). While this method is not an analytically exact calculation, it does provide a fast automated method to approximate a very complex 3 - dimensional structure, while honoring the contact with the actual shale section.

估算 SRV(三维结构)还需要估计页岩区段内每个离散面元中的裂缝网络高度。这个过程也能在选定的面元内进行，计算出特定面元和页岩段顶底的最浅和最深处的差异作为缝网高度(图 3. 36)。虽然这不是一种精确的计算方法，但它提供了一种非常复杂的三维结构的快速自动化的计算方法，同时保证了与实际页岩部分的接触。

An important aspect of the SRV measurement is the proper setup of observation well or multiple observation wells to guarantee that the entire SRV can in fact be observed. The proper design of a microseismic mapping setup takes into account maximum observation distance to ensure that the entire SRV can be imaged. Microseismic moment magnitude versus distance plots can be used to ascertain if all events were within observable range or if the SRV could in fact be larger than imaged. Other issues that affect correlations between SRV and well performance include associated formation water production, and condensate yield, which becomes relevant in less mature shales close to the oil window. Although general guidelines are applicable for SRV measurements, it is important to evaluate any given SRV measurement in conjunction with the particular reservoir setting.

测量 SRV 的一个重要方面是能正确设置一个或多个观测井，保证实际可观测到整个 SRV。正确的微震绘图设置需考虑最大观测距离，确保整个 SRV 可以成像。将微震动力矩的大小与距离曲线比较能够确定所有是否所有措施作用在可观测范围内，或者实际上的 SRV 是否可能大于成像。影响 SRV 与气井性能的其他方面包括地层产水量和凝析油产量，这些与靠近油窗的不太成熟的页岩有关。尽管 SRV 测量准则对大多储层都是适用的，但还需结合特定的储层参数来评估测量的 SRV。

It is important to note that the SRV is just the reservoir volume affected by the stimulation. It does not provide any details of the effectively producing fracture structure or spacing. Maxwell et al. (2006) introduced a concept that could eventually be used to characterize fracture density. In this approach "additional seismic signal characteristics" allow investigation of the source of the mechanical deformation resulting in the microseisms. In particular, the seismic moment, a robust measure of the strength of an earthquake or microearthquake, can be used to quantify the seismic deformation. Besides this potential geophysical approach, reservoir modeling also provides an avenue to better evaluate the effectively producing network.

值得注意的是，SRV 是增产措施所影响的储层改造体积，它并没有提供任何产生有效裂缝构造或裂缝分布的细节。麦克斯韦尔等人(2006)引入了一个最终可用于表征裂缝密度的概念。在这种方法中，“额外地震信号特征曲线”能够研究导致微震机械变形的原因，特别是地震矩，一个能反映地震或微地震强度的参数，可以用来量化地震形变。除了这种潜在的地球物理方法，油藏建模是一个更好的评估有效生产网络的途径。

3.7.2.2 FCI

Fig. 3. 36 illustrates the various types of fracture growth. Examples of each fracture growth category have been documented with direct fracture geometry measurements using tiltmeter and/or micro seismic frac mapping or inferred from fracture pressure analyses. Large-scale fracture complexity can be measured using micro seismic and/or tiltmeter fracture mapping, allowing direct detection of network fracture growth and large-scale decoupling.

3.7.2.2 FCI

图3.36显示了各种类型的裂缝的发育状态。每一类型裂缝发育的例子都被记录下来,方法包括利用测斜仪和(或)微震压裂云图直接测量裂缝几何结构或压裂压力分析推测裂缝几何结构。测斜仪和(或)微震压裂云图可以用来测量大规模复杂裂缝,能够直接检测出裂缝网络增长和裂缝网络的大规模解耦。

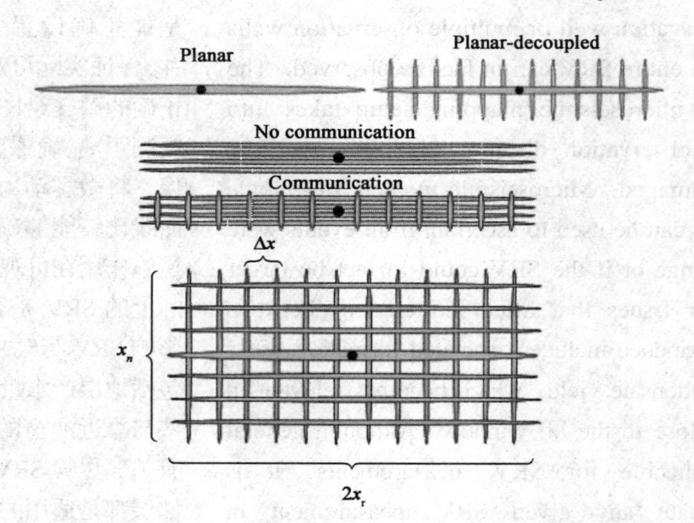

Fig. 3. 36 Fracture Growth and Complexity Scenarios

Fracture spacing; x_n—Fracture network width; $2x_r$—Fracture network length

Exercises/练习题

1. What are horizontal well completions? Discuss the main characteristics of each completion. /水平井完井方式有哪些？简述每种完井方式的特点。

2. What are the advantages and disadvantages of ball-activated completion?/投球滑套分段压裂的优缺点有哪些？

3. What are the advantages and disadvantages of plug-and-perf completion?/桥塞分段射孔压裂的优缺点有哪些？

4. Define hydrajet fracturing. Describe its operating process. /定义水力喷射压裂,简述其施工过程。

5. How does Just-in-Time Fracturing work?/即时压裂是如何实现的？

6. How does volume fracturing facilitate shale gas production? Provide a case study based on reviewing a SPE paper. /体积压裂如何开采页岩气？审阅一篇 SPE 文章后,描述一个体积压裂实例。

7. Describe the estimation of stimulated reservoir volume with a field example. /用现场实例简述储层改造体积的估算方法。

References/参考文献

[1] Bunger A P, McLennan J, Jeffrey R. Effective and sustainable hydraulic fracturing[M]. London: InTech, 2013, 9-36.

[2] Economides M J, Martin T. Modern fracturing: Enhancing natural gas production[M]. Houston: ET Publishing, 2007, 45-83.

[3] Howard G C, Fast C R. Hydraulic fracturing[J]. Society of Petroleum Engineers of AIME, 1970, 210 P.

[4] Zhang S, Wen Q, Wang F, et al. Optimization design of integral fracturing parameters for four-spot well pattern with horizontal fractures[J]. Acta Petrolei Sinica, 2004, 25(1): 74-78.

[5] Jiang T, Zhao X, Yin F, et al. Application of Multistage Hydrajet-fracturing Technology in Horizontal Wells with Slotted Liner Completion in China[C]. IADC/SPE Asia Pacific Drilling Technology Conference. Society of Petroleum Engineers, 2014.

[6] Surjaatmadja J B, Grundmann S R, McDaniel B, et al. Hydrajet fracturing: an effective method for placing many fractures in openhole horizontal wells[C]. SPE International Oil and Gas Conference and Exhibition in China. Society of Petroleum Engineers, 1998.

[7] Ekpe J, Kompantsev A, Al-Thuwaini J, et al. First Multi-Stage Swell packers Open Hole Completion in Rub Al-Khali Empty desert: Case Study[C]. SPE Middle East Unconventional Gas Conference and Exhibition. Society of Petroleum Engineers, 2011.

[8] Mu L, Ma X, Zhang Y, et al. Evaluation of multi-stage fracturing by hydrajet, swellable packer, and compressive packer techniques in horizontal openhole wells[C]. SPE Europec/EAGE Annual Conference. Society of Petroleum Engineers, 2012.

[9] Angeles R, Tolman R C, Gupta J, et al. One Year of Just-In-Time Perforating as Multi-Stage Fracturing Technique for Horizontal Wells[C]. SPE Annual Technical Conference and Exhibition. Society of Petroleum Engineers, 2012.

[10] Benish T, Angeles R, Tolman R, et al. Advancing Multi-Stage Fracturing Using Horizontal JITP and Autonomous Completion Systems[C]. IPTC 2013: International Petroleum Technology Conference, 2013.

[11] Eller J G, Garner J J, Snider P, et al. A case history: Use of a casing-conveyed perforating system to improve life of well economics in tight gas sands[C]. SPE Western Regional/AAPG Pacific Section Joint Meeting. Society of Petroleum Engineers, 2002.

[12] Lagrone K W, Rasmussen J W. A new development in completion methods-the limited entry technique[J]. Journal of Petroleum Technology, 1963, 15(07): 695-702.

[13] Al-Naimi K M, Lee B O, Bartko K M, et al. Application of a Novel Open Hole Horizontal Well Completion in Saudi Arabia[C]//SPE Indian Oil and Gas Technical Conference and Exhibition. Society of Petroleum Engineers, 2008.

[14] Li H, Deng J, Liu W, et al. Research on casing deformation failure mechanism during volume fracturing for tight oil reservoir of horizontal wells[C]//51st US Rock Mechanics/Geomechanics Symposium. American Rock Mechanics Association, 2017.

[15] Suliman B, Meek R, Hull R, et al. Variable stimulated reservoir volume (SRV) simulation: eagle ford shale case study[C]//Unconventional Resources Technology Conference. Society of Exploration Geophysicists, American Association of Petroleum Geologists, Society of Petroleum Engineers, 2013: 544-552.

[16] Mayerhofer M J, Lolon E, Warpinski N R, et al. What is stimulated reservoir volume? [J]. SPE Production & Operations, 2010, 25(01): 89 – 98.

[17] Cipolla C L, Warpinski N R, Mayerhofer M, et al. The relationship between fracture complexity, reservoir properties, and fracture-treatment design[J]. SPE production & Operations, 2010, 25(04): 438 – 452.

[18] Ahmed U, Meehan D N. Unconventional oil and gas resources: exploitation and development[M]. Boca Raton: CRC Press, 2016, 16(5) – 16(16).

Coiled Tubing 4 连续油管技术

Coiled tubing is a pipe made of low-carbon alloy steel. Due to its good flexibility, it is also called flexible tubing with several kilometers long. The coiled tubing can be used for many well operations without shutting a well and continuous lifting. The equipment is small in size, fast in operation and low in cost. To give a clear understanding of the coiled tubing technique, the chapter presents the technique from the birth, preparation, application and future market of coiled tubing.

Audio 4.1

连续油管是用低碳合金钢制作的管材,有很好的挠性,又称挠性油管。一卷连续油管长几千米,可以代替常规油管进行很多作业。连续油管作业设备具有带压作业、可连续起下的特点,设备体积小,作业周期快,成本低。为了让读者对连续油管技术有更清晰的了解,本章将从连续油管的诞生、制造过程、使用情况及未来市场等方面对连续油管进行介绍。

4.1 Overview of Coiled Tubing Technology

Coiled tubing originated from the "PLUTO" project of the Allies during the Second World War in the 1940s. The United States California Oil Company and the Bowen Oil Tools Company jointly manufactured the first coiled tubing as a light workover device with outer diameter 33.4 mm, and mainly used sand cleanout operations in oil and gas wells of the Gulf of Mexico in 1962. After 30 years of the development of coiled tubing, its value has been recognized. By the 1990s, coiled tubing technology have made rapid stride.

4.1 连续油管技术概述

连续油管源于20世纪40年代第二次世界大战期间盟军的"冥王星计划"。1962年,美国加利福尼亚石油公司和波温石油工具公司联合研制了第一台连续油管轻便修井装置,所用连续油管外径为33.4mm,主要用于墨西哥海湾油井、气井的冲砂洗井作业。在连续油管诞生30周年后,它的价值才真正被人们所认识,到20世纪90年代,连续油管技术得到了突飞猛进的发展。

4.1.1 Development of Coiled Tubing Technology

The origins of continuous-length, steel-tubing technology can be traced to engineering and fabrication work pioneered by Allied engineering teams during the Second World War. Project 99, code named "PLUTO" (an acronym for Pipe Lines Under The Ocean), was a top-secret Allied invasion enterprise involving the deployment of pipelines from the coast of England to several points along the coast of France. The 3 - in. inside diameter (ID) continuous-length pipelines were wound upon massive hollow conundrums, which were used to spool up the entire length of individual pipeline segments. The reported dimensions of the conundrums were 60 ft in width (flange-to-flange), a core diameter of 40 ft, and a flange diameter of 80 ft. These conundrums were designed to be sufficiently buoyant with a full spool of pipeline to enable deployment when towed behind cable-laying ships. Six of the 17 pipelines deployed across the English Channel were constructed of 3 - in. - ID steel pipe (0.212 - in. wall thickness). The 3 - in. - ID steel pipelines, described as "Hamel Pipe" were fabricated by butt-welding 40 - ft lengths of pipe into approximately 4,000 - ft segments of pipeline. These 4,000 - ft segments were then butt-welded together and spooled onto the conundrums. A total of 172,000,000 gallons of petrol was reported to have been delivered to the allied armies through PLUTO pipelines at a rate of more than 1 million gal/D.

Although the initial development effort of spoolable steel tubulars was reported to have occurred in the early 1940s, the first concept developed for use of continuous-length tubing in oil/gas wellbore services can be found in U. S. Patent 1,965,563, "Well Boring Machine", awarded on 10 July 1934 to Clyde E. Bannister. This approach utilized "reelable drillpipe", which was flexible enough to be coiled within a basket for storage when it was run into or out of the borehole (Fig. 4.1). This original concept used a rubber hose as the drillpipe, with the hose couplings designed to accommodate the attachment of two steel cables

4.1.1 连续油管技术发展

连续油管技术的起源可以追溯到盟军工程团队在第二次世界大战期间开创的工程制造业。项目99,代号为"PLUTO"(Pipe Lines Under The Ocean),是一个绝密的盟军入侵项目,涉及从英格兰海岸到法国沿海几个地点的管道部署。内径 3in 长度连续的管道缠绕在大型空心坩埚上,这些坩埚用于卷绕整个长度的各个管道段。据报道,这些坩埚的尺寸为60ft 宽(法兰到法兰),核心直径为40ft,法兰直径为 80ft。这些坩埚设计有足够浮力,带有一个完整的管道线轴可以在铺设电缆的船舶后面进行部署。横跨英吉利海峡部署的 17 条管道中有 6 条是 3in 内径的钢管(壁厚0.212in)。被称为"哈默尔管"的 3in 内径钢管是将 40ft 长的管道焊接到大约 4000ft 的管道段而制成的,然后将这些 4000ft 的部分对接焊接在一起并缠绕在坩埚上。据报道,PLUTO 管道以超过 1×10^6gal/d 的速度向盟军运送了总计 1.72×10^8gal 的汽油。

尽管据报道可缠绕钢管的最初开发工作始于 20 世纪 40 年代早期,但是人们在美国专利 1965563"Well Boring Machine"中第一次找到在油气井井眼服务中使用连续长度钢管的概念。这个专利于 1934 年 7 月 10 日授予克莱德·班尼斯特,采用了"可拆卸的钻杆"这一术语,当钻进或提出井眼时,该钻杆具有足够的柔韧性,可以盘绕在篮子

to provide the axial load support for the weight of the hose and bottom hole drilling assembly. The hose-coupling-cable-attachment clamps were also designed to allow removal of the steel cables as the flexible drillstring was removed from the wellbore. When pulling the flexible drillstring out of the wellbore, the separate cable lines were spooled onto drums for storage.

内进行存放(图 4.1)。这个专利使用橡胶软管作为钻杆,设计软管接头用于容纳两根钢缆的连接,以便为软管和井底钻探组件的重量提供轴向负载支撑。软管连接电缆连接夹,并且从井筒移除柔性钻柱时允许移除钢缆。将柔性钻柱从井筒中拉出时,可以将单独的电缆线缠绕在滚筒以便存放。

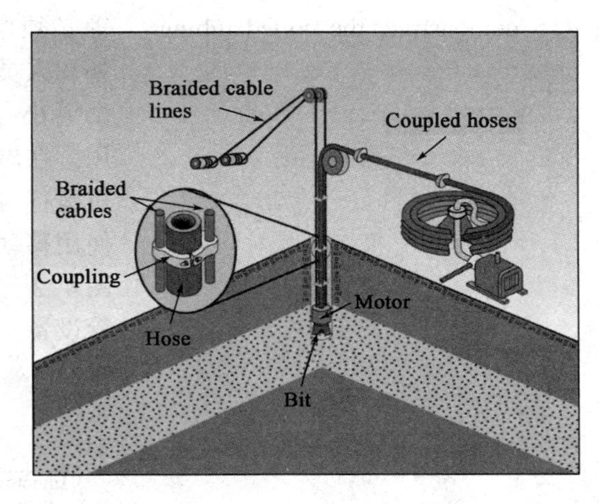

Fig. 4. 1　Illustration of Bannister Well-Boring Machine

4.1.2　Coiled Tubing System

In the modern oil and gas industries, coiled tubing (CT) refers to a very long metal pipe, normally 1 to 3. 25 in diameter which is supplied spooled on a large reel. It is used for interventions in oil and gas wells and sometimes as production tubing in depleted gas wells. Coiled tubing is often used to carry out operations similar to wirelining. The main benefits over wireline are the ability to pump chemicals through the coil and the ability to push it into the hole rather than relying on gravity. Pumping can be fairly self-contained, almost a closed system, since the tube is continuous instead of jointed pipe. For offshore operations, the "footprint" for a coiled tubing operation is generally larger than a wireline spread, which can limit the number of installations where coiled tubing can be performed and make the operation more costly. A coiled tubing operation is normally

4.1.2　连续油管系统

在现代石油和天然气工业中,连续油管指的是非常长的通过一个大卷轴卷绕起来的金属管,通常直径为 1 ~ 3. 25in。它可用于石油和天然气的井下作业,有时用作枯竭气井的生产油管。油田经常使用连续油管进行类似于测井电缆的操作。与测井电缆相比,其主要优势在于能够将化学物质通过盘管泵入,并能够将其不依靠重力推入井眼内。由于管是连续的而不是一根接一根,所以泵可以是相当独立的,几乎是一个封闭的系统。对于海上作业,连续油管作

performed through the drilling derrick on the oil platform, which is used to support the surface equipment, although on platforms with no drilling facilities a self-supporting tower can be used instead. For coiled tubing operations on sub-sea wells a Mobile Offshore Drilling Unit (MODU) e. g. semi-submersible, Drillship etc. has to be utilize to support all the surface equipment and personnel, whereas wireline can be carried out from a smaller and cheaper intervention vessel. Onshore, they can be run using smaller service rigs, and for light operations a mobile self-contained coiled tubing rig can be used. Fig. 4. 2 to Fig. 4. 5 demonstrate the coiled tubing system.

业的“占地面积”通常大于测井电缆作业距离,这会限制连续油管的安装数量,并使作业成本更高。连续油管作业通常是通过石油平台上的钻机井架进行的,该平台用于支持地面设备,在没有钻井设施的平台上可以使用自立塔。对于海上油井的连续油管作业,移动海上钻井装置(MODU),例如半潜式钻井船等必须用来支撑所有的地面设备和人员,而测井电缆可以在一个更小、更便宜的修井船上使用。在岸上,可以使用较小的钻机运行,而对于轻型作业,可以使用移动自给式连续油管钻机。图4.2 至图 4.5 展示了连续油管设备。

Fig. 4. 2 Coiled Tubing System
(source: http://www. octgproducts. com)

(a)Injector Head (b)Reel

(c)Control Cabin (d)Power Pack

Fig. 4. 3 Coiled Tubing Key Components

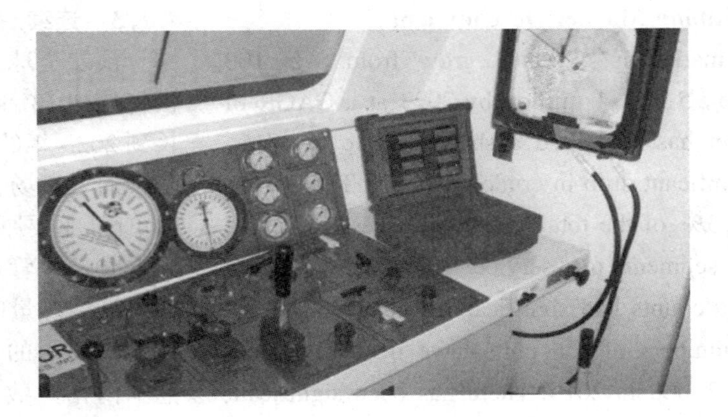

Fig. 4. 4 Inside the Control Cabin of a Modern CT Unit

Fig. 4. 5　Coiled Tubing
(source: http://www. savannaenergy. com)

The tool string at the bottom of the coil is often called the bottom hole assembly (BHA). It can range from something as simple as a jetting nozzle, for jobs involving pumping chemicals or cement through the coil, to a larger string of logging tools, depending on the operations.

油管底部的工具串通常被称为井底钻具组合(BHA)。它的作业可以像喷射嘴那样简单,也可以通过连续油管泵送化学品或水泥浆,或操作更大的测井工具串。

Coil tubing has also been used as a cheaper version of work-over operations. It is used to perform open hole drilling and milling operations. It can also be used to fracture the reservoir, a process where fluid is pressurized to thousands of psi on a specific point in a well to break the rock apart and allow the flow of product. Coil tubing can perform almost any operation for oil well operations if used correctly.

连续油管也可被用作一种更便宜的修井工具。它可以执行裸眼钻井和铣削操作,也可以用来压裂储层,在这个过程中,流体在井中的特定点上被加压到数千帕,从而破碎岩石使得油气流动。如果使用正确的话,连续油管可以执行几乎所有的油井作业。

4.1.3　Coiled Tubing Market Development

Global CT market is expected grow from $ 5, 100 million in 2013 to $5,291. 1 million by 2023 at a CAGR of 10%. This market has witnessed a sharp decline in 2015, mainly due to significant drop in crude oil price. Though CT represents only 1. 3% of the total oilfield services market, it is an important segment for service companies. Global decrease in the rig counts had significantly impacted the CT market and the numbers of CT units have decreased from 2,079 in 2014 to 2,000 in 2016. There has been significant drop in CT units in North America, Latin America, Europe, and Africa, whereas Middle East and Asia-Pacific have experienced increase in CT units. The decrease of CT units in North America, Latin America, Europe, and Africa is mainly

4.1.3　连续油管市场发展

预计全球连续油管市场规模将从 2013 年的 51 亿美元增长到 2023 年的 52.911 亿美元。连续油管市场在 2015 年出现大幅下滑,主要原因是原油价格大幅下跌。尽管连续油管仅服务油田市场总量的 1.3%,但它仍是服务公司的重要部分。全球范围内钻井数量的减少对连续油管市场产生了重大影响,连续油管装置数量从 2014 年的 2079 个减少到 2016 年的 2000 个。北美、拉丁美洲、欧洲和非洲的连

due to the weak demand for oil and decrease in oil price, which has forced many oil companies to shut down their operation temporarily. The cost of oil production in Middle East is extremely less as compared to other countries, hence the region is still able to run its operation with reasonable profit. Most of the oil production in Asia-Pacific is for the domestic use and hence decrease in oil price has not impacted the region's operation. Fig. 4. 6 represents the global coiled tubing market revenue from 2013 to 2018. Fig. 4. 7 shows the coiled tubing unit by region from 2014 to 2016.

续油管装置数量大幅下降,而中东和亚太地区的连续油管装置数量增加。北美、拉丁美洲、欧洲和非洲的连续油管装置减少主要是由于石油需求疲软和油价下跌,迫使许多石油公司暂时停产。与其他国家相比,中东的石油生产成本极低,因此该地区仍能以合理的利润运营。亚太地区的大部分石油产量都是供国内使用,因此油价下跌并未对该地区的运营产生影响。图4.6为2013—2018年世界连续油管市场收益。图 4. 7 为 2014—2016年世界连续油管市场分布。

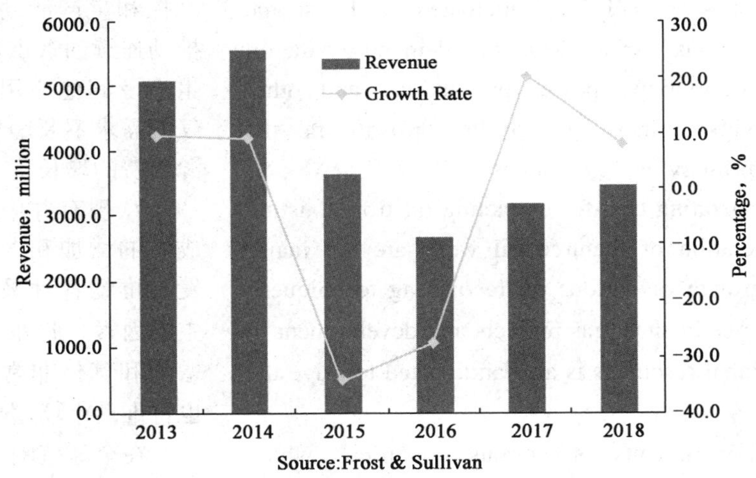

Fig. 4. 6 Global Coiled Tubing Market Revenue (2013—2018)
(source: https://ww2. frost. com)

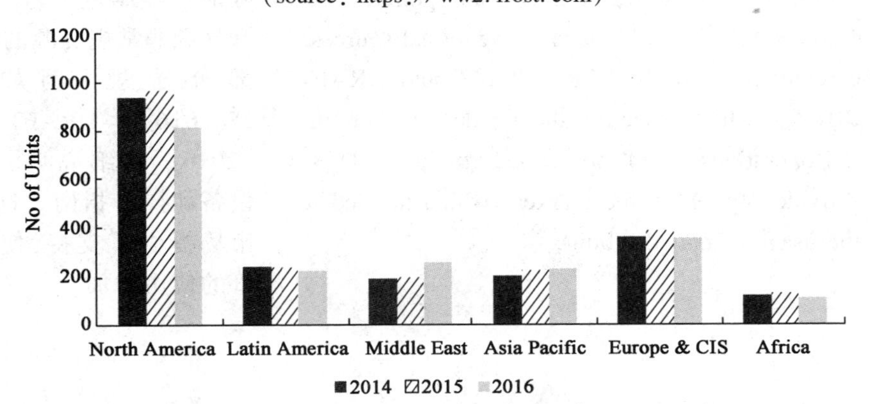

Fig. 4. 7 Coiled Tubing Units by Region (2014—2016)
(source: https://ww2. frost. com)

The global CT market has witnessed a strong decline of −34.1% in revenue in 2015, as oil price continues to be low. Maximum drop in CT business will be witnessed in North America, as companies have low bottom lines and less cash reserves for further CAPEX investments. Despite low oil price some regions such as Middle East and Asia-Pacific will have a positive growth in the CT market. This growth is mainly due to low cost of production and sufficient cash flow for capital expenditure (CAPEX) and operational expenditure (OPEX) in Middle East and high domestic oil and gas demand in Asia-Pacific.

由于油价持续走低,2015年全球连续油管市场收入大幅下降34.1%。连续油管业务的最大跌幅将在北美出现,因为进一步的资本性支出投资现金储备较少。尽管油价低,但中东和亚太等地区的连续油管市场将出现正增长。这一增长主要是由于生产成本低,中东资本支出(CAPEX)、运营支出(OPEX)高,以及亚太地区国内石油和天然气需求高、现金流充足。

4.1.3.1 Industry Insights

4.1.3.1 行业见解

The global coiled tubing (CT) market size was estimated at USD 3.25 billion in 2016. Rise in exploration & production activities globally is anticipated to boost the market over the forecast period. Mounting demand for the oil and gas in transportation, power production, and other application activities, had led to the growth in the consumption of primary energy sources.

2016年全球连续油管市场规模估计为32.5亿美元。全球勘探和生产活动的增长预计将推动连续油管市场。运输、电力生产及其他应用对石油和天然气的需求不断增加,导致一次能源消费的增长。

Increasing operating cost for extracting oil from existing wells and regeneration of matured oil wells are the major factors for the growth of various oil recovering techniques. In addition, the rise in shale gas projects and development in other unconventional resources is also anticipated to drive the market growth.

从现有井中开采石油运营成本的增加和老油井的二次开发是推动各种采油技术发展的主要因素。此外,页岩气项目的增加和其他非常规资源的开发也将推动市场增长。

Globally, governments of various countries, which supports both exploration and production activities of oil & gas, have formed several regulations and policies for extraction of conventional as well as unconventional sources. Some of the regulations are 30 CFR 250.616 and DRAFT IRP 21 (2016), which specific the guidelines for the equipment & operations used in coiled tubing. These regulations provide support to the market, which has led to increase in the usage of coiled tubing.

在全球范围内,支持石油和天然气勘探和生产活动的各国政府已经制定了若干关于开采常规和非常规资源的政策法规。部分法规[例如30 CFR 250.616和DRAFT IRP 21(2016)]详细说明了连续油管设备和操作指南。这些法规为市场提供了支持,使得连续油管的使用增加。

The market is divided into well intervention and drilling services. Rising demand for these services in oil and gas industry is expected to boost the market. A producing oil well can have a number of problems that can affect the operation and production in the well. This can result in the negative effect on the revenue generated from the well. Once the oil well goes into operation some of the problem faced are the failure of equipment and change in injection pressure. To achieve optimal production and overcome these challenges from a producing well, coiled tubing is applied.

当前连续油管的主要市场是提供修井和钻井服务。随着石油天然气行业对这些服务需求的上升将进一步推动连续油管市场发展。此外油气井生产过程中有很多问题影响油井效益,例如生产过程中一些设备出现故障及注入压力的变化。为了克服这些挑战,使得油气井达到最佳产能,工程师将会加大对连续油管的应用。

4.1.3.2 Services Insights

Well intervention was the major service segment in the year 2016, the pumping operation was major operation segment from past few years. The segment size was USD 3,250.5 million in the year 2016. CT pumping operation includes removal of fill or sand from a wellbore, acidizing or fracturing a formation, gravel packing, unloading oil well with gas, pumping slurry plugs, cutting tubulars with fluid, zone isolation, and removal of hydrocarbon, wax, and hydrate plugs. The market for the segment is expected to grow at a CAGR of 5.2% from 2017 to 2025. Fig. 4.8 shows the U.S. coiled tubing market revenue by services.

4.1.3.2 服务见解

修井作业是2016年连续油管的主要服务,泵送作业是随后几年的主要服务。2016年这一作业的规模为3.2505亿美元。连续油管泵送作业包括从井筒中清除填料或砂子,酸化或压裂地层,砾石充填,气举,泵送水泥塞,用流体切割管件,层间隔离,去除碳氢化合物、蜡和水合物堵塞。该部分的市场预计将从2017年至2025年以5.2%的复合年增长率增长。图4.8为美国连续油管市场按服务分类的收益。

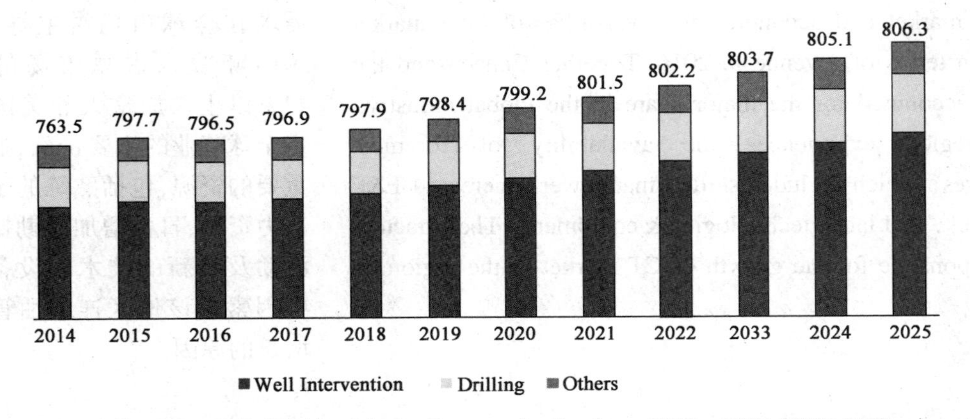

Fig. 4.8 U. S. Coiled Tubing Market Revenue by Services, 2014—2025 (USD Million)
(source: http://www.grandviewresearch.com)

4.1.3.3 Application Insights

Offshore application segment accounted for over 45% of the overall market. Coiled tubing technique is widely used in offshore extreme environments. The North Sea is an example of extreme offshore applications. Common offshore applications include well interventions or production for extending well life. As shown in the Fig. 4.9, China's coiled tubing market accounted for 48.7% of the offshore portion and 51.3% of the onshore portion.

4.1.3.3 应用见解

连续油管海上应用部分占整个市场的45%以上。连续油管技术广泛用于海上极端环境,北海是一个例子。常见的海上应用包括修井或延长井的生产寿命。如图4.9所示,中国连续油管市场收入中海上部分占比达48.7%,陆地部分占51.3%。

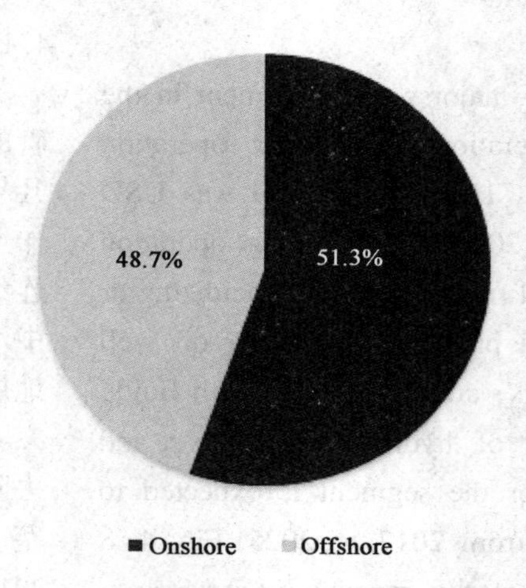

Fig. 4.9 China Coiled Tubing Market Revenue by Application, 2016
(source: http://www.grandviewresearch.com)

4.1.3.4 Regional Insights

The demand in the North America region dominated the global market and accounted for over 42% of total market share in terms of revenue in 2016. Together Canada and the U.S. accounted for the major share in the global industry. The region experiences the availability of foremost resources, which includes skilled manpower, increased E&P activities, and latest technologies & equipment. These factors are responsible for the growth of CT market in the region.

4.1.3.4 市场见解

北美地区的连续油管作业需求在全球市场占主导地位,2016年收入占总市场份额的42%以上。加拿大和美国共同占全球产业的主要份额,拥有最重要的资源,包括熟练的工程师人力资源、日益增加的勘探生产活动及最新的技术和设备。这些因素是该地区连续油管市场增长的原因。

Asia Pacific is also a potential market for CT. Countries present in the region such as China, India, Australia, and Indonesia have potential oil reserves coupled with huge investments through FDI channels in the sector. These factors are expected to generate profitable opportunities for the industry players. Coiled tubing industry is anticipated to show a significant growth in countries such as Argentina Russia, Algeria, and Poland over the forecast period.

亚太地区也是连续油管的潜在市场。中国、印度、澳大利亚和印度尼西亚等国家拥有潜在的石油储备,并通过外国直接投资渠道对这一部分大量投资。预计这些因素将为行业参与者创造更有利的机会。预计连续油管行业将在阿根廷、俄罗斯、阿尔及利亚和波兰等国家有显著增长。

4.2 Coiled Tubing Fabrication

The manufacture of coiled tubing involves a series of complex steps. The requirements for steel pipes and welding are extremely high, and the preparation process determines life span of the coiled tubing for subsequent use.

Audio 4.2

4.2 连续油管加工

制造连续油管需要一系列复杂的步骤,对钢管和焊接的要求都极高,其制备过程决定了连续油管后续使用寿命。

4.2.1 CT Manufacturing

Virtually all CT in use today begins as large coils of low-alloy carbon-steel sheet. The coils can be up to 55 in. wide and weigh over 24 tons. The length of sheet in each coil depends upon the sheet thickness and ranges from 3,500 ft. for 0.087 in. gauge to 1,000 ft. for 0.250 in. gauge.

The first step in tube making is to slice flat strips from the roll of sheet steel, and this step is usually performed by a company specializing in this operation. The strip's thickness establishes the CT wall thickness and the strip's width determines the OD of the finished CT.

The steel strips are then shipped to a CT mill for the next step in the manufacturing process. The mill utilizes bias welds to splice the flat strips together to form a single continuous strip of the desired CT string length. The mechanical properties of the bias strip welds almost match the parent strip in the as-welded condition, and the profile of the weld evenly distributes stresses over a greater length of the CT. The CT mill then utilizes a series of rollers to gradually

4.2.1 连续油管制造

实际上,目前使用的所有连续油管都是以低合金碳钢板的大盘管开始制造的。盘管宽度可达55in,重量超过24t。每个钢板中的板材长度取决于板材厚度,范围从3500ft(0.087in)到1000ft(0.250in)。

管材制造的第一步是从钢板卷上切割扁平条带,通常由专门从事该操作的公司进行。条带的厚度决定了连续油管壁的厚度,条带的宽度决定了连续油管的外径。

然后将钢带运送到连续油管轧机进行下一步的制造。轧机利用偏压焊接将扁平条带拼接在一起,形成连续油管串所需要长度的单个连续条带。在焊接状态下,偏置条焊缝的机械性能与母带相匹配,并且应力均匀

form the flat strip into a round tube. The final set of rollers forces the two edges of the strip together inside a high frequency induction welding machine that fuses the edges with a continuous longitudinal seam. This welding process does not use any filler material, but leaves behind a small bead of steel (weld flash) on both sides of the strip. Fig. 4. 10 demonstrates the manufacturing process of the coiled tubing.

分布在连续油管焊缝的轮廓上。然后连续油管轧机利用一系列滚筒逐渐将扁平条带制成圆管。最后一组滚筒将带状材料的两个边缘压在一个高频感应焊接机内，将纵向接缝熔合。这种焊接工艺不使用任何填充材料，只留下一小块焊道在条带的两侧。图 4. 10 为连续油管的制造过程。

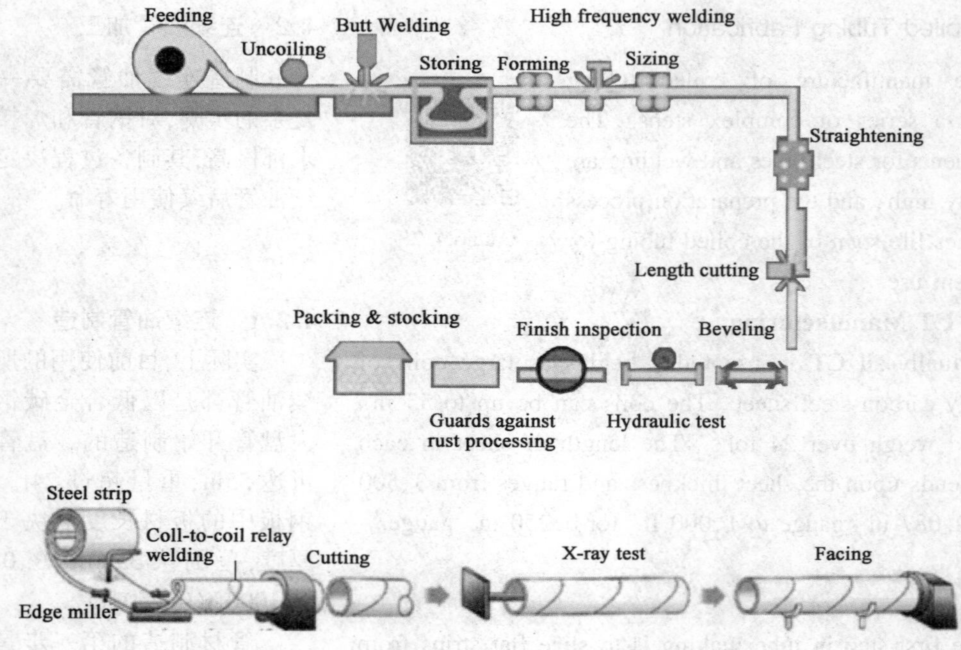

Fig. 4. 10 Coiled Tubing Manufacturing Process
(Source: http://www.tube-mill.com)

The mill removes the external bead with a scarfing tool to provide a smooth OD. The weld seam is then normalized using highly localized induction heating. Next, the weld seam is allowed to cool prior to water cooling. Full tube eddy current or weld seam ultrasonic inspection may also be performed, depending upon the mill setup. The tubing then passes through sizing rollers that reduce the tube OD slightly to maintain the specified manufacturing diameter tolerances. A full body stress relief treatment is then performed to impart the desired mechanical properties to the steel. Subsequent to

在磨机上用火焰清理工具去除外部焊道，使外表面平滑。再使用高度局部化的感应加热将焊缝标准化。接下来在水冷却之前将焊缝冷却。根据轧机设置，也可以进行全管涡流或焊缝超声波检查。之后管穿过定径滚筒，稍微减小管外径，以保持规定的制造直径公差。然后进行全身应力消除处理以赋予

the CT being wound on a shipping reel, the mill flushes any loose material from the finished CT string. Fig. 4. 11 shows the coiled tubing factory manufacturing chart.

钢所需的机械性能。在连续油管缠绕在运输卷轴上之后,磨机冲洗掉连续油管串中的所有松散材料。图 4.11 为连续油管制造工艺流程。

Fig. 4. 11 Coiled Tubing Factory Manufacturing Chart
(Source: http://www. youtube. com)

4.2.2 CT Mechanical Performance

The mechanical performance of CT is fundamentally different from all other tubular products used in the petroleum industry because CT is plastically deformed with normal use. Plastic deformation of material imparts fatigue on the CT string, and fatigue continues to accumulate over the life of the CT string, until such time as fatigue cracks develop, resulting in a CT string failure. Plastic deformation can be described as deformation that remains after the load causing it is removed. Fatigue can be defined as failure under a repeated or otherwise varying load, which never reaches a level sufficient to cause failure in a single application.

4.2.2 连续油管机械性能

连续油管的机械性能与石油工业中使用的所有其他管状产品有着本质区别,因为连续油管在正常使用时会发生塑性变形。材料的塑性变形会在连续油管串上产生疲劳损坏,并且在连续油管串的整个寿命期间疲劳将继续累积,直到产生疲劳裂纹为止,最终导致连续油管失效。塑性变形可以描述为在载荷移除后仍然存在的变形。疲劳破坏是由于反复的载荷变化引起的,单次应用不会达到疲劳破坏水平。

For standard CT operations, the tube is plastically deformed as the tube is straightened coming off the reel at point 1 as shown below in Fig. 4. 12 below. It is then bent at point 2 as it moves onto the guide arch, and is straightened again at point 3 as it travels to the injector and enters the wellbore. The CT string is then plastically deformed at the same three points during retrieval from the well.

如图 4. 12 所示,对于标准连续油管操作,管在点 1 处从卷轴上伸直时发生塑性变形,当它移动到导拱上时在点 2 处弯曲,并且当它移动到喷射器并进入井眼时在点 3 处再次伸直。然后,在从井中提出的过程中,连续油管在上面相同的三个点处发生塑性变形。

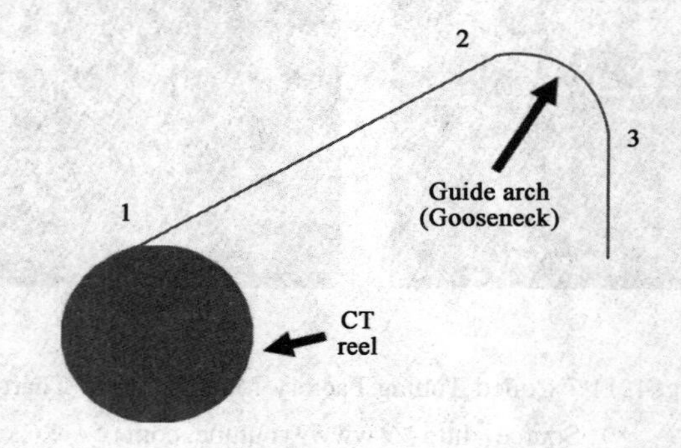

Fig. 4. 12　CT Plastic Deformation Points

(Source: http://www. icota. com/introct/introct)

CT service providers utilize sophisticated CT fatigue modeling software and field data acquisition systems to track the operating history of the CT string as it is utilized in the field. This operating history allows the CT string life to be monitored, and the string replaced prior to failure.

连续油管服务提供商利用复杂的连续油管疲劳建模软件和现场数据采集系统来跟踪连续油管现场操作历史。该操作历史记录可以监视连续油管串的寿命,并在故障之前替换管串。

4.2.3　Repairs and Splicing

4.2.3　连续油管维修

The only acceptable method of repairing mechanical or corrosion damage to a CT string is to physically remove the bad section of tubing and rejoin the ends with a temporary or permanent splice.

修复由机械或腐蚀损坏连续油管的唯一可行的方法是物理移除管道的损坏部分并用临时或永久性接头重新连接。

A temporary splice consists of a mechanical connection that is formed with a tube-tube connector. This type of connection is typically not used for prolonged operations during a CT job, but rather as an emergency repair to allow the CT string to be pulled out of the hole.

临时接头由机械连接组成,该机械连接由管—管连接器形成。这种类型的连接通常不用于长时间连续油管工作,而是用于紧急修复以使得连续油管被提出井筒。

However, connector technology continues to evolve and there are certain situations where connectors are used, such as to connect the tool string to the end of the CT. There are three general types of connectors, including the grapple, setscrew/dimple, and roll-on connector. Connector selection is based on the particular operation to be performed, as each type incorporates unique features that make it best-suited for a given application. Only butt welds are possible for field welding repair of CT strings, with TIG welding being the preferred method for permanent repair of CT work strings.

然而,连接器技术在不断发展,并且只在某些情况下使用连接器,例如将工具串连接到连续油管的末端。有三种常见类型的连接器:抓钩,膨胀螺丝(凹坑)和滚动连接器。连接器的选择基于要执行的特定操作,因为每种类型的连接器都具有其最适合某种应用的独特功能。只有接焊缝可用于现场连续油管的焊接修复,TIG 焊接是永久修复连续油管串的首选方法。

4.2.4 Alternatives to Carbon Steel CT

4.2.4 新型材料连续油管

Conventional carbon steel CT is more than adequate to meet the needs of most field operations. However, some corrosive downhole environments dictate the use of improved CT materials. QT - 16Cr is a relatively new corrosion resistant alloy (CRA) that was specifically developed for long term direct exposure to wet CO_2 environments. QT - 16Cr was commercially introduced in early 2003, and more than 30 tubing strings were in service a year later. Much of the early application was for permanent installations as a velocity string in environments containing wet CO_2 and saline conditions. It has been installed to depths greater than 18,000 ft.

传统的碳钢连续油管足以满足大多数野外作业的需要。然而,一些腐蚀性的井下环境决定了要使用改进的连续油管材料。QT - 16Cr 是一种相对较新的耐腐蚀合金(CRA),专为长期直接暴露于潮湿的二氧化碳环境而开发。QT - 16Cr 于 2003 年初在商业上推出,一年后有超过 30 个管柱在使用这种材料,早期大部分作为速度管柱在含有湿 CO_2 和盐水的环境中进行永久性安装。它的使用深度超过 18000ft。

The commercial appeal of QT - 16Cr goes beyond its favorable corrosion resistance characteristics. The material has also exhibited much improved abrasion resistance (approximately 1/4 of the material loss vs. a well-known 45 HRC low alloy steel) as well as demonstrating superior low cycle fatigue life when compared to its equivalent in carbon steel. This data indicates the grade may be an excellent candidate for future CT work string applications.

QT - 16Cr 的商业吸引力不仅在于其耐腐蚀特性,还表现出很好的耐磨性(相比于常见的 45HRC 低合金钢,该材料的损耗约为其 1/4),并且与碳钢一样,具有优异的低循环疲劳寿命。这些数据表明该材料可能是未来连续油管工作串材料的优秀候选者。

Another alternative to steel for manufacturing CT is a composite made of fibers embedded in a resin matrix. The fibers, usually glass and carbon, are wound around an extruded thermoplastic tube (pressure barrier) and saturated with a resin, such as epoxy. Heat or UV radiation is used to cure the resin as the tube moves along the assembly line. Composite CT can be manufactured with a wide range of performance characteristics by changing the mix of fibers, the orientation of their windings, and the resin matrix properties. The first commercial application for composite CT was three velocity strings deployed in The Netherlands in mid-1998.

用于制造连续油管的另一种替代材料是由嵌入树脂基质中的纤维制成的复合材料。纤维通常由玻璃和碳组成,将纤维缠绕在挤出的热塑性塑料管(压力屏障)上,并用树脂(例如环氧树脂)饱和。当管串沿组装线移动时,使用热或紫外线辐射来固化树脂。通过改变纤维的混合、绕组的取向和树脂基质特性,可以制造具有各种性能特征的复合连续油管。复合连续油管的第一个商业应用是1998年中期在荷兰部署的三个速度管柱。

The CT mills have also produced small quantities of CT made of titanium or stainless steel for highly corrosive environments, but the high cost of these materials has severely limited their use. Titanium was thoroughly explored for use in this application, but it is difficult to weld and costs approximately 10 times as much as carbon steel. As a result, only a handful of titanium strings have been manufactured.

连续油管工厂还生产少量由钛或不锈钢制成的连续油管,用于高腐蚀性环境,但这些材料的高成本严重限制了它们的使用。很多工厂深入探索了钛在这一情况下的应用,但是钛难以焊接并且成本大约是碳钢的10倍,因此只生产了少量的钛管。

4.3 Application of Coiled Tubing Technologies

Audio 4.3

The global oil and gas industry is using coiled tubing for an ever-increasing array of well intervention projects. Coiled tubing offers a number of operational and economic advantages, including: elimination of well kill and potentially damaging heavy-weight kill fluids, reduced operational footprint, horizontal intervention, and the ability to intervene without a rig.

4.3 连续油管作业

全球石油和天然气行业正在使用连续油管进行越来越多的作业。连续油管具有许多操作和经济方面的优势,包括:消除可能造成井筒破坏的大密度压井液,减少施工痕迹,水平井修井,无钻机修井。

Apart from the ones mentioned above, CT finds wide application in other areas of E&P activity as well. Some of the applications are discussed below.

除了上面提到的应用,连续油管在石油勘探与开发的其他领域也有广泛的应用。下面进行具体的讨论。

CT is routinely used as a cost effective solution for workover applications. The main advantage of CT is that it uses its own pressure to control equipment in the live well. This avoids potential formation damage associated with well killing operation.

连续油管通常作为低成本修井作业的首选方案。连续油管的主要优点是可以带压作业,这避免了进行压井操作相关的潜在地层伤害。

CT is the most common method of removing sand or fills, such as formation sand, proppants flowbacks, or fracture operation screenout and gravel-pack failures. These materials if not removed may impede fluid flow and reduce well productivity.

连续油管是最常用的冲砂或砾石充填工具(例如地层砂、返排的支撑剂,或压裂施工中脱出的砂和充填失效砾石)的方法。如果不清除这些固体颗粒可能会阻碍流体流动并降低井的产量。

In a wellbore that has developed fluid column with sufficient hydrostatic pressure, the reservoir fluid is prevented from following into the wellbore. By using CT nitrogen is pumped to displace some of these fluids and reduce the hydrostatic head.

在已形成具有足够静水压力流体柱的井筒中,连续油管可防止储层流体进入井筒。通过使用连续油管泵送氮气来置换部分流体,并减少静水压力水头。

CT can be applied when fracturing multiple zones in a single well using horizontal drilling, as it can easily move in and out of the formations. Moreover, CT has an ability to accurately spot the treatment fluid to ensure complete coverage of the zone of interest.

当用水平钻井方式在单个井中压裂多个区域时,可以使用连续油管,因为它可以轻松地进出地层。此外,连续油管能够确保处理液准确地进入目标层。

CT drilling has been commercially used for many years and can provide significant economic benefits. It can be used in both directional and non-directional well. Non-directional wells have a fairly conventional assembly as compared to directional, which requires use of orientation device.

连续油管钻井已经商业化多年,可以带来显著的经济效益。它既可用于定向井,也可用于非定向井。与定向井相比,非定向井有比较传统的组件,因此需要使用定向装置。

CT can be used for numerous pipeline applications, such as:

(1) Removing organic deposits and hydrate plugs;

(2) Removing sand or fill;

(3) Placing a patch or liner to repair minor leaks;

连续油管还可用于多种管道作业,例如:

(1)去除有机沉积物和水合物堵塞;

(2)去除砂子或填充物;

(3)放置贴片或衬管以修复轻微泄漏;

(4) Setting temporary plugs text.

4.3.1 Sand Cleanouts

A clean wellbore is not only a prerequisite for trouble-free well testing and completion, it also helps ensure optimum production for the life of the well.

CT (coiled tubing) fill cleanouts have been in existence for over four decades and today account for approximate 30% of the services per-formed with CT. Both CT and conventional jointed pipe offer two circulation modes to remove solids: either forward or reverse circulation mode. However, using conventional water-based fluids, a conventional sand cleanout method may apply excess hydrostatic pressure on the formation, resulting in some lost circulation to a sub-hydrostatic reservoir. If the losses are significant this makes sand removal impossible. Also, such losses can damage the formation. Nitrogen can be used to reduce hydrostatics, but this necessitates a very specific job design and execution, and in larger diameter wellbores, and especially in horizontal wells, can result in using large amounts of nitrogen with corresponding logistical and economic consequences.

To mitigate challenges associated with large nitrogen requirements, a sand vacuuming technology has been developed and proven in field operations worldwide. This vacuuming system consists of a specialized downhole jet pump connected to a CT string. The tool can be operated in three modes: sand vacuuming, well vacuuming and high-pressure jetting. The tool provides a localized drawdown wherever it is positioned in the wellbore and is effective in removing sand in the sand vacuuming mode and removing localized mud damage in the well vacuuming mode.

(4)设置临时桥塞。

4.3.1 清砂

干净的井筒不仅是无故障试井和完井的先决条件,而且还有助于保证油井在最佳状态下进行生产。

连续油管填充清理已经存在超过40年,现在约占连续油管服务的30%。连续油管和传统的连接管都以两种循环模式来消除固体:正向或反向循环模式。然而,使用传统的水基流体、传统的清砂方法可能对地层施加过多的静水压力,导致一些流体流失进入储层。如果滤失很大,则无法进行除砂;而且,这种滤失会损坏地层。氮气可以用来减少流体静压,但这需要非常具体的工作设计和作业,而且在较大直径的井筒中,特别是在水平井中,可能会需要大量的氮气,并因此产生相应的后勤工作和经济成本。

为了应对需求大量氮气的相关挑战,技术人员已经在全球的现场作业中开发并验证了砂吸尘技术。该真空系统由连接到连续油管的专用井下喷射泵组成,可以三种模式操作:砂吸尘,井内抽真空和高压喷射。无论位于井筒中的哪个位置,该工具都能提供局部压降,有效地在冲砂模式下去除砂子,并在井负压模式下消除局部水泥浆伤害。

For a typical sand cleanout process, the fluid could be circulated in two different directions (Fig. 4. 13): forward circulation and reverse circulation. In the forward circulation mode, the carrying fluids are pumped through the CT down through a wash tool, and they flow back to surface through the CT/completion annulus. In forward circulation, high energy jets or drill bits run on motors can be used to help break up and disperse any compacted fill in the wellbore. For reverse circulation, fluids are pumped down the CT/completion annulus with returns back up the coil. Options for breaking up compacted fill when reversing are limited. Customized nozzles are available that will deliver high energy jetting in the forward circulation mode but will allow reverse circulation without incurring any pressure drop penalty. Clearly if one encounters a compacted sand "bridge" when reversing it will be necessary to first empty any solids inside the CT string and then switch to for-ward jetting mode to break up the "bridge" and then subsequently switch back to reversing mode. For safety reasons, certain limitations apply when reversing up the coiled tubing.

对于典型的清砂过程,流体可以在两个不同的方向上循环(图4.13):正向循环和反向循环。在正向循环模式中,液体通过连续油管向下通过清洗工具,并且通过连续油管(完井)环流回到地面。在正向循环中,在发动机上运行的高能喷射或钻头可帮助破碎和分散井筒中的任何压实填料。对于反向循环,液体沿连续油管(完井)环空泵送并返回油管。反向循环时分解压实填充的选择有限。定制喷嘴可在正向循环模式下提供高能喷射,反向循环时不会产生任何压降损失。显然,如果在反向循环时遇到压实的砂“桥”,则必须首先清空连续油管工具串内的固体,然后切换到正向循环模式以打破砂“桥”,再切换回反向循环模式。出于安全考虑,在倒转连续油管时会有一些限制。

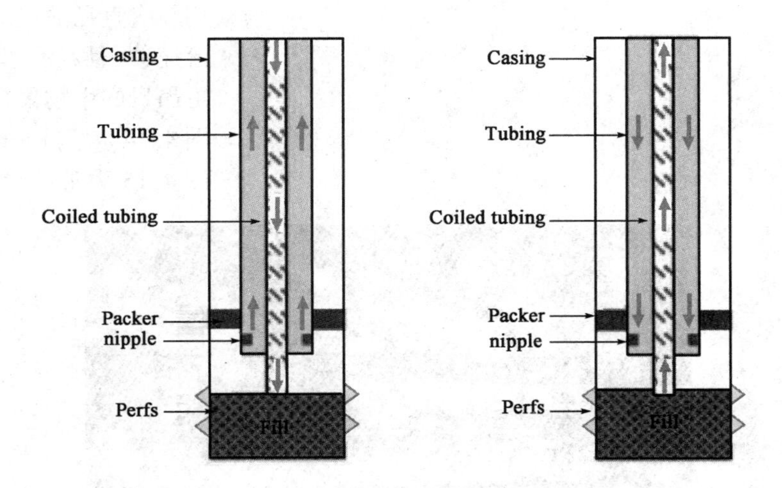

Fig. 4. 13 Two Typical Sand Cleanout Circulation Modes

The traditional approach to remove sand has been running in with CT to the top of fill, increasing the flow rate and then beginning penetration into the fill. The cleanout might be carried out by continuous slow penetration into the fill followed by stationary circulation or else penetrating, say, 15 ~ 50 ft of the fill followed by moving the CT uphole over the last penetration interval and continuing circulation while running back in hole. The operation might continue taking "bites" of the fill and reciprocating the CT until the target depth (TD) or bottom of the wellbore is reached. At TD, the sand in transit to surface is intended to be circulated out with one to two hole volumes of circulated fluids, with the CT generally remaining stationary at TD followed by pulling out of the hole. In some cases, the traditional cleaning method might involve pumping a certain amount of fluid for every penetration interval, then circulating a bottoms-up volume and then penetrating another interval and repeating the process until TD is reached. These stationary circulation or multi-bite methods often involve the use of larger diameter coiled tubing, higher flow rates, and in some cases, using more costly bio-polymer fluid systems. Fig. 4. 14 shows the cleaning on fracturing sands. Fig. 4. 15 is a schematic of typical sand cleanout process.

传统除砂方法是将连续油管运行到填充顶部,增加流速,然后开始渗透到填充物中。清理过程是通过连续油管缓慢穿过填充物,随后固定循环冲砂液或者穿透比如15 ~ 50ft的填充物,再穿透最后一个填充物段后向上移动连续油管,在回到井眼的过程中不间断循环。该操作可以连续对填充物进行"咬合"并使连续油管往复运动直到达到目标深度或井筒底部。在目标深度,运送到地面的砂子一般需要用一到两个孔隙体积的循环流体循环,连续油管通常在目标深度处保持静止,然后从井眼中拉出。在某些情况下,传统的清洁方法可能包括为每个穿透间隔泵送一定量的流体后循环一个自下而上体积的流体,然后穿透另一个间隔并重复该过程直到达到目标深度。这些固定循环或多咬合方法通常涉及使用较大直径的连续油管、较高的流速,并且在一些情况下,使用更昂贵的生物聚合物流体系统。图4.14为清洁压裂产生砂砾。图4.15为常见清砂过程。

Fig. 4. 14 Carry out Cleaning on Fracturing Sands or Stratum Sands
(Source: http://www. oilservice. keruigroup. com)

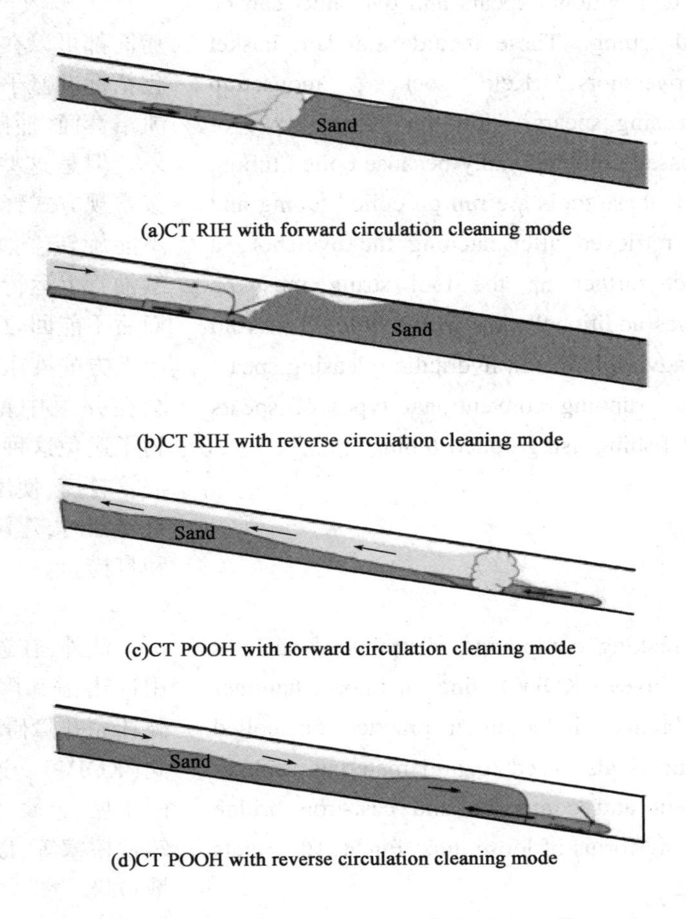

Fig. 4. 15 Typical Sands Cleanout Methodology/Process

4.3.2 Downhole Accident Handling

4.3.2.1 Coiled Tubing Fishing

When fishing on coiled tubing for devices stuck in hole, hydraulic jars are typically used as means of providing impact force to free the device. The drawback is that the pipe must be cycled over the gooseneck multiple times to fire and reset the jar. Down-hole vibration technology places impact energy right where the device is stuck and frees it quickly, even from deep or deviated wells.

4.3.2 井下事故处理

4.3.2.1 连续油管打捞

当使用连续油管打捞卡在井眼中的装置时，通常使用液压缸提供冲击力来释放装置。其缺点是管必须在鹅颈管上多次循环才能点火和复位激震器。井下振动技术可以将冲击能量直接对准设备卡住的位置并快速释放，甚至可以应用于深井或斜井。

Most types of conventional spears and overshots can be conveyed on coiled tubing. These include standard basket and spiral-type overshots, Kelo sockets, mousetrap overshots, and releasing spears. However, these types of tools cannot be released conventionally because coiled tubing cannot be rotated. If these tools are run on coiled tubing and the fish cannot be retrieved after latching the overshot, a hydraulic disconnect farther up the tool string must be activated. This leaves additional tools in the hole. To avoid this situation, it is advisable to run hydraulic releasing spears and overshots before running conventional types of spears and overshots when fishing using coiled tubing.

大多数类型的常规矛和打捞筒都可以在连续油管上输送，包括标准篮子和螺旋式打捞筒、凯洛套筒、捕鼠器打捞筒和释放矛。但是这些类型的工具不能按常规方式释放，因为连续油管不能旋转。如果这些工具在连续油管上运行，并且在锁定打捞筒后不能回收，则必须激活工具串上方的液压断开连接，但这就会在井眼中留下额外的工具。为了避免这种情况，在使用连续油管打捞，使用传统类型的矛和打捞器时，建议使用液压释放矛和打捞筒。

In addition, breaking completion obstructions such as knock-out isolation valves (KOIV) using an impact hammer and ceramic disk breaker is common practice on coiled tubing. Coiled tubing is also used to mill materials such as scale, metal, cement and composite and cast-iron bridge plugs, as well as many forms of loose junk. Fig. 4. 16 depicts coiled tubing fishing.

此外，在连续油管上通常使用冲击锤和陶瓷圆盘破碎机来破坏完井障碍物，例如分离隔离阀(KOIV)。连续油管也用于磨削水垢、金属、水泥及复合材料、铸铁桥塞等，以及各种形式的松散垃圾。图 4. 16 所示为连续油管打捞。

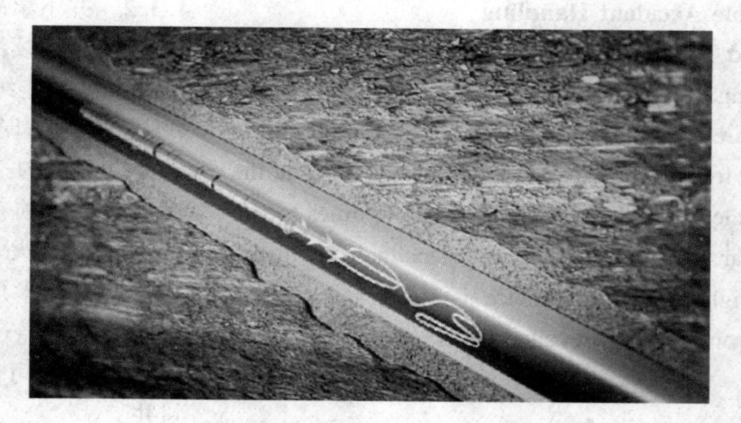

Fig. 4. 16 Coiled Tubing Fishing
(Source: http://www. oilservice. keruigroup. com)

4.3.2.2 Cutting of Coiled Tubing

(1) Technical characteristics: a motor is started to drive a cutting-off tool/carry a cutting projectile by hydraulic action, and cutting is carried out on a target tubing or drill rod.

(2) Mainly solved problem: downhole accidents are solved, and it is a method to enable a drilling tool to be pulled out when the tubing or drill rod is clamped.

(3) Technical advantages: the outer diameter of a pipe column and tool combination is small, so that it can pass through the tubing to rapidly solve the problem of encountering clamping. A regular fish head is maintained, and conditions for fishing in the next step are created.

Fig. 4.17 shows the coiled tubing cutting.

4.3.2.2 连续油管切割

(1)技术特征:通过液压作用启动电动机驱动切割工具或者携带切割抛射物,并在目标油管或钻杆上进行切割。

(2)主要解决的问题:解决了井下事故,是一种在油管或钻杆被卡紧时能够拔出钻具的方法。

(3)技术优势:管柱外径和工具组合小,可以通过管道快速解决钻杆被卡的问题。保持常规打捞头,并为下一步打捞创造条件。

图4.17为连续油管切割。

Fig. 4.17 Coiled Tubing Cutting
(Source: http://www.oilservice.keruigroup.com)

4.3.2.3 Coiled Tubing Perforation

(1) Technical characteristics: by utilizing a differential pressure ignition type perforation gun carried by the coiled tubing, after being delivered to the target layer by electronically and mechanically positioned depth, a target layer is perforated via compressing to ignite the perforation gun by the tubing and the casing.

4.3.2.3 连续油管射孔

(1)技术特征:在输送到电子和机械定位的目标深度后,通过油管和套管下入由连续油管携带的压差点火式射孔枪对目标层进行射孔。

(2) Mainly solved problems: handling for problems of the first shoot before fracturing in the horizontal section or purposeful downhole accident handling.

(2)主要解决的问题:处理水平段压裂前的第一次射孔问题或有目的的处理井下事故。

(3) Technical advantages: compared with tubing conveyed peroration, there are two guns in a single round, and delivery in the horizontal section is rapid, the period is short etc.

(3)技术优势:与油管输送相比,单次可携带两支枪,水平段输送速度快,周期短等。

Fig. 4. 18 shows coiled tubing perforation.

图4. 18为连续油管射孔。

Fig. 4. 18 Coiled Tubing Perforation
(Source: http://www. oilservice. keruigroup. com)

4. 3. 2. 4 Pigging and Scraping by Coiled Tubing

4. 3. 2. 4 通过连续油管进行清管和刮削

(1) Technical characteristics: by adopting the coiled tubing with a pigging gauge or a scraper, handling operation is carried out on a casing diameter and a well wall.

(1)技术特征:采用带有清管规或刮刀的连续油管,可以在一个套管直径内和井壁上进行操作。

(2) Mainly solved problems: operation in the horizontal section, and operation under pressure of the production well.

(2)适用情况:在水平段操作,以及生产井内带压操作。

(3) Technical advantages: the period is short, it can be carried out in the horizontal well, and pollution from snubbing to the stratum is low.

(3)技术优势:周期短,可在水平井内进行,从缓冲到地层的污染较低。

4.3.3 Reservoir Stimulation

4.3.3.1 Fracturing

Continuing low prices for oil and gas stimulate new technologies improve the production of low permeability reservoirs. Hydrajet-fracturing with coiled tubing, a unique technology for low-permeability horizontal and vertical wells, uses fluids under high pressure to initiate and accurately place a hydraulic fracture without packer, saving operating time and lowering operating risk. In addition, Conveying stimulation and fracturing systems on coiled tubing allows these operations to be carried out in a "live" well, so there is no need for a costly and time consuming well kill and no chance of irreparable wellbore damage from the use of kill-weight fluids. Fig. 4. 19 is the schematic of the hydraulic coiled tubing unit.

4.3.3 储层改造

4.3.3.1 水力压裂

石油和天然气的持续低价刺激新技术的发展从而提高了低渗透油藏的产量。使用连续油管进行水力压裂,是一种低渗透水平井和垂直井的独特开采技术,使用高压启动流体可以精确进行水力压裂而且无需封隔器,从而节省了操作时间并降低了操作风险。此外通过连续油管输送压裂工具使得这些作业可以带压进行,而且不需要耗费大量时间和成本,也不会因使用压井液而造成不可修复的井筒损坏。图4.19为液压连续油管装置。

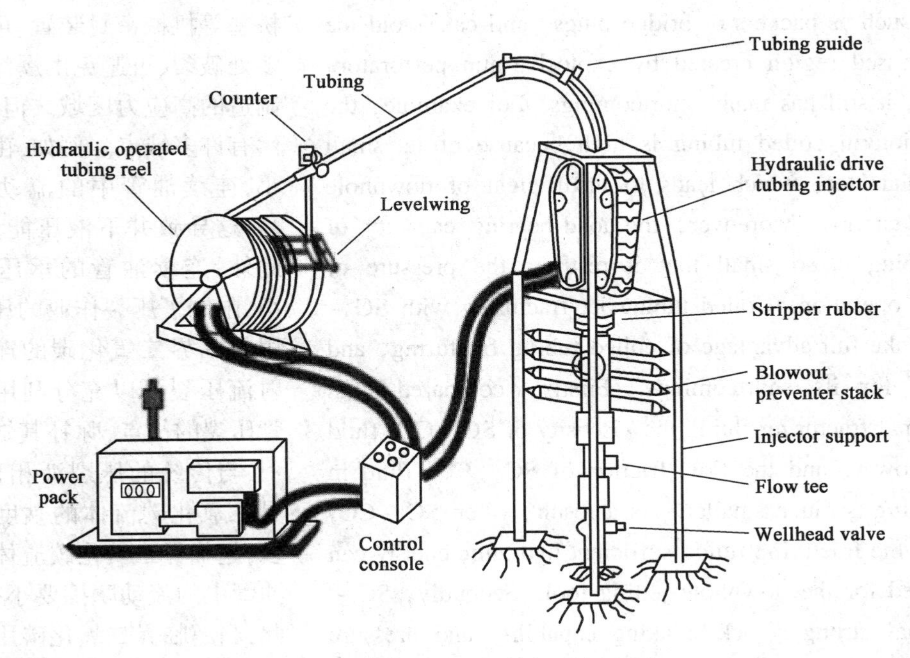

Fig. 4. 19 Hydraulic Coiled Tubing Unit
(source: http://www. slideshare. net)

Hydrajet-fracturing technique, a relatively new stimulation technology, combines hydrajetting, hydraulic fracturing, acidizing, and a dual-paths pump fluid technique. This fracture stimulation method has primarily used jointed pipe to achieve hydraulic fracturing injection rates for horizontal completions. Applying hydrajet-fracturing technique with larger-OD coiled tubing can initiate and accurately place fractures without mechanical packer, saving job time and reducing operation risk. This adaptsto multi-layer fracturing in vertical wells and multi-stage stimulation in open hole horizontal wells. Hydrajet-fracturing technique is more important and promising to exploit low-permeability reservoirs.

水力喷射压裂技术是一种相对较新的增产技术,它结合了水力喷射、水力压裂、酸化和双路泵流体技术。这种压裂增产方法主要使用连续油管来实现水力压裂需要的注入速率。将水力喷射压裂技术应用于较大直径的连续油管时,可以在没有机械封隔器的情况下精确控制裂缝位置,从而节省工作时间并降低操作风险,适用于垂直井中的多层压裂和裸眼水平井中的多级压裂。水力喷射压裂技术对于开发低渗透油藏更为重要且有很大潜力。

Coiled tubing hydrajet fracturing is an effective stimulation method, and plays an important role in the development of unconventional oil and gas. It can place many fractures in a well without using mechanical-sealing devices, such as packers or bridge plugs, and can avoid the highly-stressed region created by explosive-gun perforator. However, it still has many shortcomings. For example, the flow friction in coiled tubing is high because of its small internal diameter, which leads to insufficient of downhole hydraulic energy. Moreover, the load-bearing capacity of coiled tubing is so small that it restricts the pressure of fracturing operation. Coiled tubing jet fracturing with SC - CO_2 can take full advantage of coiled tubing fracturing, and make up for its shortcomings. Firstly, compared with conventional fracturing fluid, the viscosity of SC - CO_2 fluid is much lower, and the flow friction of SC - CO_2 fluid in coiled tubing is much smaller. As a result, when SC - CO_2 is used as the fracturing fluid, sufficient hydraulic energy can be provided for the downhole jetting tool. Secondly, SC - CO_2 jet has stronger rock-breaking capability and pressure boosting effect than water jet. Therefore, the two operations, SC - CO_2 perforating and fracturing, can be carried out within the load-bearing capacity of coiled tubing. Thirdly, since SC - CO_2 can improve the stimulation effect

连续油管水力喷射压裂是一种有效的增产方法,在非常规油气开发中发挥着重要作用。其优点包括无需使用封隔器或桥塞等机械密封装置、可在井内多处破裂、可避免由爆炸射孔器制造的高应力区域。但是,它仍然有许多缺点。例如,由于内径小,连续油管中的流动摩擦很高,这导致井下液压能量不足。此外,连续油管的承压能力很小,限制了压裂作业的压力。采用超临界二氧化碳的连续油管射流压裂可以充分利用连续油管压裂的优点,弥补其缺点。第一,与传统的压裂液相比,超临界二氧化碳流体的黏度要低得多,超临界二氧化碳流体在连续油管中的流动摩擦要小得多,因此,当超临界二氧化碳用作压裂液时,可以为井下喷射工具提供足够的液压能量。第二,超临界二氧化碳射流比水射流具有更强的破岩能力和增压效果。因

instead of doing damage to the formation, SC – CO_2 fluid doesn't have to return to the surface, which shortens the operating time and reduces the costs. To sum up, coiled tubing jet fracturing with SC – CO_2 combines the advantage of SC – CO_2 jet and coiled tubing fracturing, and is expected to become an efficient, safe and environmentally friendly stimulation method of developing unconventional oil and gas resources. Fig. 4. 20 demonstrates the schematic of coiled tubing jet fracturing with SC – CO_2.

此,超临界二氧化碳射孔和压裂两种操作可以在连续油管的承载能力范围内进行。第三,由于超临界二氧化碳可以改善增产效果而不会对地层造成伤害,因此超临界二氧化碳流体不必返回地面,这缩短了操作时间并降低了成本。综上所述,采用超临界二氧化碳的连续油管喷射压裂结合了超临界二氧化碳喷射和连续油管压裂的优点,有望成为开发非常规油气资源的高效、安全、环保的增产方法。图4.20为采用超临界二氧化碳的连续油管喷射压裂示意图。

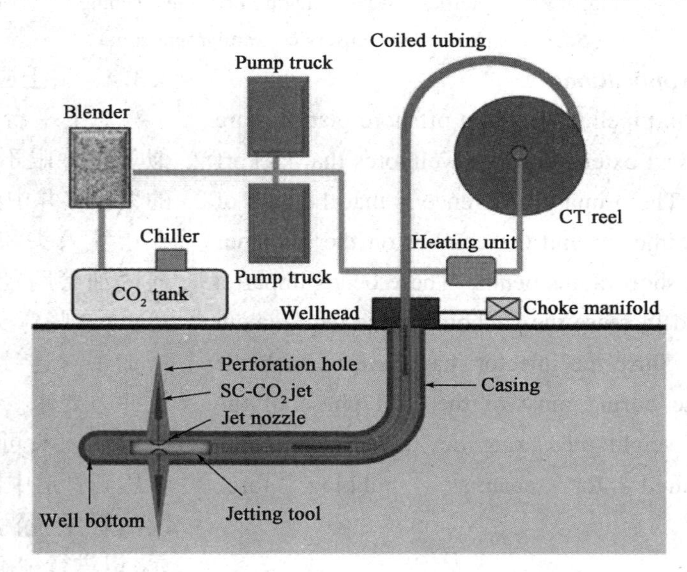

Fig. 4. 20 Schematic of Coiled Tubing Jet Fracturing with SC – CO_2

4.3.3.2 Acidification by Dragging of Coiled Tubing

(1) Technical characteristics: by utilizing belt pressure and ductility of the coiled tubing, acid is uniformly distributed on the target layer.

(2) Mainly solved problem: quantitative acid distribution cannot be realized on acidification of the target layer.

4.3.3.2 通过拖拽连续油管进行酸化

(1)技术特点:利用连续管的带压作业和延展性,使酸液均匀分布在目标层上。

(2)主要解决的问题:目标层酸化不能实现定量酸液分布。

(3) Technical advantages: acid distribution at a fixed point is implemented by dragging on the target stratum, so that stratum reform is carried out.

(3)技术优势:通过拖动至目标层实现固定点的酸分布,从而进行地层改造。

Fig. 4. 21 shows the acidification by dragging of coiled tubing.

图 4. 21 为拖拽连续油管进行酸化。

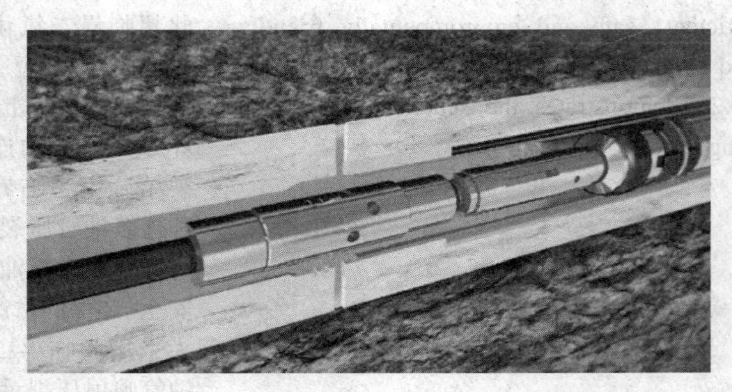

Fig. 4. 21 Acidification by Dragging of Coiled Tubing
(Source: http://www.oilservice.keruigroup.com)

4. 3. 4 Offshore Applications

CT operations in pipelines from an offshore platform are similar to operations in extended reach wellbores that kickoff at a shallow depth. The primary difference is that the path of the CT between the injector and the conduit on the sea floor may include several short radius bends. These bends impart a high drag force, and increase the snubbing force requirement on the CT injector. Since the injector may have to snub the CT into the pipeline during most of the RIH phase of the operation, the CT weight measuring device (weight cell) must be configured for accurate snubbing force measurement.

4. 3. 4 海上应用

海上平台连续油管操作类似于在井眼不深时就开始偏移的大位移井中的操作,主要区别在于注入头和海底管道之间的连续油管路径会有几个短半径弯头,这些弯头产生高阻力,并增加了对连续油管喷射器上的缓冲力要求。由于大部分下钻阶段注入头可能必须将连续油管压到管道中,因此必须配置连续油管重量测量装置(重量单元)以进行精确的缓冲力测量。

In addition, CT operations on many offshore platforms are constrained by the lifting capacity of the crane, as well as deck loading and space limitations. A loaded CT reel is typically the heaviest component of the CT system. Various solutions to address this issue have been successfully implemented in the field, including:

此外许多海上平台的连续油管操作受起重机的起重能力及甲板载荷和空间的限制,装载的连续油管卷轴通常是连续油管系统中最重的部件。对于该问题的各种解决方案已成功实施,包括:

(1) Disassembling the CT equipment into the smallest, lightest lifts possible, and reassembling the equipment on the platform.

(1)将连续油管设备尽可能拆卸成最小、最轻的组件,并在平台上重新组装设备。

(2) Cut the CT string into sections, spool the sections onto lightweight shipping reels, lift the reels onto the platform, then reconnect the sections on the platform.

(3) Use a barge or jackup with a heavy-lift crane to hoist all of the CT equipment onto the platform.

(4) Lift the CT unit, minus the CT string, onto the platform. Then spool the CT string onto the work reel from a loaded reel on a floating vessel.

(5) Install only the CT injector on the wellhead, leaving the CT reel and other CT unit components on a barge, workboat, or jackup, positioned alongside the platform.

The first four options can be applied where crane lift capacity is the controlling factor. Option (2) has been applied successfully numerous times in the North Sea, and requires high quality CT welding services to be available. Options (3) ~ (5) require more equipment and personnel versus that of typical CT operations, with an associated increase in the cost of the CT operation. Option (3) is rarely used, due to the high cost and scarcity of floating cranes.

In permanent installations, CT may be used as a flowline between offshore structures. The CT installation costs in this application are normally much less than for conventional barge-lay installations of welded line pipe. Prior case studies have documented savings in excess of 50 percent. In addition, the lower internal surface roughness of CT flowlines provides for lower frictional pressure loss than equivalent size jointed pipe. This provides additional economic benefit in the form of lower operating and maintenance costs.

(2)将连续油管串切割成多个部分,将部分卷绕到轻型运输卷轴上,将卷轴提升到平台后重新连接平台上的部分。

(3)使用带有重型起重机的驳船或自升式起重机吊起所有连续油管设备到平台上。

(4)将连续油管单元(取下管串)提升到平台上,然后将连续油管串从浮动容器上装载的滚筒上卷到工作滚筒上。

(5)仅在井口安装连续油管注入头,将连续油管滚筒和其他连续油管装置组件放在平台旁边的驳船、工作船或自升式平台上。

在起重机提升能力是控制因素的情况下,可以应用前四个选项。选项(2)已在北海成功应用多次,但需要高质量的连续油管焊接服务。与典型的连续油管操作相比,选项(3)~(5)需要更多的设备和人员,并且相关的连续油管操作成本会增加。由于浮动起重机的高成本和稀缺性,选项(3)很少使用。

在永久性安装中,连续油管可用作海上设施之间的流线。该应用中连续油管安装成本通常远低于焊接管线的传统驳船安装成本。之前的案例研究表明连续油管节省了超过50%的成本。此外,连续油管流线内表面粗糙度较低,因此比相同尺寸的连接管具有更低的摩擦压力损失。这样以较低的操作和维护成本提供了额外的经济效益。

In 2012, Baker Hughes designed and built coiled tubing systems specifically designed for deepwater down-line applications. The design brief for the customized system was as follows:

2012年,贝克休斯石油公司设计并制造了深水下线应用的连续油管系统。定制系统的设计简介如下:

(1) Capable of operating in water depths up to 3,000 m (9,842 ft);

(1)能够在水深达3000m(9842ft)的水中作业;

(2) Designed for large diameter pipe of 2 ⅞ in or 3½ in;

(2)设计用于2⅞in或3½in的大直径管道;

(3) DNV-certified to allow offshore lifting;

(3)DNV认证,允许海上提升;

(4) Road transportable in two loads;

(4)可在两个负载中运输管线;

(5) Standard basic components giving easy access to spare parts and trained mechanics/operators;

(5)标准基本组件,可轻松获取备件及联系到经过培训的机械师或操作员;

(6) Flexible frame to allow use on a wide variety of vessels, either through a moonpool or over the side.

(6)灵活的框架可在各种船只上使用(通过船井或侧面)。

Historically, the main use of the coiled tubing downline has been as a conduit for supplying air or nitrogen to dewater the subsea pipelines. This also typically requires that MEG or another pipeline hydrate-inhibiting fluid be pumped as part of a conditioning pig train. To date, coiled tubing has been used in water depths of around 2,200 m (7,217 ft), but with exploration already taking place in water depths down to 3,000 m (9,842 ft), this was selected as the target water depth.

从历史上看,连续油管的主要应用是供应空气或氮气以使海底管道脱水。这通常还要求连续油管泵送乙二醇或其他管道水合物抑制流体。到目前为止,连续油管的使用深度约为2200m(7217ft),但已有实例将其用在水深至3000m(9842ft)的地方。因此,以上两深度通常被选为连续油管海底施工水深区间。

Exercises/练习题

1. List the main components of a coiled tubing system. /列出连续油管的主要组成部件。
2. Describe briefly the applications of the coiled tubing technology. /简述连续油管的应用。
3. What are the major advantages of the coiled tubing conveyed fracturing?/连续油管水力压裂的主要优点有哪些?

References/参考文献

[1] Long N, Raj R, Srisa-Ard S, et al. Coiled tubing operations from a work boat[J]. Journal of Petroleum Technology, 2011, 63(6).

[2] Li J, Misselbrook J, Sach M. Sand cleanouts with coiled tubing: choice of process, tools and fluids[J]. Journal of Canadian Petroleum Technology, 2010, 49(08): 69 - 82.

[3] Tian S, Li G, Huang Z, et al. Investigation and application for multistage hydrajet-fracturing with coiled tubing [J]. Petroleum Science and Technology, 2009, 27(13): 1494 - 1502.

[4] Wilson A. Targeted Fracturing Using Coiled-Tubing-Enabled Fracture Sleeves[J]. Journal of Petroleum Technology, 2012, 64(6): 77 - 79.

[5] Cheng Y, Li G. Feasibility analysis on coiled-tubing jet fracturing with supercritical CO_2[J]. Oil Drilling & Production Technology, 2013, 35(6): 73 - 77.

[6] Zhang Y, Hao Y, Samuel R. Analytical Model to Estimate the Drag Forces for Microhole Coiled Tubing Drilling[J]. Journal of Energy Resources Technology, 2013, 135(3): 033101.

[7] Xuejun H, Huikai Z, Jian Z, et al. Analysis of Circulating System Frictional Pressure Loss in Microhole Drilling with Coiled Tubing[J]. The Open Petroleum Engineering Journal, 2014, 7(1):22 - 28.

[8] Perry K. Microhole coiled tubing drilling: a low cost reservoir access technology[J]. Journal of Energy Resources Technology, 2009, 131(1):849 - 860.

[9] Li X, Li G, Wang H, et al. A coupled model for predicting flowing temperature and pressure distribution in drilling ultra-short radius radial wells[C]. IADC/SPE Asia Pacific Drilling Technology Conference. Society of Petroleum Engineers, 2016.

[10] Bybee K. Coiled-tubing underbalanced through-tubing drilling[J]. Journal of petroleum technology, 2004, 56(6): 49 - 50.

[11] Bybee K. Coiled-Tubing Underbalanced Drilling in the Lisburne Field, Alaska[J]. Journal of Petroleum Technology, 2008, 60(6): 79 - 82.

[12] Van Venrooy J, van Beelen N, Hoekstra T, et al. Underbalanced drilling with coiled tubing in Oman[C]. SPE/IADC Middle East Drilling Technology Conference. Society of Petroleum Engineers, 1999.

[13] Denney D. Wired BHA for underbalanced coiled-tubing drilling[J]. Journal of petroleum technology, 2000, 52(10): 54 - 55.

[14] Graham R. Underbalanced drilling with coiled tubing: A safe, economical method for drilling and completing gas wells[J]. Journal of Canadian Petroleum Technology, 1997, 36(8):19 - 27.

[15] Dawson P G, Gerald P. Well drilling: U. S. Patent 2,548,616[P]. 1951 - 4 - 10.

[16] 傅阳朝,李兴明,张强德.连续油管技术[M]. 北京:石油工业出版社, 2000, 12 - 33.

[17] Bannister C E. Well boring machine: U. S. Patent 1,965,563[P]. 1934 - 7 - 10.

[18] Ledgerwood Jr L W. Efforts to Develop Improved Oilwell Drilling Methods[J]. Journal of Petroleum Technology, 1960, 12(4): 61 - 74.

[19] Calhoun G H, Herbert A. Equipment for inserting small flexible tubing into high-pressure wells: U. S. Patent 2,567,009[P]. 1951 - 9 - 4.

[20] Slator D T, Hanson Jr W E. Continuous-String Light Workover Unit[J]. Journal of Petroleum Technology, 1965, 17(1): 39 - 44.

[21] Martin J R, van Arnam W D, Normoyle B. QT - 16Cr Coiled Tubing: A Review of Field Applications and Laboratory Testing[C]. SPE/ICoTA Coiled Tubing Conference & Exhibition. Society of Petroleum Engineers, 2006.

Intelligent Well Completion 5
智能完井技术

Audio 5.1

Intelligent completion systems are also called intelligent Wells. Intelligent wells are wells which have equipment which can be controlled automatically or manually. Intelligent well completions or intelligent wells enable operators to acquire data, monitor and remotely control well operations for maximum productivity. Moreover, intelligent well technology plays a crucial role in maintaining financial stability and drilling activity by providing details regarding reservoirs characteristics. It is equipped with downhole sensors, which monitor well and reservoir conditions as well as it monitors valves to control the inflow of fluids from the reservoir. Additionally, the intelligent well technology allows measuring flows from each producing formation in real time and regulating water injection rates in injection wells. Besides, intelligent well technology creates additional economic and environmental benefits, since it allows a reduction in the number of well pads and associated field infrastructure, and also helps meet the requirements of environmental regulations.

智能完井系统也称智能井(Intelligent Well,IW),是具有可以自动控制或手动控制的设备的井。智能完井系统或智能井使操作人员能够采集数据、监测和远程控制油气井生产,从而最大程度地提高油井产能。此外,智能完井技术通过提供有关储层特性的详细信息,在维护经济稳定方面和钻井作业方面发挥着至关重要的作用。智能完井系统配备了可以监测油井、井下储层信息及井下阀门的传感器,以对油藏流体的流入进行控制。智能完井技术能够实时测量各产层的流量并且调节注水井注水量。由于智能完井技术可以减少井场及相关基础设施的数量,智能完井技术带来额外的经济效益和环境效益,同时也有助于满足有关环境保护法律法规的要求。

North America is currently the largest market for intelligent well followed by Europe and Asia-Pacific. In North America, the growth of the intelligent well market is attributed to factors such as shale revolution and presence of large drilled wells in the U. S. which drives the market for intelligent well.

目前,北美是最大的智能完井市场,紧随其后的是欧洲和亚太地区。在北美,智能完井市场的增长可归因于美国页岩气革命及美国原油开采活动增多等因素,这些因素推动了智能完井市场的发展。

A recent trend is that of IT integration and digitization of oil fields. Integrating IT with software allowed the continuous monitoring of performance of equipment that can be fixed immediately. This tracking method has substantially reduced maintenance time and devices failures, which resulted in monitory losses and delay in the completion of projects. Digital oilfield technologies are gaining importance, as they allow decision making and remote executions with the objective to maximize production, improve capital efficiency, and minimize safety hazards.

油田产业发展的新趋势是信息技术集成化和数字化。将信息技术与软件相结合,可以对能够立即修复的设备的性能进行持续监测。这种监测方法能够大大缩短设备维修时间,减少设备故障,从而避免项目中完井部分的监测损失及工程延期。数字油田可以最大程度地提高油井产能,提高资本配置效率,同时最大限度地消除安全隐患,有利于开发技术人员分析决策和远程控制提供,发挥着越来越重要的作用。

Factors such as increasing drilling activities in oil fields and rising focus on shale gas and tar sand reserves are driving the onshore well type. The exploration of new reserves, and creation of growth prospects for the exploration and production activities in the U. S. , Canada and Mexico are expected to trigger the demand for intelligent well market.

油气田钻井活动增加及页岩气、焦油砂越来越受到关注推动着油气井发展。在美国、加拿大和墨西哥国内进行的探明新储量、开拓勘探与生产活动发展前景预计将触发智能井市场需求。

5.1 Overview

5.1 概述

An intelligent well completion enables you to monitor, evaluate, and actively manage production in real time without any intervention. Intelligent well completion is very valuable in that it provides real time zonal downhole monitoring, enables surface controlled production, reduce production of undesirable gas, increase recovery and extends the economic life of a well and allow production testing of individual zones without interventions.

智能完井系统能够在无需进行任何井筒干预作业的情况下实时监测、评估、主动管理生产。智能完井技术具有很高的实用价值,它提供了井下层段实时监测,实现了地面控制生产,减少了不需要的气体的生产,提高了采收率,延长了油井的经济寿命,并且能够在不进行任何井筒干预作业的情况下在各层段进行生产测试。

5.1.1 History and Technology Development

The demand of oil and gas resources is high and the forecasts show a trend for higher requirements in the future. More unconventional resource exploitation along with an increase in the total recovery in current producing fields is required. At this pivotal time the role of emerging technologies is of at most importance. Fig. 5.1 shows the increase of the global oil and gas resource demand under the rapid growth of the world's economy.

5.1.1 智能完井技术的产生和发展

目前,石油和天然气资源的需求很高,根据预测,未来较长一段时期,油气资源需求总体将呈较大的增长趋势。伴随着当前油田最终采收率提高,需要加强页岩气、煤层气、油页岩、天然气水合物等非常规油气资源的勘查开发。新兴技术在这一关键时刻发挥着至关重要的作用。图5.1展示了世界经济高速增长下的全球油气资源需求增长。

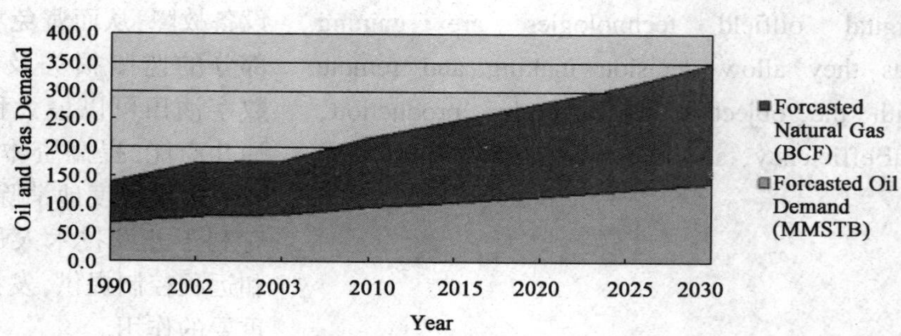

Fig. 5.1 Hydrocarbon World Demand under High World Economic Growth (EIA)

Sources of hydrocarbons are still abundant around the globe. Many of these resources are harder to produce than the reserves being produced currently. Another important challenge in this current situation is to maximize recovery at a profitable rate.

在全球范围内,烃类化合物仍然有十分丰富的来源。但相较于石油资源,其他资源的有效开采比较困难。此外,在目前的形势下,如何根据给定的利润率最大程度地提高采收率是另一个挑战。

The positive aspect of this situation is that the technology is progressing as those reserves get more challenging to produce. Many petroleum engineering technologies have been developed in order to ease the production of new reservoir.

Slanted and horizontal well technology was developed in the early 1920's but was rarely used until the 1980's. However with the technology advancement in the industry horizontal wells are not uncommon anymore.

针对上述情况,从积极的方面看,由于开采难度不断加大,与之相关的技术将持续进步。目前,已经发展了多种石油工程新技术以开发新油藏。

斜井和水平井技术于20世纪20年代被提出,40年代付诸实施,80年代相继在美国、加拿大、法国等国家得到广泛的工业化应用。随着工业技术的发展,水平井已是一种常见的技术。

The industry has a tendency of being careful with new technologies till the technology is proved both theoretically and operationally. The change usually takes place in more than one aspect i. e. production, drilling and reservoir strategies. However, adapting some of the new technologies will be a must in order to produce the resources in the best manner that will yield profit to these companies.

石油行业内的新技术在理论上和操作上被证明具有可行性之前,石油行业倾向于对其保持谨慎的态度。通常,新技术需要受到多方面的认可,包括生产、钻探、油藏管理。为了以能够为公司(企业)创造利润的最佳方式开采资源,必须推动一些新技术的应用。

One of the new technologies that have emerged in the past decades years is what is called "Intelligent Well Technology" or "Intelligent Well Completion Technology". The main driver for this technology is the emergence of horizontal and multi-lateral wells around the world. The main aspect of the intelligent well technology is the ability to control flow from many laterals or zones utilizing down-hole control valves.

智能完井技术是过去几十年里出现的新技术之一,发展的主要动力是世界范围内长水平段水平井和多分支井的增多。智能井的主要特征是利用井下层间控制阀控制多个分支井眼或储层的流体流量。

Schlumberger defines an intelligent well as a well equipped with monitoring equipment and completion components that can be adjusted to optimize production, either automatically or with some operator intervention. WellDynamics defines an intelligent well as a well that combines a series of components that collect, transmit and analyze completion, production and reservoir data, and enable selective zonal control to optimize the production process without intervention. The Intelligent Well Reliability Group (IWRG) defines an intelligent well as a well equipped with means to monitor specified parameters (e. g. fluid flow, temperature, pressure) and controls enabling flow from each of the zones to be independently modulated from a remote location (e. g. at the wellhead, or a nearby offshore platform, or a distant facility). Baker Hughes defines an intelligent well as implementation of fundamental process control downhole. Intelligent wells enable surveillance, interpretation and actuation in a continuous feedback loop, operating at or near real-time.

斯伦贝谢公司定义智能井是配备了监测设备和完井组件的井,可以自动地或在人工干预的情况下进行调整,以达到优化产能的目的。WellDynamics 公司定义智能井为结合了一系列能够采集、传输和分析完井综合数据、井眼生产数据及油藏数据,并且能够在无需进行人工干预的情况下实现选择性层段控制以优化生产过程的组件的油井。智能井可靠性小组认为智能井配备有能够监测井下流体流量、温度、压力等指定参数的多种装置并且能够控制他们,使得各个生产层段的流量能够受到远程遥控(例如井口、附近的海上平台或远距离设施)而独立调整。贝克休斯公司定义智能井为实现了井下基本过程控制的油井,能够在一个连续信息反馈闭环系统中实现监测、信息分析,并主动实施控制;能完成几乎实时地操作控制。

In summary, an intelligent well completion can be defined as a system capable of collecting, transmitting and analyzing completion, production, and reservoir data without physical intervention, and taking action to better control well and production processes. Fig. 5. 2 demonstrates the element of an intelligent well completion system.

总的来说,可以将智能井定义为一种系统,该系统能够在无需外界(人工)干预的情况下采集、传输和分析完井综合数据、井眼生产数据及油藏数据,并能够执行相关操作以更好地控制油井和生产过程。图 5. 2 为智能完井系统的组成。

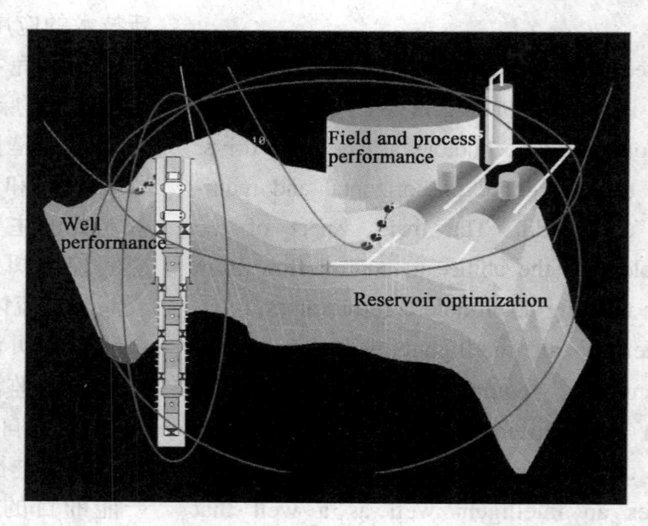

Fig. 5. 2　Intelligent-Well Completion Concept

Until the late 1980s, remote monitoring was generally limited to surface transducers around the tree and choke, remote hydraulic control of subsurface safety valves (SSSVs), and (electro -) hydraulic control of tree valves. The first computer-assisted operations optimized gas lift production by (remote) choke control near the tree and assisted with well monitoring and control of pumped wells. With the development, successful implementation, and improved reliability of a variety of permanently installed sensors, operators began to consider direct control of wellbore inflow to provide significant economic benefit. The service industry responded with high-level systems designed to provide full monitoring and control functionality.

20 世纪 80 年代前期,远程监测技术的应用通常只局限于采油树与节流器周围的地面传感器、远程液压控制地下安全阀(SSSV)及(电动)液压控制采油树相关阀门。第一套计算机辅助作业系统通过(远程)控制采油树附近的节流器优化了气举采油,并且辅助对油井进行监测与控制。随着多种永久性传感器的不断发展、成功安装应用及可靠性的改善,操作人员开始考虑直接对井筒内流体流入动态进行控制以获取显著的经济效益。油田服务行业设计出了具有全面监测及控制功能的高级作业系统。

Initially, intelligent-completion flow-control devices were based on technology used by conventional wireline-operated sliding sleeve valves. These valves were reconfigured to provide on/off and variable-position choking by use of hydraulic, electrical, or electrohydraulic actuation systems. Further development resulted in choke devices resistant to erosion and configured for high-differential-pressure service. Additional equipment based on conventional SSSV technology provided in-line ball valves for on/off closure.

最初,智能完井流量控制装置是基于传统的钢缆操作滑套所使用的技术而设计的。滑套经过重新配置,通过使用液压驱动系统、电动驱动系统或电动液压驱动系统而具备了井下开关和变位节流功能。进一步的发展使节流装置能够抵抗腐蚀并且承受较高的压差。基于传统地下安全阀技术设计的辅助设备中还有可用于井下开关的直通式球阀。

Initially, these fully integrated systems were not widely accepted because of the high incremental capital cost and perceived low possibility of success resulting in a high risked cost, which, at the time, did not meet project-screening criteria. To counter this challenge, lower-cost hydraulic systems were offered to provide some of the functionality of the initial high-end systems. These budget systems permitted packaging a variety of sensors together hydraulic control devices to provide a composite intelligent well completion.

最初,因为其投资成本较高且被认为能够成功应用的可能性较低,有可能成为高风险投资,不符合项目筛选标准,这些全集成系统并没有被广泛接受。针对这一问题,推出了低成本的液压系统以提供最初的高端系统的部分功能。液压系统将多种传感器连同液压控制装置封装在一起形成了一种复合式智能完井系统。

Often, data-handling and transmission procedures left much to be desired and reflected the ad hoc nature of early installations with the proliferation of stand-alone PCs, basic production-monitoring systems, and consequential data overload. Permanent downhole pressure and temperature gauges and intelligent completions were combined with some form of intranet or Internet data transmission, increasing speed and use of the data. Sensors were developed to measure flow rates by use of either nonintrusive systems or venturi meters. Combinations of these devices can be linked with additional fiber-optic systems to measure distributed temperature profiles, multipoint pressures, and acoustic signals. The performance lifetime of such systems was variable but is moving toward acceptable levels as suppliers invest in increased reliability engineering.

通常,数据处理及传输程序在许多方面需要完善,并且随着独立的 PC 机、基本的生产监测系统和相应的数据过载的普及,数据处理及传输程序也能反映早期安装设备的有关特性。永久性井下压力、温度测量仪器及智能完井系统与某种形式的内联网或互联网数据传输装置相结合,提高了数据传输速度及数据利用率。早先研制出了通过利用非侵入式系统或文丘里流量计来测量流速的传感器,这些设备可以与其他光纤系统相连接测量分布式温度剖面(曲线)、多点压力及声学信号。这类系统的有效工作寿命虽然还不完全稳定,但随着供应商在改善系统可靠性方面持续投资,其有效工作寿命已接近达到可接受的水平。

Intelligent-completion technology medium-term goals are summarized as follows.

(1) Prevent routine intervention for reservoir-management purposes;

(2) Leverage systems giving multiple horizon or reservoir penetrations per well;

(3) Self-optimize and automate wells and process facilities;

(4) Design processes on an optimum system rather than component basis (e. g., downhole/subsea vs. surface facilities and infrastructure).

智能完井技术的中长期研究目标为:

(1)出于对油藏管理的考虑减少日常井筒干预作业;

(2)利用杠杆系统使每口井可以水平或垂直穿透多个储层;

(3)油井自身优化及油井、工艺设备自动化;

(4)对生产系统进行优化设计而不仅对基本组件(例如井下、海底设施与地面设施及基础设施)进行优化设计。

Intelligent-completion system reliability should exceed 95% operability 10 years after installation.

成功安装10年后,智能完井系统的可靠性应超过95%。

5.1.2 Value of Intelligent Well Completion

5.1.2 智能完井技术的价值

The value-added menu of intelligent completions includes the elements listed below.

智能完井技术具有的价值主要体现在以下几个方面。

5.1.2.1 Quantifiable (Hard)

5.1.2.1 可量化(硬件)

(1) Reduction in well count to drain reserves or to drain more reserves per well;

(2) Savings on intervention cost;

(3) Well's ability to respond immediately to (un) expected changes in the production or injection performance in all operating environments. (This translates into intervention-cost savings and minimal deferment);

(4) Increased ultimate recovery from improved well management.

(1)减少井数来开采现有储量或每口井可开采更多储量;

(2)降低人工干预成本;

(3)油井在任何作业环境下都能对生产能力或注入能力预期的或突然的变化做出响应(这可以转化为降低人工干预成本和避免工程延期);

(4)通过加强对油井的管理提高最终采收率。

5.1.2.2 Difficult to Quantify (Soft)

5.1.2.2 不可量化(软件)

(1) Early data acquisition to enhance the probability of success of infill wells;

(2) Identification of key variables to be measured to optimize reservoir-management options;

(1)井下传感器采集到的前期数据有助于加密井井位的确定;

(2)通过识别需要测量的关键变量优化油藏管理;

(3) Means to mitigate the downside that is so often difficult to envision in new developments;

(3)减少在新油田开发过程中难以预见的不利因素的影响;

(4) Health, safety, and environmental dividends from unmanned operations;

(4)通过实现无人化操作提高健康、安全与环境绩效;

(5) Smaller environmental footprint from a reduction in the number of wells;

(5)通过减少井数降低对环境的破坏;

(6) Opportunity to acquire relevant data in wells to be abandoned.

(6)采集即将被废弃的油井的相关数据。

The value of the intelligent-well technologies is derived from the ability to actively modify the well configuration and performance through flow control and to monitor the response and performance through downhole data acquisition. Analysis of these data combined with predictive reservoir simulations enable realization of greater asset value by the utilization of this virtual feedback control system.

智能完井技术的价值体现在通过控制井下流量而使得井身结构和油井生产状态得到积极的改进与提高,并且通过采集井下数据而监测油井的生产。将这些数据的分析结果与预测性油藏模拟相结合,就可以通过利用虚拟反馈控制系统(的数据资料)实现储层效益最大化。

It is estimated that as much as 10% of accelerated or incremental recovery is an achievable target for intelligent-well completion applications. The applications and benefits of remote completion monitoring and control depend on the type of well considered in each development. In particular, multizone or multilateral wells (both injectors and producers) may benefit greatly from remote control. This benefit will tend to accelerate the trend toward investment in fewer, more highly productive wells. Incremental initial-well capital costs for intelligent-completion systems vary from U. S. $200,000 for a permanent downhole gauge system to U. S. $2,500,000 for a fully specified multizone remote controlled completion.

据估计,应用智能完井技术可使油藏的采收率提高10%。智能完井技术的远程完井监测、控制技术的应用和效益取决于作业井的类型。特别是对于多层开采井和多分支井(包括注水井和生产井),远程控制可使其经济效益显著提高,具体体现在有效降低生产投资的同时提高油井产能。智能完井系统的初始投资成本各不相同,从一套永久性井下参数测量系统的200000美元到特定的多层遥控完井系统的2500000美元不等。

In general, intelligent-well system candidates may be identified when economic evaluations indicate benefits from optimizing zonal/manifold water cuts, sweep, or reserves access. Other benefits may result from downhole production allocation, which may be less complex and less expensive than at the seabed or surface; improved lift efficiency and pressure maintenance; production acceleration and reduced project life; efficient use of available well slots; reduced intervention costs; and selective zonal stimulation treatments from the surface.

通常,经济评估结果表明对储层(管汇)的产液含水程度、注入流体波及范围、储量进行优化后可获得经济效益的油井,具备安装智能井的资格,可作为安装智能井的候选井。此外,智能井还可以从以下几方面获得经济效益:采用与海底设备或地面设备相比更为简单且投资更小的井下产液调节系统;提高举升效率并且改进产层的压力保持状况;提高采油效率并且缩短采油作业施工时间;充分发挥井下割缝衬管的泄油作用;降低人工干预成本;可在地面对产层进行选择性增产处理。

5.1.3 Main Applications

Intelligent well completion system include a battery of completion equipment designed to do the following:

(1) Monitor well operating conditions downhole (e.g., flow, pressure, temperature, phase composition, and water pH);

(2) Image the distribution of reservoir attributes away from the well (e.g., resistivity and acoustic impedance);

(3) Control the inflow and outflow rates of segregated segments of the well.

Combined with quality readings at surface of total rates and other non-well mapping technologies, such as time-lapse seismic, intelligent wells also provide the tools to manage wells, identify undrained oil, and make informed decisions that optimize hydrocarbon recovery.

5.1.3 主要用途

智能完井系统包括一系列的完井装置,可以应用这些装置来完成以下工作:

(1)监测井下油气生产信息(如流量、压力、温度、相态组成、地层水的pH值);

(2)对距离井筒较远的储层的特征分布进行成像(如电阻率和声阻抗);

(3)控制隔离井段的流入流速和流出流速。

通过与地面流量监测和其他非油井测绘技术(如时移地震技术)相结合,智能完井技术也可以实现油井管理,确定未泄油区,形成可提高采收率的决策信息。

5.1.3.1 Optimal Sequential Production

Wells often intercept more than one hydrocarbon-bearing zone. The decision to produce these commingled or sequentially is driven by many regulatory and reservoir-management concerns. When only one zone is produced at a time, in a typical bottom-up sequence, producers are required to deplete the current zone to the economic limit before plugging it back and perforating the next zone uphole. This results in a very suboptimal production profile with long interludes of declining rates until the next zone is perforated. An intelligent completion that can open the most prolific zone at any time from surface brings forward otherwise-deferred oil, without sacrificing reserves. In cases in which weak aquifers cannot provide adequate pressure support, shutting off one zone temporarily to be opened again at a later time can result in incremental ultimate recovery of that zone.

5.1.3.1 优化生产顺序

油气井通常穿透多个含油气区。这些含油气区的动用顺序(是混合生产还是按一定顺序生产)是基于许多油藏管理工作的考虑得以确定的。当一次只对一个含油气区域进行开采时,按照自下而上的顺序依次对多个含油气区进行开采,当前含油气区域开发至经济极限产量后将其封堵,然后将射孔器下至下一个含油气区域的上部进行射孔。但是,在对下一个含油气区域进行射孔前的很长一段时期,产量将呈递减趋势,这样导致不能达到最理想的生产状态。智能井可以随时从地面开启产能最高的层位,在不浪费油气储量的情况下开采原本需要延后开采的原油。在弱含水层无法提供足够压力支持的情况下,智能井可以暂时关闭一个含油气区域并在一定时间后再次开启,这可使得该区域的最终采收率增加。

5.1.3.2 Fluid Transfer for Sweep or Pressurization

When a high pressure gas or water zone lies over or under production intervals, a well can be used to transfer fluids to support the producing interval in a controlled manner. The practice of uncontrolled "dumpflooding" has been employed for many years to dump water from a high-pressure aquifer into oil-producing zones. Intelligent completions, including metering and controlling flow rates of the transferred fluids, widen the range and lower the risks associated with such a practice.

5.1.3.2 用于驱油和加压的流体输送

当高压气层或高压水层处于产层之上或产层之下时,井筒可输送流体,从而通过控制流体流动辅助产层生产。之前一直采用不受控制的"自流注水"法将高压含水层中的水引流至产油层,智能井能够测量并且控制流体流量,扩大了"自流注水"法的适用范围同时降低了与之有关的风险。

5.1.3.3 Drive-Recovery Processes

The success of water or gas drives into multiple zones, or employing horizontal wells, is tied to the sweep efficiency of injected fluids. The open literature is full of secondary- and tertiary-recovery examples on the effects of heterogeneity on sweep. Intelligent completions are ideally suited for wells producing from, or injecting into, layer-cake-type reservoirs or for long horizontal wells with significant contrast in rock properties along their trajectory.

5.1.3.3 提高采收率

水驱、气驱开采多个层段或采用水平井能否成功与注入流体的波及效率有关。公开的文献中研究了大量关于油层的非均质性影响驱油效果的二次采油和三次采油的实例。智能井非常适用于千层饼状储层采油或注水的井,或沿井眼轨迹岩性差异明显的长水平井,提高其采收率。

5.1.3.4 Flow Profiling

Inflow and outflow along wells is modeled on the basis of kh. Often, production logs provide inflow profiles that do not match expectations, and significant segments of the well are found to not be contributing to production. This information is not only key to the ensuing stimulation, but also plays an important role in understanding unswept or undrained oil. In many cases, production logs are not even run because of factors such as cost, risk of the intervention, and wells on pump.

5.1.3.4 流量分析

以地层系数为基础对沿井筒的流入和流出动态剖面进行建模。通常,生产测井(生产动态测井)确定的流入剖面与预期不符,并且发现主力产层没有产量贡献。这些信息不仅对后续的增产措施起着关键作用,而且对于了解油藏未泄油区域、未波及区域也有着重要的作用。在许多情况下,受成本、干预风险等诸多因素的影响,生产测井无法进行。

Distributed temperature sensing (DTS) is a fiber-optic technology that provides a temperature profile that can, in many cases, be translated into a flow profile along the well. DTS has been extensively employed in steamdrives in California and Canada, in which producers are to be steam free and temperature logs are to provide key information to operate both injectors and producers.

油气井光纤分布式温度传感技术(DTS)是一种能提供温度分布(曲线)的光纤技术,在很多情况下,这种温度分布(曲线)可以转化为沿井筒的流动剖面。该技术已广泛应用于加利福尼亚州和加拿大的蒸汽驱井组,其中生产井是无蒸汽的,温度测井将提供关键的数据信息来操作注入井和生产井。

The use of point-temperature data in a warm-back mode at injectors is a well-established methodology to determine injectivity. DTS widens its application to the full well.

利用节点温度数据对注水井进行井温测试是确定吸水能力的行之有效的方法。油气井光纤分布式温度传感技术可使其应用范围扩大到全井。

5.1.3.5 Downhole Reservoir Imaging

5.1.3.5 井下油藏成像

The rapid uptake across the industry of time-lapse seismic to monitor key data such as pressure depletion, water fronts, and steam chests is testimony to the value that this type of information provides to ultimate recovery. Downhole acquisition results in very repeatable seismic surveys and adds in quality because near-surface distortion is eliminated.

油气行业采用时移地震技术以监测压力降、水驱前缘等关键数据证明这些信息对于最终采收率而言具有重要价值。井下采集到的数据可以形成可重复利用的地震勘测调查报告,并且,由于消除了近地数据失真,可以提高报告质量。

Permanently deployed geophones have been used and should be considered for passive listening in cases in which fracture orientation and growth and caprock integrity are concerns, and for active listening in waterfloods and other enhanced oil-recovery processes, to better define oil/water/gas/steam maps to assist in infill drilling and sidetracking opportunities. Reservoir management decisions can be significantly improved with the knowledge of reservoir features and fluid distribution. Borehole seismology can provide higher resolution and become, if deployed cost-effectively, a powerful monitoring tool.

永久性部署的地震检波器已经投入使用,可以应用地震检波器在裂缝方向、裂缝发育及盖层完整性受到关注的情况下进行被动监测,并且考虑应用地震检波器主动监测水驱过程及其他提高原油采收率过程,以更好地确定油、水、气、蒸汽分布从而为加密钻井和侧钻提供帮助。通过了解和掌握储层特征及流体分布,可以显著提高油藏管理水平。井中地震勘探技术可以提高地震勘探分辨率,部署恰当的话,能有效监测。

Dynamic reservoir drainage imaging (DRDI) provides another, albeit more limited, approach. An array of electrodes is permanently located across a formation, potentially allowing determination of the saturation field at some distance away from the wellbore, with an extremely high acquisition frequency when compared with conventional 4D seismic.

另一种井下油藏成像技术是储层泄油动态成像技术(DRDI)。它将一系列电极永久性地分布在地层中,可以确定距离井筒一定距离的流体饱和度场的分布,与常规的4D地震相比具有极高的采集频率。

5.2 System Classification and Characteristics

The Intelligent Well system is broken down into major mini systems working together which are monitoring system, analytic system, and control system. Intelligent Well Completion maybe higher in costs but it has benefits such as increased reservoir information, increased options for placement of chemicals, better understanding of flood-fronts. However, the system has its disadvantages such as, difficulty in measuring flow rate zone-independently, high volumes of information and data with increased complexity in handling and processing data presented at the surface.

Audio 5.2

5.2.1 Elements of Intelligent Well Completion

An intelligent well completion is a system capable of collecting, transmitting and analyzing completion, production, and reservoir data, and taking action to better control well and production processes without physical intervention. The value of the intelligent well technologies comes from their capability to actively modify the well zonal completions and performance through downhole flow control, and to monitor the response and performance of the zones through real time downhole data acquisition, thereby maximizing the value of the asset. An Intelligent Completion combines a series of components that collect, transmit and analyze completion, production and reservoir data, and enable selective zone control to optimize the production process.

The system will be broken down into three major mini-systems working together, namely: monitoring system; analytic system; control system.

(1) Monitoring system. This consists of downhole gauges, sensors, flowmeters, densiometers. Downhole flow measurements of pressure, temperature, flowrate, density, phase properties, etc. as required can be obtained in real time via the monitoring component. The data collected proves useful in making production allocation decisions.

5.2 智能完井系统组成及特征

智能完井系统由监测系统、分析系统和控制系统这三个相互合作的微型系统组成,成本可能更高,但也具有可获取更多油藏信息、增加了注入化学试剂的选择、能够更好地追踪驱替前沿从而准确预测油藏特征等优势。但是,智能完井系统也有其缺点,例如,难以独立地测量各层段流速,信息量和数据量大从而增加了在地面进行数据处理的复杂性。

5.2.1 系统组成

智能完井系统是一种能够收集、传输、分析完井、生产和油藏数据,并采取行动以更好地控制油井和生产过程且无需物理干预的系统,其价值来源于它们可以通过井下流量控制主动改变井区完井数据和性能,以及通过实时井下数据采集监测层段的生产情况,从而实现最优生产。智能完井结合了一系列组件用于收集、传输、分析完井、生产和油藏数据,并启用选择性层段控制来优化生产过程。

智能完井系统由相互合作的三个主要微型系统组成,即:监测系统、分析系统和控制系统。

(1)监测系统。监测系统包括井下测量仪器、传感器、流量计和密度计。监测系统能够实时获取压力、温度、流速、密度、相位特征等数据。这些采集到的数据有助于制定产量配置决策。

(2) Analytic system. The analytic system ordinarily consists of the designated database and analysis computer that collects and evaluates received dowhhole data. With the increasing sophistication of sensors, larger volumes of data will be collected and the increased need for prompt data processing and decision making arises. Data sorting and filtering in real time also assumes increased importance. Computing systems of hybrid efficiency and more powerful processors will apparently provide the flexibilities of real time data sorting, processing and decision making, if necessary.

(2)分析系统。分析系统通常由指定的数据库及采集并评价已有井下数据的分析计算机组成。随着传感器技术日趋成熟,分析系统能够采集到更多的数据,对即时数据进行处理及决策功能的需求日益增加的同时,数据分类和过滤功能也变得越来越重要。拥有混合效率且具有更强大功能处理器的计算系统能够更加灵活地进行实时的数据分类与处理及决策制定。

(3) Control System. Interval control valves (ICVs set of variable chokes along the tubing), mainfold valves, wellhead choke and lift mandrels. The control system's main component for an IWC is the interval control valve. The interval control valve makes use of actuators and controllers in receiving commands from the control server. These actuators and controllers could however be improved by introducing applications of nanotechnology. In these respects, further research work need to be done to adapt such innovations to the completion system.

(3)控制系统。控制系统包括层间(流动)控制阀、汇流阀、井口节流器和举升工作筒。其中层间(流动)控制阀是一套沿着油管部署的可调节节流器,是智能完井系统中控制系统的重要组成部分。来自控制服务器的指令通过执行器和控制器接收传递给层间(流动)控制阀。这些执行器和控制器的性能可以通过引入纳米技术加以改进,但还需要开展进一步的研究工作以使其更好适应智能完井系统。

(1) Flow Control Devices. Most current downhole flow control devices are based on or derived from sliding sleeve or ball-valve technologies. Flow control may be binary (on/off), discrete positioning (a number of preset fixed positions), or infinitely variable. The actuating motive force for these systems may be provided by hydraulic or electric systems. Current generation hydraulically operated flow control devices have evolved to be more reliable, more resistant to erosion, provide greater flow control, and generate greater opening and closing forces.

(1)流量控制装置。当前大多数井下流量控制装置是基于或源自滑动套管或球阀技术。流量控制方式可以是二进制(开/关)、离散定位(多个预设固定位置)或无级调速。这些系统的驱动动力可以由液压或电动系统提供。目前的液压操作流量控制装置已经更可靠、更耐腐蚀、能提供更大的控制流量,并产生更大的打开和关闭力。

(2) Feedthrough Isolation Packers. To realize individual zone control, each zone must be isolated from each other by packers incorporating feedthrough systems for control, communication, and power cables.

(2)直通式隔离封隔器。为了实现单独层段的控制，每个层段必须利用封隔器隔离，其中包含用于控制、通信和电力电缆的直通系统。

(3) Control, Communication and Power Cables. Current intelligent well technology requires one or more conduits to transmit power and data to downhole monitoring and control devices. These may be hydraulic control lines, electric power and data conductors, or fiber optic lines. For additional protection and ease of deployment, multiple lines are usually encapsulated and may be armored.

(3)控制、通信和电力电缆。目前的智能完井系统需要一个或多个管道将电力和数据传输到井下并监测和控制设备，这可以通过液压控制线、电力和数据导体或光纤来完成。为了提供额外的保护和易于部署，通常封装多条线路，或者可以铣装。

(4) Downhole Sensors. A variety of downhole sensors are available to monitor flow performance parameters from each zone of interest. Several single-point electronic quartz crystal pressure and temperature sensors may be multiplexed on a single electric conductor, thus allowing very accurate measurements at several zones. Fig. 5. 3 is a schematic of a typical intelligent well completion system.

(4)井下传感器。有多种井下传感器可用于监测目标层段的流量性能参数。几个单点电子石英晶体压力和温度传感器可以在单个电导体上多路复用，因此可以在多个层段进行非常精确的测量。如图 5. 3 所示为典型智能完井系统示意图。

Intelligent Well Completions systems possess flexibility in reacting to detrimental conditions. It also allows for data management, with optimization and control techniques in place, in turn, reservoir management. Over the years, real time remote control and monitoring of wells using IWC has proved more advantageous in maintaining well productivity, improved sweep, controlled production of stacked reservoirs, managing water or gas production (if unwanted) and minimized interruptions from conventional well testing procedures.

智能完井系统能够灵活地应对不利条件，也能运用优化技术和控制技术进行数据管理进而改善油藏管理。多年来，使用智能完井系统对油井进行实时远程调控及监测的实践已经证明其在保持油井产能、改善油层波及程度、控制错叠油藏生产、管理不需要的水或气体的产出及最大程度地减少常规试井测试过程中的中断作业等方面具有更大的优势。

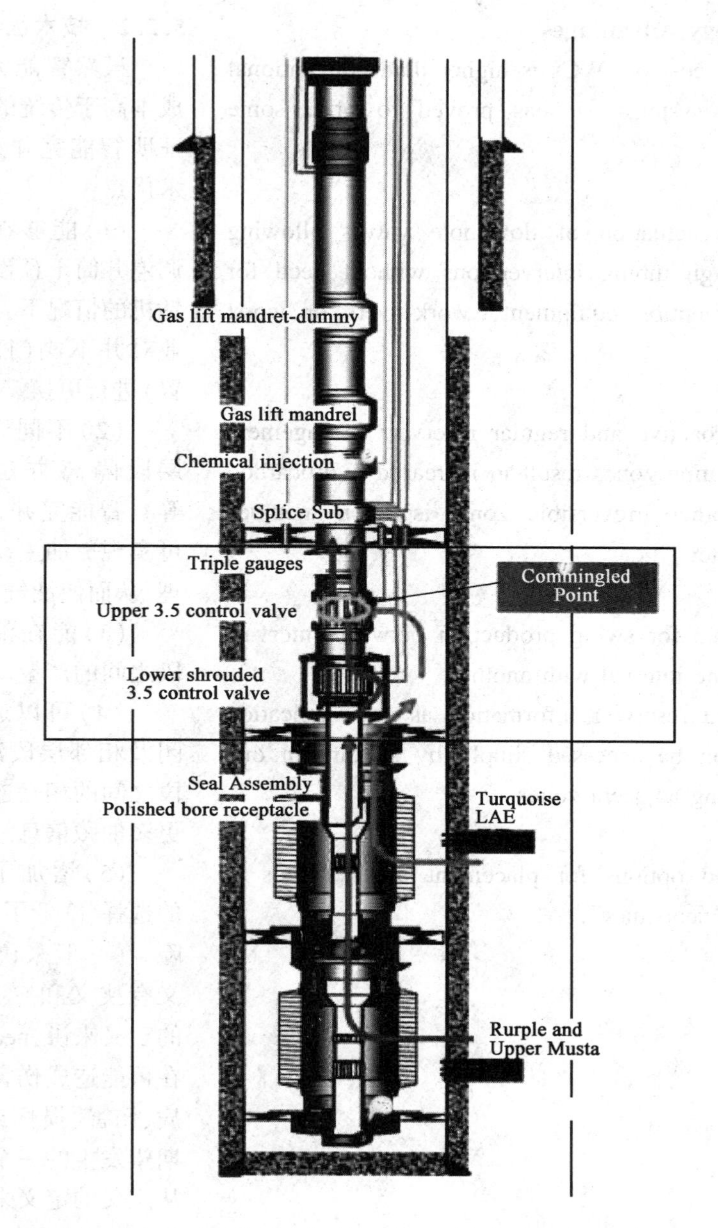

Fig. 5. 3 Schematic of a Typical Intelligent Well Completion System

Among the various systems installed, hydraulic motive power is dominant, even though a variety of electric and hybrid electro-hydraulic completions have been deployed successfully. Among the various providers, WellDynamics has been dominating the market.

在已安装的各种智能完井系统中,尽管已经成功安装部署了多种电动及电动液压混合智能完井系统,但液压动力智能完井系统仍然占据主导地位。在不同的智能完井系统供应商中,WellDynamic 公司一直在市场上占据着主导地位。

5.2.2 Technology Advantages

Although the cost of IWCs is higher than conventional well completions systems, it has proved to offer some benefits such as:

(1) Remote actuation of downhole valves allowing replacement through tubing interventions without need for rigless well intervention equipment, work-overs or worst case, a rig.

(2) More proactive and regular reservoir management from systems isolating zones result in increased hydrocarbon reserves unlike other irreversible zonal isolation methods (cement plugs, etc.).

(3) Allowance for swing production between intervals i. e. alternating one interval with another.

(4) Increased reservoir information as communication between zones can be accessed simply by shutting-in one interval and flowing adjacent zones.

(5) Increased options for placement of chemicals at reduced risks of "near-miss".

(6) Better understanding of flood-fronts allow for accurate forecast of reservoir performance.

Intelligent completion functionality contributes to the production process as a result of the following.

5.2.2 技术优点

虽然智能完井系统的投入成本高于传统的完井系统,但已证明智能完井系统具有如下技术优点:

(1)能够在不需要无钻机调停井筒干预设备、修井作业或钻机的情况下,通过油管干预作业对井下阀门的远程启动(装置)进行更换。

(2)不同于其他不可逆的层段隔离方法(如注水泥塞等),智能完井系统的层段隔离可实现更加主动、规范的油藏管理,从而使油气可采储量增加。

(3)能在油藏中调整各层段之间的产量。

(4)可以通过层间间隔关闭及相邻层段流体流出实现层段之间的相互连通,因此可获取更多油藏信息。

(5)增加了注入化学试剂的选择,降低了"near-miss"的风险。(一般来讲,near-miss 的定义有狭义和广义之分。从狭义的定义来讲,near-miss 是指有潜在可能造成伤害,但却未产生疾病、伤害、损坏或其他损失的不期望发生的一个或一系列事件。从广义的定义来讲,near-miss 是指有潜在可能造成更严重伤害的事件或情况,通过它可以增加改善业绩的机会。)

(6)能够更好地追踪驱替前缘,从而准确预测油藏性能。

智能井对于油气生产过程较为有利的原因有:

(1) Removing or reducing the frequency of intervention required for reservoir and production monitoring/optimization and enabling tuning of production, which will no longer be limited by control of the surface process.

(2) Increasing ultimate recovery and production by zonal/branch or inflow-profile optimization facilitated by timely remote-control inputs.

(3) Reducing gross fluid handling, waste product, surface hardware costs (e. g., lines, separation, and metering), manpower, and support services.

(1)不需要地面流程的控制可以免除或减少油藏与生产监测、优化所需的干预作业并且实现生产调整。

(2)通过及时的远程遥控输入促进层段、分支优化或流入剖面优化,从而提高最终采收率及产量。

(3)减少流体、废弃物处理,降低地面设备成本(如地面管线、分离设备和计量设备)、人力和后勤服务投入。

5.2.3 Defect of System

Even with the above identified benefits associated with IWCs, the following shortfalls were identified:

(1) Difficulty in measuring flow rate zone-independently due to unavailability of annular flowmeters and relatively expensive fibre optics used.

(2) High volumes of information and data with increased complexity in handling and processing of the data presented at the surface.

(3) Increased risks and additional costs associated with its installation due to the complexity from handling control cables.

Difficulty in obtaining downhole measurements without hampering the flow of alternate or adjacent zones i. e. zone-independent flow measurement resulting in production loss from shut-in zones and possibility of losing considerably useful formation pressure (reservoir pressure).

5.2.3 技术缺陷

尽管智能完井系统具有上述已明确的优势,但也存在以下问题:

(1)由于无法利用环形流量计,并且使用的光纤相对昂贵,难以独立地测量各层段流速。

(2)信息量和数据量大,增加了在地面进行数据处理的复杂性。

(3)处理控制电缆的复杂性增加了与其安装过程相关的风险和额外成本。

在不限制交替的或相邻层段内流体流动(即与层段无关的流量测量)的情况下难以获得井下测量结果,这将导致在关闭的储层层段内造成生产损失并且可能损失相当有用的地层压力(油藏压力)。

The long-term vision for the intelligent completion, however, is as a subsystem within an integrated intelligent production system, in which several wells, having the capability for individual and/or collective automated self-control, are linked by field, process, and reservoir-management systems. Currently, this vision implies closed-loop linkage between monitoring and flow-control equipment that is driven by feedback generated by the comparison of reservoir-performance sensors with the output from reservoir simulations. It is recognized that there are three distinct control challenges inherent in this objective:

智能完井系统的长期目标是作为集成化智能生产系统中的一个子系统投入应用,在集成化智能生产系统中,具有独立性和(或)集体性自动化自身控制能力的油井通过油田现场、工艺流程及油藏管理系统相互连接。目前,这种设想意味着监测设备和流量控制设备之间实现闭环连接,这是通过对比油藏性能传感器采集到的数据与油藏数值模拟输出结果而实现的。要实现这一目标,需要应对三个控制方面的挑战:

(1) Well-performance optimization in which intelligent artificial lift completion systems enable classic real-time control to optimize production as a function of controlled inputs (e.g., gas lift injection, electrical submersible pump speed, and surface choke settings). The response of wells to these control optimization inputs will be seen in minutes.

(1)油井性能优化。在油井性能优化方面,智能完井系统能够根据受到控制的输入信息(如气举注入量、电潜泵速度及地面节流塞设置参数)进行实时控制以优化生产,油井在短时间内就可对这些优化过程中输入的信息做出响应。

(2) Field optimization in which the overall control system optimizes output within constraints imposed by the field production infrastructure. This optimization is driven by direct monitoring of production parameters, but the control-time constants change in that the overall field response to process-control inputs is measured in hours or days.

(2)现场优化。在现场优化方面,整个控制系统在现场生产基础设施的约束下对输出结果进行优化。这种优化是通过对生产参数直接监测而实现的,由于过程控制输入信息的整体现场响应是以小时或以天为单位进行测量的,控制时间常数因此会发生变化。

(3) Reservoir optimization in which field control inputs are driven by the output of reservoir simulators, which are validated (history matched) with data provided by the intelligent production system. Optimization requires removing the direct linkage between sensor output and control commands to account for reservoir heterogeneity and to match control functionality with reservoir responses, which

(3)油藏优化。在油藏优化方面,油藏数值模拟输出结果与智能生产系统提供的数据进行验证(历史拟合)得到现场控制输入信息。考虑到储层非均质性,并且要使控制功能与在几个月内甚至几年内测量得到的

may be measured in months or even years. It may be expected that as sensors capable of providing accurate and detailed reservoir characterization and response are developed and deployed, the requirement for reservoir simulation for control purposes will diminish.

油藏响应相匹配,油藏优化需要消除传感器输出结果和控制指令之间的直接联系。可以预计,随着能够提供准确、详细的油藏特征及油藏响应的传感器的开发和部署,对以控制为目的的油藏数值模拟的需求将减少。

5.3 Application of Intelligent Well Completion

5.3 智能完井技术的应用

5.3.1 Application Case 1-Dump Flooding in Western Kuwait

5.3.1 应用实例1——科威特西部自流注水

By completing an injector well, dump flooding permit water taken from an overlying aquifer to dump into a depleted zone using inflow control and monitoring to provide pressure support.

Audio 5.3

自流注水是指通过智能井,把一个地层的水直接注入另一个油层,提供油层压力支持。

In early 2007, a West Kuwait well was completed as a controlled dump flood well utilizing intelligent well technology. Water from the Zubair aquifer formation flew to the Minagish Oolite oil formation in a controlled and monitored dump flood process. Utilizing a variable interval control valve, the amount of injection fluid was regulated, while permanent down hole monitoring devices transmit pressure data to surface, enabling evaluation of the flow rate. The intelligent well also permits "soft starts" of the dump flood to avoid borehole destabilization.

2007年年初,在西科威特的一口井利用智能井完成了可控自流注水的完井作业。来自Zubair含水层的水在控制和监测下自流至Minagish Oolite油层。利用可变间隔控制阀,注入流体的量可以进行调节,井下永久监测装置可以将压力数据传输到地面,从而评估流速。智能井还可以做到自流注水的"软启动"以避免井眼不稳定。

In the Kuwait dumpflood intelligent well application, an interval control valve (ICV) was deployed on the production tubing to control the rate of flow of water from the Zubair formation to the Minagish Oolite formation, along with a permanent dual pressure monitoring system (PDHMS) to transmit pressure and temperature data to a surface acquisition unit for display and recording (Fig. 5.4).

科威特自流水井智能井的生产油管上部署了间隔控制阀(ICV)以控制从Zubair层到Minagish Oolite层的水流量,同时永久性双压力监测系统(PDHMS)将压力和温度数据传输到表面采集单元并显示和记录(图5.4)。

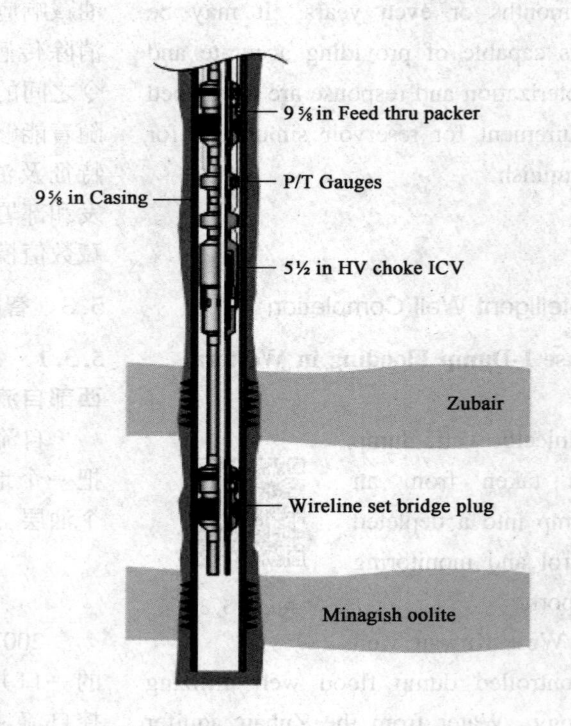

Fig. 5.4 Dumpflood Completion

The dump flooding has the following benefits:

(1) Ability to monitor wellbore producing and injection pressure in real time.

(2) Ability to monitor crossflow rate and quantify flow rate in real time.

(3) Ability to determine productivity index and injectivity index for the producing zone and injection zone respectively.

(4) Ability to perform pressure transient analysis independently on both the production zone and the injection zone to determine each reservoir pressure, *KH*, skin, and potential for productivity/injectivity improvement through stimulation.

(5) Ability to independently stimulate and clean-up both the production zone and the injection zone.

(6) Ability to pre-produce the injection zone to reduce regional wellbore pressure and benefit from pre-production of hydrocarbon.

自流井中智能井的好处具体如下：

(1)能够实时监控井筒生产和注入压力；

(2)能够实时监控横流流速并量化流速；

(3)能够分别确定生产区和注入区的生产指数及吸水指数；

(4)能够在生产层段和注入层段独立地执行压力瞬变分析以确定每个储层的压力、*KH*、表皮系数及通过增产提高采收率或注入能力的潜力；

(5)能够独立地进行增产作业和清理生产区、注入区；

(6)能够预先生产已注入区以降低层段井筒压力并从预生产的油气中获益；

(7) Ability to "soft-start" the dumpflood process, initially and after shut-in, to reduce the drawdown pressure transient and geomechanical shock to the wellbore of the producing zone.

(8) Ability to monitor and control drawdown pressure and flow flux rate to maintain flow conditions within wellbore stability and sand control guidelines.

(7)能够在初始和关井后阶段"软启动"自流注水,以减少生产井筒的瞬态压降和地层岩石冲击;

(8)能够监测、控制压降和流量以维持可以保证井筒稳定性及防砂要求的流动条件。

5.3.2 Application Case 2-Intelligent-Well Completions in Agbami

5.3.2 应用实例2——Agbami油田智能完井的应用

The Agbami field, discovered in 1998, is 70 miles offshore Nigeria at a water depth of approximately 5,000 ft. First production from the Agbami field was 29 July 2008.

1998 年发现的 Agbami 油田位于尼日利亚海上 70mi 处,水深约 5000ft。Agbami 油田的首次生产是在 2008 年 7 月 29 日。

The field is a northwest/southeast-trending doubly plunging anticline, with a significant thrust fault through the crestal axis of the structure. As Fig. 5.5 shows, the field has four main reservoirs: 17MY, 16MY, 14MY, and 13MY. Approximately 80% of the field in-place volume and reserves are in the deepest sand (17MY), having average porosity of 18% and average permeability of 270 mD.

该油田是一个西北—东南向的双倾伏背斜,该结构的脊轴有明显的逆冲断层。如图 5.5 所示,该油田有四个主要储层:17MY,16MY,14MY 和 13MY。大约80%的已采量和储量都位于最深的砂体(17MY),其平均孔隙度为 18%,平均渗透率为 270 mD。

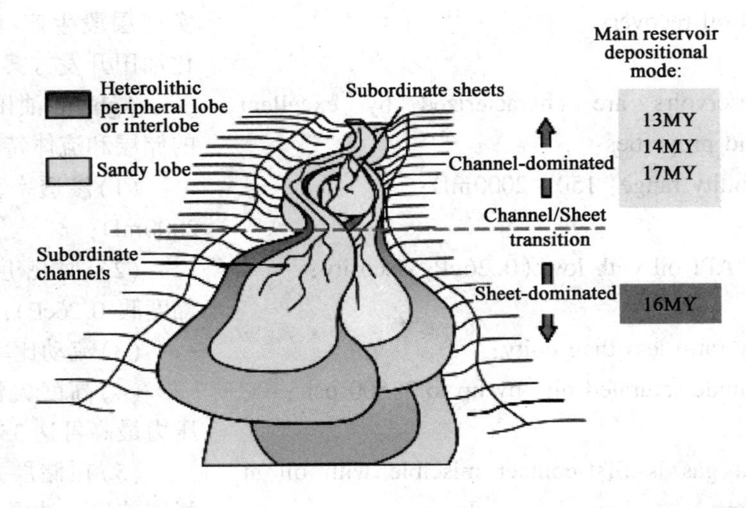

Fig. 5.5 Agbami-Reservoirs Depositional Model

The field-development plan has 38 subsea wells (including 20 producers, 12 water injectors, and six gas injectors) tied back to a floating production, storage, and offloading (FPSO) vessel through subsea flowlines. The drilling-and-completion program is spread over three development-drilling stages. The project is in Stage1, with eight producers and two gas injectors on line. Pressure maintenance will be by peripheral water injection and crestal produced-gas reinjection to ensure compliance with a "no flare" policy.

该油田计划开发38口海底油井(包括20口生产井、12口注水井和6口注气井),通过海底管线连接回浮式采油、储存和卸载(FPSO)船。钻完井计划在三个钻井开发阶段展开。该项目处于第一阶段,目前已有8口生产井和2口注气井。将通过周边注水和顶部产气再注入来维持压力确保符合"无明火"政策。

5.3.2.1 Geologic and Engineering Considerations

5.3.2.1 地质和工程考虑

The field has multiple lobes within each major reservoir. Although these sand lobes, or zones, are in pressure equilibrium geologically, the vertical/lateral connectivity under dynamic conditions is uncertain. Therefore, waterflood and gasflood fronts are likely to advance through the reservoirs at different rates, leading to conformance issues. To optimize field performance and recovery, IWCs with downhole control valves are installed in the Agbami wells. The system will provide zonal information and control of production from and injection into the completed subreservoir zones. IWCs will enable zonal production/injection management for optimal field development and oil recovery.

该油田的每个油藏内有多个朵叶体。尽管这些砂舌或层段在地质上处于压力平衡状态,但动态条件下的垂直、横向连通性是不确定的。因此,水驱和气驱前沿可能以不同的速率通过储层,导致前沿不一致。为了优化现场生产及提高采收率,在Agbami井中安装了带有井下控制阀的智能完井系统。该系统将提供层段信息,并控制生产和注入子储层。智能完井系统将实现层段生产/注入管理,以优化油田开发方案和提高采收率。

Agbami reservoirs are characterized by excellent reservoir and fluid properties:

(1) Permeability range: 150 ~ 2000mD;

(2) 45 ~ 47°API oil with low (0.26cP) viscosity;

(3) Mobility ratio less than unity;

(4) Highly undersaturated oil, by up to 3,500 psi;

(5) Injected gas is first-contact miscible with oil at reservoir conditions.

Agbami油田具有以下优良的储层和流体特性:

(1)渗透率范围:150 ~ 2000mD;

(2)油相密度为45 ~ 47°API,黏度低(0.26cP);

(3)流动比率小于1;

(4)高度欠饱和油藏,饱和压力最高可达3500 psi;

(5)在储层条件下,注入的气体首先与油混溶。

With favorable reservoir and fluid properties, high net-/gross-pay ratio, and structural dip (10° ~ 30°), gravity-stable gas and water fronts should result in good oil recovery. However, uncertainty in key subsurface parameters (e. g., sand connectivity, fault location/sealing capacity, and injection conformance) requires a proactive approach to Agbami field development by application of IWC and other recovery-enhancing technologies.

由于有利的储层和流体性质、较高的总产层和净产层之比及结构倾角(10° ~30°)、重力稳定的气体和水前沿,应该能够得到良好的采收率。然而,关键地下参数(例如砂体连通性、断层定位、密封能力和注入一致性)的不确定性需要通过应用智能完井系统和其他提高采收率技术来更好地开发 Agbami 油田。

5.3.2.2 Agbami IWC

5.3.2.2 Agbami 油田智能完井系统

The Agbami IWC system collects and transmits completion, production, and reservoir data and enables selective zonal control to optimize the production process without physical intervention. The IWC-system control is provided by use of interval-control valves (ICVs) in the multizone-completion wells. The system has three components: sensors (permanent downhole gauges, flowmeters, and densitometers) at points along the production network from sandface to separator (i. e., monitor); an integrated database and analysis computer to collect and evaluate sensor output (i. e., analysis); and up to two downhole variable chokes and wellhead chokes, manifold valves, and a riser network (i. e., control).

Agbami 油田智能完井系统收集并传输完井、生产和油藏数据,而且能实现选择性层段控制,以优化生产过程,无需物理干预。该油田通过在多层段完井井中使用间隔控制阀(ICV)来提供智能完井系统控制。该系统有三个部件:传感器(永久性井下计量、流量计和密度计),沿着生产网络从砂面到分离器(即监测)分布;一个集成的数据库和分析计算机,用于收集和评估传感器输出(即分析);最多两个井下可调油嘴和井口油嘴、管汇阀和提升管网(即控制)。

The IWC system assists in mitigating short-term production problems without well interventions. In the long term, an IWC system helps reduce the effect of geological uncertainties through management of gasflood/waterflood fronts. The monitor, analysis, and control process in combination with the IWC system can identify operating ranges for wells and locate excess capacity in the well and subsea system. Relevant-time identification of well performance has improved reaction time to execute mitigation plans when needed since field startup.

智能完井系统有助于在没有井干预的情况下缓解短期生产的问题。从长远来看,智能完井系统通过管理注气、注水前沿从而减少地质不确定性的影响。与智能完井系统结合的监测、分析、控制过程可以识别井的操作范围并确定井和海底系统中的过剩容量。井性能相关时间的识别减少了井场运作以来需要执行计划的反应时间。

5.3.2.3 Value Added from IWCs

The Agbami field-development plan originally proposed IWCs only for the water and gas injectors to enable injection-profile modification and control. However, the deployment of IWC has found increasing use in the field by improving the understanding of compartmentalization and heterogeneity from development-drilling results.

(1) Incremental recovery. Reservoir modeling indicates most wells have direct incremental recovery by use of ICVs because of the ability to manage individual zones. The ability to optimize zonal contributions by use of zonal-drawdown, productivity-index, and production data helps to maximize well production. It will also help to sustain plateau production and minimize decline rates.

(2) Prudent reservoir management. A clear understanding of produced or injected zonal volumes enables effective voidage replacement and pressure maintenance with gas and water injection. This information enables optimizing reservoir production and displacement processes to maximize sweep efficiency and recovery. Zonal pressure and production data from IWCs enhance simulation history matching of fluid breakthroughs and the ability to manage production and injection for optimal oil recovery.

(3) Optimize infill drilling. Zonal volumes facilitate model history matching better than use of total-reservoir volumes because flood fronts are understood and managed better. Understanding flood fronts supports the ability to forecast reservoir performance accurately and identify bypassed volumes for infill drilling.

5.3.2.3 智能完井系统的附加价值

Agbami油田最初仅计划为注水和注气井安装智能完井系统以实现注水剖面的优化和控制。之后通过安装智能完井系统,基于开发钻井结果得到了对油藏分区化和非均质性更深入的理解,使得智能完井系统的应用更加广泛。

(1)提高采收率。油藏数模表明,由于能够管理各个层段,大多数井通过使用间隔控制阀可以直接提高采收率,通过调控层段压降、生产指数和生产数据来优化层段贡献的能力从而最大化井产量。智能完井系统还有助于维持平稳生产并最大限度地降低产量下降。

(2)精细油藏管理。通过智能完井系统能够清楚地了解生产或注入的层段体积,从而可以通过注气和注水达到有效的孔隙置换和压力维持。这些信息可以优化油藏生产和驱替过程,从而最大限度地提高波及效率和采收率。来自智能完井系统的层段压力和生产数据优化了流体突破的历史拟合,更好地管理生产和注入以达到最佳采收率。

(3)优化加密钻井。层段体积比总油藏体积更适用于模型历史拟合,因为这样可以更好地理解和管理水驱前沿。了解水驱前沿有助于准确预测油藏性能并识别加密钻井的绕流体积。

(4) Reduced expenses. IWCs provide capital-expense savings by eliminating or greatly reducing workover/ sidetrack activities. There also are significant operating-expense savings that will accrue by eliminating production logs, which otherwise would be required to evaluate zonal injection/production contributions to/from individual sand lobes. Operationally, ICV action will reduce costs of handling excess gas and water. It will also prevent FPSO gas and water capacity from becoming a production bottleneck.

(4)减少开支。智能完井系统通过消除或大大减少修井、侧钻作业来节省支出;同时取消生产日志记录还可以节省大量的运营费用,否则将需要生产记录来评估层段注入、生产对(来自)单个砂质朵叶体的贡献。在操作上,间隔控制阀的使用可以降低处理多余气体和水的成本,同时防止 FPSO 气体和水成为生产瓶颈。

5.4 Global Intelligent Well Market

North America is currently the largest market for intelligent well followed by Europe and Asia-Pacific. Factors such as increasing drilling activities in oil fields and rising focus on shale gas and tar sand reserves are driving the onshore well type. The exploration of new reserves, and creation of growth prospects for the exploration and production activities in the U. S., Canada and Mexico are expected to trigger the demand for intelligent well market.

Audio 5.4

5.4 全球智能完井市场

目前,北美洲是最大的智能完井市场,紧随其后的是欧洲和亚太地区。油田钻井作业增加及页岩气和焦油砂越来越受到关注等因素,正在推动陆上油井发展。在美国、加拿大和墨西哥等国的新探明储量、勘探与生产活动发展前景预计将触发智能完井市场需求。

5.4.1 Market Status

The growth in the demand for oil and gas translates into the increased demand in the intelligent well market. Advancement in technology will increase the efficiency, control and monitoring of the well. Additionally it also increases the overall recovery and also facilitates the production from unconventional resources like shale gas and tar sands reserves. The development of the efficiency of the intelligent well, advancement in equipment technology and the increase in investments by market vendors are pushing the market towards growth. The growth of intelligent well market is directly linked to the growth in drilling and completion activities across the globe. Moreover, increased use of horizontal and multilateral wells acts as a major driver for the market.

5.4.1 智能完井市场现状

石油和天然气需求的增长将转化为智能完井市场需求的增加。技术进步将改善油井的(生产)效率及对油井的监测和控制。此外,技术进步还提高了总采收率,并且促进了对页岩气、焦油砂等非常规(油气)资源的开发。智能完井(生产)效率的提高、设备技术的进步及市场供应商投资的增加正在推动智能井市场增长。智能完井市场的增长也与钻完井活动在全球范围内的增长直接相关。此外,水平井和多分支井的增加也是智能完井市场增长的主要推动力。

However, uncertainties associated with low crude oil prices is further projected to hinder the industry growth for intelligent well.

但是,与低油价相关的不确定性会阻碍智能完井市场的增长。

Currently, North America dominates the global intelligent well market share. Major value generation activities of the market are recorded in North America and the other regions are growing at a higher growth rate with the exploration of new reserves, creating growth prospects for the exploration and production (E&P) activities due to technological developments.

目前,北美地区在全球智能完井市场中占主导地位。智能完井市场的主要价值创造活跃在北美地区,这得益于新储量的探索,其他区域的智能完井市场正在以较高的增长率增加,技术进步为勘探和生产活动创造了增长前景。

Asia-Pacific is expected to grow significantly in the forecast period. The growth is primarily due to shale activity in China and increasing offshore oil and gas exploration in countries such as India and Indonesia. Moreover, rise in offshore oil and gas exploration and enhanced oil recovery techniques will make APAC the fastest growing region in the global market.

预计亚太地区的智能完井市场在预测期内(2018—2022)将会显著增长。这种增长主要是因为中国页岩气的开发,以及印度和印度尼西亚等国家不断增加的海上油气勘探(活动)。此外,海上油气勘探(活动)的增加及提高原油采收率技术将使亚太地区成为全球智能井市场中增长最快的区域。

5.4.2 Major Service Company

5.4.2 主要服务公司

The key players of global intelligent well market are Baker Hughes, Halliburton, Schlumberger, Weatherford International, National Oilwell Varco, Superior Energy Services, Inc. Trican Well Service Ltd., RPC Inc., Nabors Industries Ltd, Salym Petroleum Development N. V. and others.

全球智能完井市场的主要服务公司有贝克休斯公司、哈里伯顿公司、斯伦贝谢公司、威德福国际有限公司、国民油井华高公司、RPC 公司、纳伯斯工业公司等。

5.4.2.1 Baker Hughes

5.4.2.1 贝克休斯公司

1. Accelerate Production, Increase Ultimate Recovery and Reduce Total Cost of Ownership

1. 推动生产,提高最终采收率并且降低总成本

Our intelligent well systems are part of our world-class intelligent optimization systems that can help you collect and monitor downhole data and remotely control reservoir zones to optimize reservoir efficiency.

贝克休斯公司的智能完井系统是其具有国际水平的智能优化系统的一部分,该系统能够采集和监测井下数据并对储层特性和油井动态进行远程控制,从而优化生产效率。

This reservoir surveillance and control technology, combined with our completion and production expertise, will reduce your total cost of ownership. Our intelligent well products and services will also increase the ultimate recovery from your reservoir.

这种油藏动态监测和控制技术结合了贝克休斯公司的完井技术及生产技术的专业知识，能够降低总成本。贝克休斯公司有关智能完井技术的产品和服务也能够提高油藏最终采收率。

Several components make up an intelligent wellbore. They include well monitoring, intelligent completion tools and chemical automation.

智能完井的井筒是由一系列模块组成的,这些模块包括油井监测模块、智能完井工具模块和化学自动化模块。

2. Monitoring Eliminates Well Conditions Guesswork

2. 通过实时监测准确掌握油井及油藏的动态变化

Well monitoring instrumentation measures pressures, temperatures, flow rates, water-cut, and density in the wellbore with both electronic and fiber optic gauges.

监测仪器能够实时测量温度、压力、流量、含水率,通过使用电子传感器和光纤传感器可测量井筒中流体密度。

3. Cut Down Well Intervention with Remote Flow Control

3. 通过远程流量控制减少人工干预

Intelligent completion technologies, such as zonally isolated, hydraulically adjustable valves and chokes, let you adjust product inflow from any zone, without well intervention. If you're injecting water or gas, our tools can also remotely control the flow to individual zones.

智能完井系统使用了层段隔离、液压调节阀和节流塞等设备,可在不进行井筒干预作业的情况下在油藏中调整各层段之间的产量。贝克休斯公司的智能完井工具也能远程控制向各个产层注入水或注入气的过程。

4. Simplify Chemical Applications through Automation

4. 通过自动化改造简化化学试剂的使用

Chemical automation products save you money by replacing manual application of chemicals needed to keep your wells flowing. Our SENTRYNET chemical automation tools make it easy to control your chemical regimen at remote oil and gas production facilities, including unmanned satellite platforms, remote well locations, and pipelines.

化学试剂(使用)自动化装置取代人工操作来保持井内(油气)流动,从而节约了成本。贝克休斯公司的 SENTRYNET 化学试剂(使用)自动化装置能够在无人卫星平台、远程井场和油气输送管道等远程油气生产设施上简化化学试剂操作并使其按计划完成注入。

5.4.2.2 Halliburton

1. Overview

Halliburton is the world's leading provider of intelligent completion technology to the upstream oil industry. SmartWell system technology, introduced in 1997, was the industry's first intelligent well completion system(Fig. 5.6).

5.4.2.2 哈里伯顿公司

1. 概述

哈里伯顿公司是世界领先的为油气行业上游领域里智能完井技术提供产品及服务的供应商。1997 年推出的 SmartWell（智能完井）系统是油气行业第一套智能完井系统(图 5.6)。

Fig. 5.6 Intelligent Well Systems

(Source: www.halliburton.com/en-US/ps/well-dynamics/well-completions/intelligent-completions/default.page? node-id = hfqel9vs&nav = en-US_completions_public)

Halliburton's SmartWell system intelligent completion technology helps operators to optimize production without costly well intervention. Reliable and fit-for-purpose SmartWell systems enable operators to collect, transmit and analyze downhole data; remotely control selected reservoir zones; and maximize reservoir efficiency by:

哈里伯顿公司的 SmartWell 系统可在不需要成本高昂的井筒干预作业下帮助操作人员优化生产。SmartWell 系统允许使操作人员采集、传输和分析井下数据，远程控制特定的储层层段。它主要通过以下几种方式最大限度地提高生产效率：

(1) Increasing production. Commingling of production from different reservoir zones increases and accelerates production and shortens field life.

(1) 提高产量。储层不同层段混合生产，提高生产产量，加快采油速度，缩短油田开发年限。

(2) Increasing ultimate recovery. Selective zonal control enables effective management of water injection, gas and water breakthrough and individual zone productivity.

(2) 提高油田最终采收率。选择性层段控制可以有效地管理注水过程，避免气窜和水窜，提高各个产层的产能。

(3) Reducing capital expenditure. The ability to produce from multiple reservoirs through a single wellbore reduces the number of wells required for field development, thereby lowering drilling and completion costs. Size and complexity of surface handling facilities are reduced by managing water through remote zonal control.

(3)减少资本支出。通过在一个井眼内控制多个储层的开采,减少了全面开发油田所需的油井数量,从而降低了钻井和完井成本。通过远程层段控制关闭或节流含水率较高的产层,可以减少地面(产出水)处理设施的规模和复杂程度。

(4) Reducing operating expenditure. Remote configuration of wells optimizes production without costly well intervention. In addition, commingling of production from different reservoir zones shortens field life, thereby reducing operating expenditures.

(4)减少作业支出。远程控制可在不需要成本高昂的井筒干预作业下优化生产。此外,储层不同层段混合生产缩短了油田开发年限,从而减少了作业支出。

A SmartWell system completion consists of some combination of zonal isolation devices, interval control devices, downhole control systems, permanent monitoring systems, surface control and monitoring systems, distributed temperature sensing systems, data acquisition and management software and system accessories.

SmartWell系统由产层隔离装置、层间(流动)控制装置、井下(流体)控制系统、永久性监测系统、地面控制和监测系统、分布式温度传感系统、数据采集和管理软件及系统附属设备组成。

2. Well Type/Service Offering

2. 产品与服务

1) Chemical Injection Systems

1)化学试剂注入系统

The Halliburton chemical injection system provides operators with precise wellbore chemical management to help optimize production as well as help reduce the need for costly interventions. Fig. 5. 7 shows the Halliburton chemical injection mandrel.

哈里伯顿公司的化学试剂注入系统能够给操作人员提供精准的化学试剂(注入)管理方法,有助于优化生产及减少油井生产期间所需进行的成本高昂的井筒干预作业。图5.7为哈里伯顿公司的化学试剂注入装置。

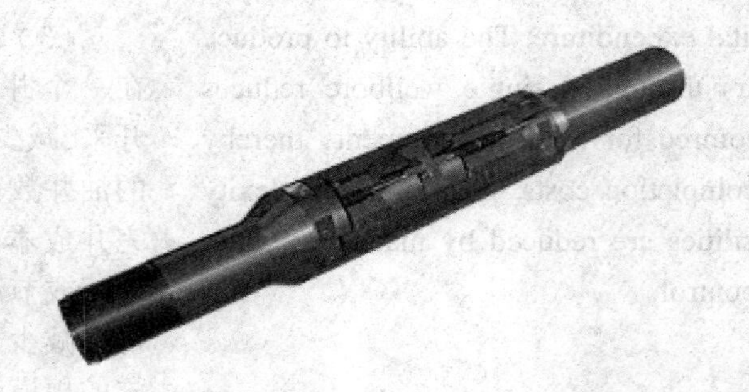

Fig. 5.7 Chemical Injection Mandrel
(Source: http://www.halliburton.com/en-US/ps/well-dynamics/well-completions/intelligent-completions/chemical-injection/Chemical-Injection-System.page? node-id = ijb921z5&nav = en-US_completions_public)

The Halliburton chemical injection system enables operators to address asphaltines, corrosion, emulsions, foaming, hydrates, paraffin, scale, and scavengers. Two types of chemical injection mandrels are available on the system: a robust, machined, non-welded, one-piece mandrel for all applications, including deepwater; or a welded pup-joint mandrel for low-risk environments like shallow, non-deviated wells or land applications.

哈里伯顿公司的化学试剂注入系统使操作员能够解决沥青质处理、腐蚀、乳化、起泡、天然气水合物、石蜡处理、结垢和净化处理等问题。该系统具有两种类型的化学试剂注入工作筒:一种是稳固、经机械加工、非焊接式的一体化工作筒,适用于包括深水环境在内的各种作业环境;另一种是焊接式的短节工作筒,适用于如浅井、非斜井或陆地井等风险较低的作业环境。

(1) Features:

①Custom mandrels for high-or low-profile applications;

②Dual check valves available with different cracking pressures;

③Injection lines of corrosion-resistant 316 stainless steel or Incoloy alloy 825;

④Standard cast or pressed-steel cable protectors;

⑤Installation services, including multi-line spooling;

⑥Subsea, platform and land applications.

(1)特点:

①具有可应用于不同作业环境的定制工作筒;

②具有不同开启压力的双止回阀;

③具有由耐腐蚀的316不锈钢和Incoloy 825合金制成的(化学试剂)注入管线;

④具有符合行业标准的铸钢或压制钢电缆保护套;

⑤提供安装服务,包括多线绕线;

⑥可应用于海底、海上平台和陆地。

(2) Key benefits:

①Maintains flow assurance;

②Optimizes production;

③Helps reduce costly interventions;

④Corrosion-resistant injection lines;

⑤Redundant checks.

2) Interval Control Valves

Halliburton interval control valves (ICV) allow operators to control flow into or out of an isolated reservoir interval, wherever selective control over production or injection is required.

(1) eMotion-HS remotely operated circulating valve. Provides high circulation rates without the need for any interventions during completion deployment-saving time and helping to reduce risk.

(2) HS circulating valve. The Halliburton HS circulating valve is designed for use in SmartWell and conventional completion systems.

(3) HS interval control valve (HS - ICV). HS - ICV (Fig. 5. 8) is designed for deepwater environments, and can withstand pressures up to 15,000 psi and temperatures up to 325℉ (163 ℃).

(2)优点:

①保持流道通畅;

②优化生产;

③有助于减少成本高昂的井筒干预作业;

④(化学试剂)注入管线耐腐蚀;

⑤冗余检查。

2)层间(流动)控制阀

哈里伯顿公司的层间(流动)控制阀(ICV)使操作人员能够在任何需要对生产过程或者注入过程进行选择性控制的时候控制(流体)流入或流出隔离的储层层段。

(1)eMotion-HS 远程遥控循环阀。eMotion-HS 远程遥控循环阀在完井作业期间能够在无需开展任何井筒干预作业的情况下提供较高的循环速率,缩短了施工工期并且降低了作业风险。

(2)HS 循环阀。哈里伯顿公司的 HS 循环阀适用于 SmartWell 系统和常规完井系统。

(3)HS 层间(流动)控制阀。HS - ICV(图 5. 8)适用于深水作业环境,可承受 15000 psi 的高压和 325℉ (163℃)的高温。

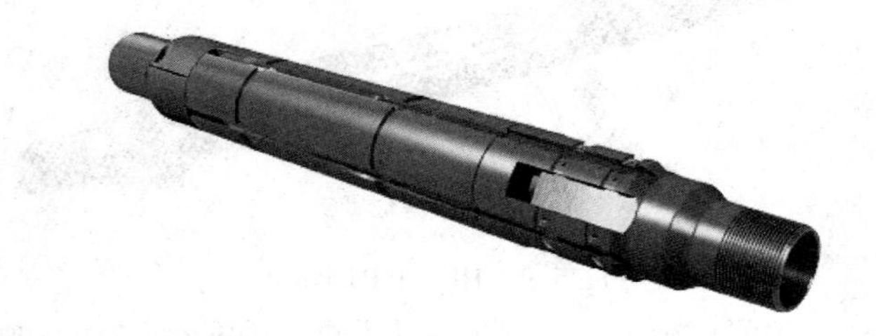

Fig. 5. 8 HS - ICV

(Source: http://www. halliburton. com/en-US/ps/well-dynamics/well-completions/intelligent-completions/interval-control-valves. page? node-id = hfqel9vt&nav = en-US_completions_public)

(4) MC interval control valve (MC - ICV). MC - ICV is a multi-position interval control valve that provides incremental flow control over individual reservoir zones, allowing optimization of reservoir architecture downhole and maximizing recovery.

(4) MC 层间(流动)控制阀。MC - ICV 是一种可为储层各个层段提供增量流量控制的多位置层间(流动)控制阀,能够优化井下储层位置管柱结构并且最大限度地提高采收率。

3) Zonal Isolation Devices

3)产层隔离装置

Individual reservoir zones must be isolated in order to control flow. Halliburton offers a range of high-performance, field-proven packers and isolation devices with control line bypass capabilities specifically designed for SmartWell system applications, as well as a line of packers for marginal assets.

为控制不同产层的流量,必须将储层各个生产层段隔离。哈里伯顿公司提供一系列专为 SmartWell 系统的应用而设计的具备控制管线旁路(管理)功能、性能好、经过现场试验证明的封隔器和隔离设备;也可提供一系列用于改善边际成本的封隔器。

(1) HF - 1 packer. HF - 1 packer (Fig. 5.9) is a single-string, retrievable, cased-hole packer that features a facility for bypass of multiple electrical and/or hydraulic control lines. Available for use as both the top production packer and as one of many lower packers isolating adjacent zones, the HF - 1 packer can operate under higher loads and greater pressures than standard production packers.

(1) HF - 1 封隔器。HF - 1 封隔器(图 5.9)是一种单管柱可回收式套管井封隔器,是具有多条电力控制管线和(或)液压控制管线的旁路(管理)设施。HF - 1 封隔器既可用作生产封隔器,又可作为用来隔离相邻产层的常规封隔器,相比于标准生产封隔器,HF - 1 封隔器能够在更高的负载和更大的压力下使用。

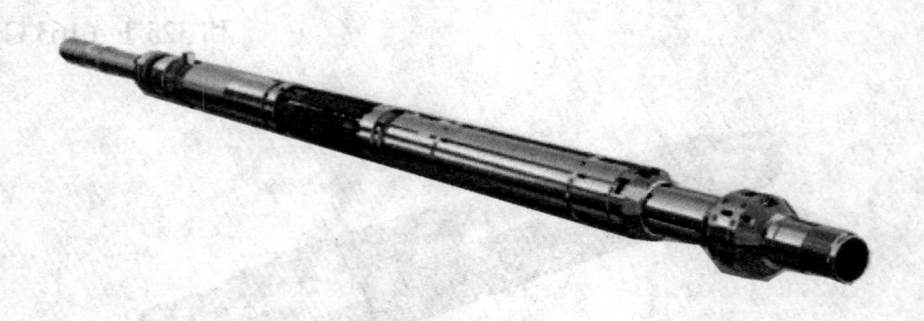

Fig. 5.9 HF - 1 Packer

(Source: http://www.halliburton.com/en-US/ps/well-dynamics/well-completions/intelligent-completions/zonal-isolation-devices.page? node-id = hfqel9vu&nav = en-US_completions_public)

(2) MC packer. MC series packers, available as both production packers or isolation packers (no slips), are single-string, cased-hole, retrievable packers primarily designed for use in SmartWell system completions. In some instances, the MC production packer has been used as an isolation packer below the primary HF – 1 production packer.

(2)MC 封隔器。可用作生产封隔器或隔离封隔器(无卡瓦)的 MC 封隔器是单管柱套管井可回收式封隔器,主要是为 SmartWell 系统而设计的。在某些情况下,MC 生产封隔器被用作 HF – 1 生产封隔器下面的隔离封隔器。

(3) Seal stack assembly. Seal Stack Assembly isolates individual zones in SmartWell intelligent completions in applications where it is not possible or desirable to use packers for isolation.

(3)密封组件。密封组件可以隔离 SmartWell 系统中那些不能够或不希望使用封隔器进行隔离的产层。

4) Downhole Control Systems

4)井下(流体)控制系统

Halliburton intelligent downhole control systems provide a method of integrating the surface control system (either manual or automated) with downhole SmartWell equipment.

哈里伯顿公司的井下(流体)控制系统提供了将地面控制系统(手动或自动)与井下 SmartWell 系统的设备相结合的方法。

(1) Accu-Pulse incremental positioning module. The Accu-Pulse incremental positioning module provides incremental opening of a multiple-position interval control valve (ICV) allowing operators to control produced or injected fluid rates, greatly enhancing reservoir management capabilities.

(1)Accu-Pulse 增量位置定位模块。Accu-Pulse 增量位置定位模块为允许操作人员控制生产流体速率或注入流体速率的多位置层间(流动)控制阀(ICV)提供增量开启,极大地提高了油藏管理水平。

(2) Digital hydraulics downhole control system. The Digital Hydraulics system allows for control of multiple downhole devices, with the minimum number of control lines, using hydraulic pressure sequencing and a mechanical decoder downhole.

(2)Digital Hydraulics 井下(流体)控制系统。Digital Hydraulics 井下(流体)控制系统允许利用液压排序和一个井下机械解码器(即数字液力解码器),以最少数量的控制管线对多个井下设备进行控制。

(3) Direct hydraulics downhole control system. The Direct Hydraulics system utilizes two hydraulic lines, one each for the open and close functions, connected directly to the device that is to be controlled. When networking multiple devices, a common close line can be used to minimize the number of lines needed.

(3)直接液压式井下(流体)控制系统。直接液压式井下(流体)控制系统利用两条液控管线进行控制,直接与需要控制的设备相连,分别控制打开和关闭作业。需要连接多个设备时,可以用一条公共的闭合线路来使所需管线数量最小化。

(4) SmartPlex downhole control system. The SmartPlex downhole control system (Fig. 5. 10) is an electro-hydraulic multi-drop system that provides simple and reliable zonal control of multiple valves (12) in a single wellbore with a minimum number of the control lines.

(4) SmartPlex 井下(流体)控制系统。SmartPlex 井下(流体)控制系统(图 5. 10)是一种多产层电动—液压控制系统,在单个井筒中用最少数量的控制管线就能够为井下多个(12 个)产层控制阀提供简易和可靠的控制。

Fig. 5. 10 HAL40695 SmartPlex

(Source: http://www. halliburton. com/en-US/ps/well-dynamics/well-completions/intelligent-completions/downhole-control-systems/default. page? node-id = hfqel9vv&nav = en-US_completions_public)

5) SmartWell System Accessories

5) SmartWell 系统附属设备

Halliburton offers a full complement of auxiliary components, from hydraulic disconnects, to control line clamps and protectors, to flatpack and connectors.

哈里伯顿公司提供从水力断开装置,到控制管线线夹和控制管线保护器,再到封装设备和连接器等各种作业所需的全套辅助设备。

(1) CLCS. The Halliburton Control Line Cut Sub (CLCS) incorporates reliable and TEC Cutter field-proven methods for effective removal of the tubing string and control lines in contingency situations.

(1) 控制管线切削短节。哈里伯顿公司的控制管线切削短节(CLCS)与可靠且经过现场试验证明的 TEC 切削齿相结合,可以在紧急情况下有效切割油管柱和控制线。

(2) Flatpack. Flatpack is the physical link between surface control systems and downhole equipment, and is available in a variety of different configurations to suit the application and well environment.

(2) 封装设备。封装设备是地面控制系统和井下设备之间的物理连接,可应用于各种不同的组合配置,以适应现场应用及井筒环境。

(3) FMJ connector. FMJ Connector (Fig. 5. 11) is a high-performance, fully-testable, triple ferrule metal-to-metal seal for use with ¼ in hydraulic control line.

(3) FMJ 连接器。FMJ 连接器(图 5. 11)是一种高性能、经充分测试、三套圈金属对金属密封形式的连接器,适用于¼in 液压控制管线。

Fig. 5. 11 FMJ Connector

(Source: http://www.halliburton.com/en-US/ps/well-dynamics/well-completions/intelligent-completions/smartwell-system-accessories.page? node-id = hfqel9vw&nav = en-US_completions_public)

(4) SmartWell electro-hydraulic disconnect tool and hydraulic disconnect tool. SmartWell Electro-Hydraulic Disconnect tool and hydraulic disconnect tool facilitate the removal of the upper completion from the lower completion without any destructive or mechanical intervention, leaving the intelligent completion lower assemblies such as packers and ICVs in place.

(4) SmartWell 电动—液压断开装置和水力断开装置。SmartWell 电动—液压断开装置和水力断开装置有助于在不进行任何破坏性干预或机械干预的情况下将上部完井(生产完井)设备从下部完井(油藏完井)设备中移除,从而使封隔器和层间控制阀等智能完井系统中的下部组件处于适当的位置。

(5) Splice sub connector. The Splice Sub Connector tool secures the flatpack to the tubing and is a reliable and field-proven method of splicing two ends of flatpack, thereby facilitating repairs or lengthening the flatpack.

(5) 铰接短节连接器。铰接短节连接器可将封装设备固定在油管上,是一种经过现场试验证明的可靠铰接封装设备两端的装置,有助于对封装设备进行维修或延长其使用寿命。

6) Permanent Monitoring

Encompassing the latest in cutting-edge downhole pressure, temperature, flow, and density sensing technology, the DataSphere permanent monitoring suite seamlessly integrates Halliburton Intelligent Completions′ set of permanent monitoring tools. The DataSphere architecture is designed for versatility and modularity, providing customized solutions with superior performance and enhanced

6)永久性监测

DataSphere 永久性监测系统包含最新的井下压力、温度、流量和密度传感技术,能够很好地与哈里伯顿公司智能完井系统中一系列的永久性监测工具相结合。DataSphere 永久性监测系统的体系结构旨在实现功

reservoir monitoring capabilities.

能性和模块化，提供增强油藏监测水平能力的卓越定制解决方案。

Through the use of the some of the best and most reliable downhole gauge systems, array systems, and wireless through-wellbore systems, the DataSphere permanent monitoring suite helps operators meet industry challenges associated with increased reservoir contact. Real time downhole data is conveyed through our most advanced surface systems, capable of remotely transmitting crucial wellbore information to operators. Ultimately, the DataSphere permanent monitoring suite allows operators to obtain long term wellbore data and make informed decisions that can increase hydrocarbon recovery for the life of the well.

DataSphere 永久性监测系统通过使用一些最好的、最可靠的井下测量系统、阵列系统和无线传输系统来帮助操作人员应对增加的与储层有关的行业挑战。利用最先进的地面系统传输井下采集到的实时数据，能够将关键的井筒信息远程传输给操作人员，DataSphere 永久性监测系统使操作人员能够获取长期井筒数据并做出正确的决定，从而在油井的开采寿命内提高油气采收率。

(1) DataSphere array system. The DataSphere Array system is a reliable, permanent, multi-point reservoir monitoring system that provides distributed pressure and temperature sensing in single-zone and multi-zone land and subsea applications.

(1) DataSphere 阵列系统。DataSphere 阵列系统是可靠、永久性的多点储层监测系统，可以在单产层、多产层陆地及海底应用中提供分布式压力、温度传感。

(2) DataSphere FloStream venturi flow meter. The FloStream flowmeter is designed specifically to meet the requirements of the well in relation to material selection and expected flow rates. FloSteam flowmeter system operation is based on the Venturi principle.

(2) DataSphere FloStream 文丘里流量计。DataSphere FloStream 文丘里流量计是为满足油井在材料选择和预期流量上的要求而专门设计的。FloSteam 文丘里流量计系统基于文丘里原理进行工作。

(3) DataSphere LinX monitoring systems. The LinX monitoring system provides real-time, well integrity monitoring in subsea wells without the need to halt production.

(3) DataSphere LinX 监测系统。DataSphere LinX 监测系统可在不需要停止生产的情况下实时监测海底油井的完整性。

(4) DataSphere ROC permanent downhole gauges. ROC permanent downhole gauges provide reliable, real-time permanent monitoring of downhole conditions, thereby increasing productivity through the life of the well or reservoir.

(4) DataSphere ROC 永久性井下测量仪器。DataSphere ROC 永久性井下测量仪器能够对井下情况进行可靠、实时的永久性监测，从而提高油井或油藏开采寿命内的油气生产能力。

(5) DataSphere SmartLog permanent downhole gauge. The DataSphere SmartLog downhole gauge system provides reliable and economical downhole pressure, temperature and vibration measurements for optimized reservoir and production management.

(5) DataSphere SmartLog 永久性井下测量仪器。DataSphere SmartLog 永久性井下测量系统可以经济可靠地测量井下压力、温度及振动，从而优化油藏（经营）管理和生产（运作）管理。

7) Surface Control and Monitoring Systems

7)地面控制和监测系统

Halliburton digital infrastructure surface control and monitoring system enables operators to monitor permanent downhole gauges (PDGs) and control downhole interval control valves (ICVs) from surface. Consisting of both electrical and hydraulic systems, Digital Infrastructure also enables operators to remotely configure ICVs, expand deployed systems, and interpret and model data acquired by the system.

哈里伯顿公司的数字化基础设施——地面控制和监测系统使操作人员在地面就能够监测永久性井下测量仪器(PDG)并且控制层间(流动)控制阀(ICV)。操作人员还可以由电路系统和液压系统组成的数字化基础设施进行远程操控 ICV，扩展已部署的系统，对系统采集的数据进行分析和建模。

The digital infrastructure system consists of the SmartWell Master supervisory application, as well as surface hydraulic systems, standalone monitoring systems, subsea interface cards, and components provided by third-parties.

数字化基础设施系统由 SmartWell Master 监控应用(系统)、地面液压系统、独立监测系统、水下接口卡及第三方提供的组件组成。

(1) Land and platform control systems. The surface hydraulic system (SHS) supplies pressurized hydraulic fluid to downhole interval control valves (ICVs) located in the well. It also retrieves pressure and temperature data from permanent downhole gauges (PDGs).

(1)陆地和海上平台控制系统。地面液压系统(SHS)向处于井筒内的层间(流动)控制阀(ICV)提供加压流体。该系统也可从永久性井下测量仪器(PDG)中获取压力和温度数据。

(2) Portable control systems and ancillary equipment. The digital infrastructure system includes portable manual hydraulic units, portable data acquisition units, SHS simulators, PDG simulators and downhole cable simulators.

(2)便携式控制系统和辅助设备，包括便携式手动液压单元、便携式数据采集单元、SHS 模拟器、PDG 模拟器和井下电缆模拟器。

(3) SCADA and software applications. The SmartWell Master software application is the supervisory application for Digital Infrastructure system. Designed to provide a central point of control, the SmartWell Master application integrates field control system peer connectivity with the engineer's computer control and data acquisition activities.

(3) SCADA 和软件应用。SmartWell Master 软件应用系统是数字化基础设施系统的监控应用系统,旨在提供控制中心,将油田控制系统对等连接与工程师的计算机控制和数据采集活动相结合。

(4) Standalone Permanent Monitoring Systems. The XPIO 2000 data acquisition and control system allows operators to monitor and control downhole gauges and topside instrumentation.

(4)独立永久性监测系统。XPIO 2000 数据采集和控制系统使操作人员能够监测和控制井下仪表仪器。

(5) Subsea control and monitoring systems. Halliburton offers a range of subsea interface cards that enable operators to monitor PDGs and to control Halliburton's Surface Controlled Reservoir Analysis and Management System (SCRAMS) completions.

(5)水下控制和监测系统。哈里伯顿公司提供一系列的水下接口卡,这些水下接口卡使操作人员能够监测永久性井下测量仪器和控制油藏分析及管理系统(SCRAMS)。

8) Remote Open Close Technology

8)远程(遥控)打开或关闭技术

Halliburton's range of Remote Open Close Technology (ROCT) products are designed to simplify well operations by removing wireline runs-making the operation more efficient. The field-proven technology replaces traditional wireline plug and prong equipment, dramatically reducing the number of, or in some cases eliminating all wireline runs from an operation. As a consequence, the technology helps reduce risk and save time by reducing the need to rig-up wireline and all the associated pressure control equipment.

哈里伯顿公司的远程(遥控)开启或关闭技术(ROCT)系列产品旨在通过取代钢缆作业模式来简化井下作业使作业效率更高。经过现场试验证明的远程(遥控)开启或关闭技术取代了传统的钢缆作业,大大减少了完成一次井下作业时进行电缆作业的次数,在某些情况下甚至能够免除电缆作业。因此,远程(遥控)开启或关闭技术能够减少对装配钢缆及所有相关压力控制设备的需要,有助于降低风险并且缩短作业工期。

(1) eMotion remotely operated downhole control unit. A computer-controlled, downhole hydraulic power unit. It is permanently deployed as part of the tubing and is used to remotely open and close a slave valve such as a sliding sleeve or ball valve.

(1) eMotion 远程遥控井下控制单元。eMotion 远程遥控井下控制单元是一种由计算机终端控制的井下液压动力单元,可作为油管的一部分永久性部署在油管上,通过计算机终端远程遥控打开关闭滑套或球阀等液压自控换向阀。

(2) eMotion-HS remotely operated circulating valve. Provides high circulation rates without the need for any interventions during completion deployment-saving time and helping to reduce risk.

(2) eMotion-HS 远程遥控循环阀。eMotion-HS 远程遥控循环阀在完井作业期间能够在无需开展任何井筒干预作业的情况下提供较高的循环速率，缩短了施工工期并且降低了作业风险。

(3) eMotion-LV remotely operated isolation barrier valve. Helps save time, money and reduce risk by eliminating all wireline runs from completion placement operations.

(3) eMotion-LV 远程遥控隔离阀。eMotion-LV 远程遥控隔离阀够免除在完井作业过程中进行的钢缆作业，有助于缩短作业工期、减少资金投入及降低作业风险。

(4) eRED ball valve. A computer-controlled ball valve that can be repeatedly opened and closed by remote command. It is deployed below either a lock or bridge plug and can be used as a downhole barrier or flow control device.

(4) eRED 球阀。eRED 球阀是一种由计算机终端控制的球阀，可以通过远程指令反复开启和关闭。eRED 球阀安装在桥塞等装置下方，可用作井下隔离装置或流量控制装置。

(5) Evo – RED bridge plug. Evo – RED bridge plug helps save time and reduce risk by eliminating interventions during well operations.

(5) Evo – RED 桥塞。Evo – RED桥塞能够免除在井下作业期间进行井筒干预作业，有助于缩短作业工期以及降低作业风险。

5.4.2.3 Schlumberger

5.4.2.3 斯伦贝谢公司

1. Overview

1. 概述

Intelligent completions incorporate permanent downhole sensors and surface-controlled downhole flow control valves, enabling you to monitor, evaluate, and actively manage production (or injection) in real time without any well interventions. Data is transmitted to surface for local or remote monitoring.

斯伦贝谢公司的智能完井（系统）包括永久性井下传感器和地面控制的井下流量控制阀，能够在无需进行任何井筒干预作业的情况下实时监测、评估、主动管理井下流体生产（或注入），采集到的数据可传输到地面以便于进行本地或远程监测。

These completions:

(1) Provide real-time zonal downhole monitoring of pressures and temperatures;

(2) Enable surface-controlled production from each zone or lateral to optimize production and reservoir management;

(3) Reduce production of undesirable water or gas;

(4) Increase recovery and extend the economic life of the well.

Initially used in subsea wells, where intervention is expensive and high-risk, intelligent completions have since proven their value in managing production from multilateral wells, horizontal wells with multiple zones, wells in heterogeneous reservoirs, and mature reservoirs.

这样的完井作业能够:

(1)井下层段压力和温度数据的实时监测;

(2)能够从地面对各个层段或各个分支井眼的油、气生产进行控制以优化生产及油藏管理;

(3)减少不需要的水或气体的生产;

(4)提高采收率,延长油井经济寿命。

最初应用于高作业成本、高风险的海上油井的智能完井系统现在已经在管理多分支井、多产层水平井、非均质油藏井及其油藏井的生产方面体现了其价值。

2. Well Type/Service Offering

1) Production and Reservoir Management System Manara

Manara production and reservoir management system provides downhole permanent monitoring and in-lateral flow control of multiple zones in real time—for the first time, even in multilateral wells.

(1) Optimize production and ultimate recovery, from single wellbores to multilaterals.

Unlimited zones and compartments, full production information, and infinite control in each enables unprecedented production and reservoir management in heterogeneous (e. g., carbonate) or multilayered reservoirs, extended-reach developments, and extreme reservoir contact (ERC) wells.

(2) Monitor and control more zones on a single control line.

2. 产品与服务

1)生产和油藏管理系统

Manara生产和油藏管理系统能够实时进行多个选择性层段的井下(情况)永久性监测和分支井眼流量控制流动控制。

(1)从单个井筒到多分支井筒,优化生产和提高采收率。

无数的选择性层段和层间封隔层段、完整的生产信息及对各个层段的无限控制有助于在非均质(例如碳酸盐岩)油藏或多层油藏井、大位移开发井和极大储层接触井中实现前所未有的生产(运作)管理及油藏(经营)管理。

(2)在单个控制管线上监控和控制更多选择性层段。

The revolutionary Manara system provides in-situ measurements of pressure, temperature, flow rate, and water cut across the formation face in each zone of each lateral. All sensors are packaged in one compact station, together with an electric flow control valve (FCV) that has infinitely variable settings controlled from surface through a single electrical control line. Use multiple stations to maximize hydrocarbon sweep and recovery with fewer wells, reducing capex, opex, and surface footprint.

Manara 生产和油藏管理系统可以对每个分支井眼的不同目的层的井底地层的压力、温度、流速和含水量进行现场测量。所有传感器和一个可通过单个电力控制管线由地面控制的电动流量控制阀都封装在一个紧凑型组件中。使用多功能的组件可以用更少的井数来最大限度地提高油气的波及程度和采收率,同时减少资本支出、作业支出和井场占地面积。

(3) Enhance drainage without intervention.

(3)不进行井筒干预的情况下增强泄油。

Data reach the surface almost instantly, where wellbore and reservoir surveillance software facilitates understanding of flow behavior in real time and effective use and analysis of the data through clear displays and FCV setting recommendations. Data workflows embedded in the software reduce the amount of data gathering and manipulation needed, freeing up production and reservoir engineers to manage the asset and focus on priority wells. Decisions that used to take days or weeks can now be made in hours.

采集到的数据快速地传输到地面,操作人员通过井筒和油藏监测软件及流量控制阀实时了解井下各油层油气水的流动状况,有效利用和分析采集到的数据。嵌在软件中的工作流程减少了需要采集和操作的数据量,使生产工程师和油藏工程师能够自由地进行资产管理并且专注于需要优先考虑的油井。过去需要数天或数周才能做出的决定现在只需几个小时。

2) IntelliZone Compact Modular Multizonal Management System IntelliZone

2)紧凑型模块化多产层管理系统

IntelliZone Compact modular multizonal management system is the first fully integrated intelligent flow control system for multizone wells. It provides an efficient, reliable, and cost-effective means of controlling wells on land and offshore.

紧凑型模块化多产层管理系统是第一个用于多产层井的全集成智能流量控制系统,它提供了一种有效、可靠且具有成本效益的对陆地和海上油井进行控制的方法。

(1) Flow control technology for multizone wells.

IntelliZone Compact modular multizonal management system is the first fully integrated flow control technology for multizone wells. It brings together as one compact unit an advanced design and production modeling engine, a fully integrated completion module, and a user-friendly remote operating system.

(2) Preassembly and pretesting minimize installation rig time.

The system is assembled and tested at manufacturing and delivered as a single, integrated unit ready to install out of the box, minimizing field connections and saving rig time compared with conventional intelligent completions.

(3) Multidrop capability reduces deployment complexity.

The system's integrated design helps reduce the number of hydraulic lines required without reducing the number of valves the operator can control downhole. The multidrop capability simplifies the wellhead interface and eliminates the requirement for custom wellheads.

(4) Two-click computer control simplifies valve control.

With control logic programmed into the surface control system, valve operation is a simple two-click computer command. Unlike sliding sleeves, no rig-based intervention is required to modify valve position.

3) Permanent Monitoring

Remotely monitor wells and reservoirs in real time.

(1)多产层井流量控制技术。

紧凑型模块化多产层管理系统是第一个用于多产层井的全集成智能流量控制系统。它将一个紧凑型单元集成为一个先进的设计和生产建模引擎、一个全集成的完井模块、一个用户友好型远程遥控系统。

(2)预组装和预(性能)测试可最大限度地缩短(安装过程中的)钻机作业时间。

该系统是在生产制造过程中进行组装和(性能)测试的,作为一种随时都可以安装在井底的一体化、集成化单元,与传统的智能完井(系统)相比,能够最大限度地减少现场安装时间并且节省钻机作业时间。

(3)多节点能力降低了系统安装的复杂程度。

该系统的集成化设计有助于在不减少井下控制阀门数量的情况下减少所需的液压(控制)管线数量。多节点能力简化了井口设备,并且免除了对定制井口设备的需要。

(4)双击计算机控制简化了阀门控制。

利用地面控制系统的控制逻辑程序,可将阀门操作简化为双击计算机发送指令。与滑套不同的是,它不需要基于钻机的干预来调整阀门的位置。

3)永久性监测

可对油气井和油气藏进行实时远程监测。

Schlumberger is the industry-recognized permanent monitoring leader, having deployed more than 12,000 permanent downhole pressure and temperature gauges over the past four decades.

斯伦贝谢公司是业内公认的永久性监测领域领导者,在过去的四十年里已经部署了12000多个永久性井下压力、温度测量仪器。

WellWatcher permanent monitoring systems integrate the most advanced permanent downhole measurement technology with surface acquisition and data communication systems to allow remote monitoring of wells and reservoirs in real time. Pressure, temperature, density, and flow rate data are transmitted to remote locations via satellite, the Internet, or cable.

WellWatcher 永久性监测系统将最先进的永久性井下测量技术与地面数据采集系统和数据通信系统相结合,能够对油气井和油气藏进行实时远程监测。井下压力、温度、密度和流速等数据可以通过卫星、互联网或电缆远程传送。

The data provide long-term reservoir and production monitoring without the cost of well interventions, helping you optimize well productivity and hydrocarbon recovery by identifying trends throughout the producing life of your well or field.

这些数据能够在不需要进行成本高昂的井筒干预作业的情况下提供长期的油藏监测和生产监测,有助于通过确定油井或油田在油田开采寿命内的生产趋势来优化油井生产、提高油气采收率。

4) Downhole Interval Control Valves

4)井下层间(流动)控制阀

Interval control valves are used in multizone intelligent completions, commingled-flow completions, auto (natural) gas-lift wells, and well environments with scale deposition, severe erosion, or high temperatures.

层间(流动)控制阀用于多产层智能井、混合流动井、自动(自然)气举井,以及具有水垢沉积、严重侵蚀或高温等环境中的油井。

Schlumberger installed the first interval control valve (ICV) in May 2000 for Norsk Hydro offshore Norway. Since then, we have successfully installed more than 1,400 flow control valves in 24 countries in a range of different environments, including subsea.

斯伦贝谢公司于2000年5月为挪威近海的Norsk Hydro公司安装了第一个层间(流动)控制阀。从那以后,斯伦贝谢公司已经成功地在24个国家安装了1400多个包括海底在内的各种不同环境中的层间(流动)控制阀。

Schlumberger flow control valves are widely used around the globe.

斯伦贝谢流量控制阀在世界范围内被广泛应用。

Interval or flow control valves can be operated automatically, manually, or remotely as part of an intelligent completion. Used to control multiple zones selectively, they reduce water cut and gas cut, minimize well interventions, and maximize well productivity.

作为智能完井(系统)的一部分,层间(流动)控制阀(或流量控制阀)可以自动调节、手动调节或远程遥控。层间(流动)控制阀(或流量控制阀)用于选择性地控制多个产层,能够降低含水率和含气率,最大限度地减少井筒干预,并且最大限度地提高油井产量。

Schlumberger provides on/off and multiposition, annular and inline flow control valves for both producer and injector applications. Valves are available in a range of sizes (2⅞ in, 3½ in, 4½ in, and 5½ in) and metallurgies to suit your reservoir parameters.

斯伦贝谢公司为生产井和注水井的应用提供开关和多位置环形直通式流量控制阀。阀门有各种尺寸(2⅞in、3½in、4½in和5½in)和材质以满足不同的油藏参数。

5) Multitrip Connectors

5)多行程连接器

Essential to multitrip intelligent completion installations, these connectors enable retrieval of the upper completion without disturbing the lower completion and its zonal isolation packers, flow control valves (FCVs), chemical injection mandrels, isolation valves, and monitoring sensors.

多行程连接器对于多行程智能完井装置至关重要,可以在不影响下部完井(油藏完井)设备及产层封隔器、流量控制阀(FCV)、化学试剂注入工作筒、隔离阀、监测传感器正常工作的情况下回收上部完井(生产完井)设备。

(1) Deploy intelligent completions in the lower completion.

(1)在下部完井(油藏完井)设备中安装智能完井组件。

Schlumberger multitrip connectors make it possible to install intelligent completion assemblies in the lower completion by providing communication with the upper completion. In addition, they minimize the risks and costs associated with workovers in these wells. Remedial operations (e.g., replacement of an ESP) can be conducted in the upper completion without the need to retrieve the lower completion.

斯伦贝谢公司的多行程连接器可以连通上部完井(生产完井)设备和下部完井(油藏完井)设备,从而在下部完井(油藏完井)设备中安装智能完井组件。此外,多行程连接器还可以最大限度地降低与井中进行的修井作业有关的风险和成本。修井作业(例如更换电潜泵)可以在上部完井(生产完井)设备中进行而不需要回收下部完井(油藏完井)设备。

(2) Disconnect and reconnect upper completion multiple times.

(2) 多次断开和重新连接上部完井(生产完井)设备。

The concentric design of these connectors enables multiple connect and disconnect operations during the life of a well without the need for any alignment downhole. This key feature makes the connectors reliable and easy to use in highly deviated to horizontal, multizone, multilateral, and extended-reach (ERD) wells.

这些连接器的同心同轴设计使其能够在无需进行任何井下校准的情况下在油井的开采寿命内多次连接和断开。这一特性使其在大斜度井、水平井、多产层井、多分支井和大位移井中变得可靠可用。

(3) Handle completion design changes simply.

(3) 简单地应对完井设计变更。

AGILITI modular digital completions give you maximum flexibility in intelligent completions, which simplifies well design and planning to accommodate unexpected downhole conditions. Predesigned modules and downhole wet-mate electric and electrohydraulic connectors simplify and expedite design changes that arise from reviewing your drilling data.

AGILITI 模块化数字化完井系统最大限度地提升了智能完井系统的灵活性,这能够简化油井的设计和规划以适应难以预计的井下情况。预先设计的模块及井下插拔电连接器、电动液压连接器简化并推进了因复查钻井数据而引起的设计更改。

6) Packers for Intelligent Completions

6) 智能完井用封隔器

(1) Multiport packers that allow hydraulic and electric communication with devices below.

(1) 能够与下部设备进行水力连通和电力连通的封隔器。

Multiport packers isolate fluids and pressure in wells with intelligent completions that rely on hydraulic and electric lines for control and data transmission. They allow passage of electric and hydraulic conduits for fiber-optic cables, tubing-mounted reservoir monitoring equipment, subsurface flow-control valves, and other equipment that requires connection to the surface. The packers enhance safety, protect against formation damage, reduce the need for costly workovers, and protect reservoir integrity and productivity.

多端口封隔器在应用依靠液压管线和电力管线进行控制和数据传输的智能完井系统的井中隔离流体和压力。光缆、装在油管上的油藏监测设备、地下流量控制阀及其他需要连接到地面的设备的电力管线和液压管线可以通过多端口封隔器。多端口封隔器可以提高安全性,减小地层伤害,减少进行成本高昂的修井作业的需要,并且保护油藏完整性和产能。

(2) Modular multiport packers.

The MRP – MP packer series comprises retrievable, tubing-conveyed, hydraulically set production and isolation packers with differential pressure ratings up to 5,000 psi. Multiple configuration options—accommodate a range of applications. Up to five ports are available to enable communication with devices installed below the packer.

(2)模块化多端口封隔器。

MRP – MP 系列封隔器包括可回收油管输送液压式生产封隔器和隔离封隔器,其额定压差高达 5000psi。MRP – MP 系列封隔器包括多种配置选项,分别适用于各种应用需要。最多有五个端口可用于与安装在封隔器下部的设备进行连通。

(3) High-pressure multiport packer.

XMP premium multiport production packers are tubing-conveyed, hydraulically set, retrievable packers designed and tested to ISO 14310 V0. Differential pressure ratings up to 10,000 psi are available.

(3)高压多端口封隔器。

XMP 优质多端口生产封隔器是以 ISO 14310 标准中的 V0 等级为标准设计和测试的,是油管输送液压式可回收封隔器,其压差额定值高达 10000psi。

(4) Ultrahigh-pressure multiport packer.

BluePack Ultra RH – MP ultrahigh-pressure retrievable hydraulic-set, multiport production packers are critical elements of deepset and deepwater wells. They are ISO 14310 V0 validated to 15,000 psi and 300 degF. Up to seven feedthroughs increase lower-completion flexibility.

(4)超高压多端口封隔器。

BluePack Ultra RH – MP 超高压可回收液压式多端口生产封隔器是深井和深海油井的关键部件,经 ISO 14310 标准 V0 等级验证,其压差额定值为 15000psi,温度额定值为 300degF,多达七个端口增加了下部完井(油藏完井)设备的灵活性。

(5) Gravel-pack multiport packer.

The QUANTUM MultiPort gravel-pack multiport packer is a hydraulically set, retrievable production packer used as the upper packer in stacked configurations for gravel-packed intelligent completions. Designed and tested in accordance with ISO 14310 V3, these packers are available for a variety of service applications, including H_2S and CO_2 environments.

(5)砾石充填多端口封隔器。

QUANTUM MultiPort 砾石充填多端口封隔器是一种液压式可回收的生产封隔器,在砾石充填智能完井系统中用作上部封隔器。QUANTUM MultiPort 封隔器以 ISO 14310 标准 V3 等级为依据进行设计和测试,可在包括含 H_2S 和 CO_2 环境在内的各种应用环境中使用。

5.5 Future Development of Intelligent Well

5.5 智能完井技术的未来发展

Nowadays, economic scenario imposes a new challenge for cost reduction. Well designs and technologies to achieve another level of efficiency are based on the following drivers: complexity, flexibility, HSE, CAPEX/OPEX and availability. Fig. 5. 12 illustrates the drivers for new intelligent well completion technologies.

Audio 5.5

如今,经济形势对降低成本提出了新的挑战。实现更高效率水平的井筒设计和智能完井技术将基于以下驱动因素:复杂性、灵活性、HSE 管理体系、资本支出(作业支出)和可用性。图 5. 12 展示了驱使新型智能完井技术诞生的动力源。

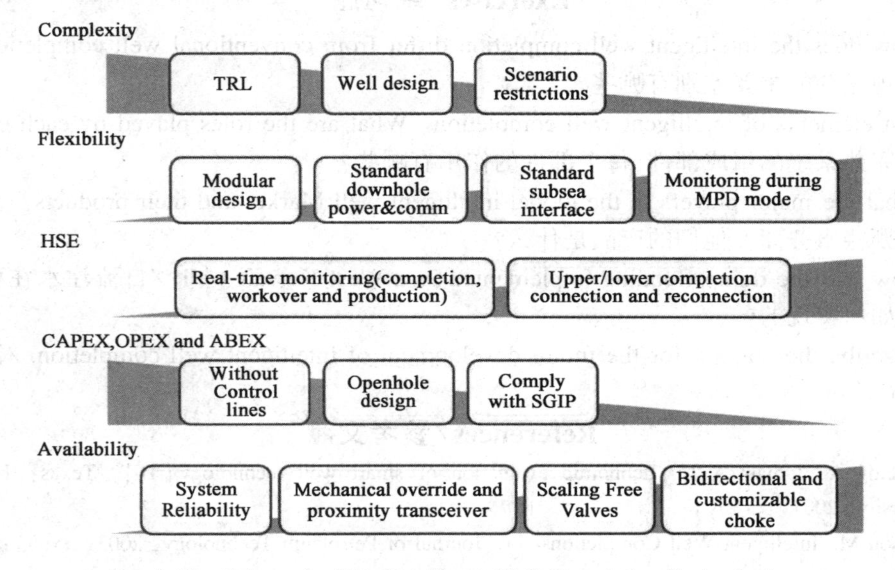

Fig. 5. 12 Drivers for New Intelligent Well Completion Technologies

The oil and gas industry is climbing the learning curve rapidly toward recognizing the ultimate potential of intelligent-well technology—to contribute to efficiency and productivity. Beyond the attraction of interventionless completions in the high-cost arena of subsea and deepwater wells, intelligent-well technology has already demonstrated the ability to deliver improved hydrocarbon production and increased recovery with fewer wells. Intelligent-well technology can improve the efficiency of waterfloods and gasfloods in heterogeneous or multilayered reservoirs when applied to injection wells, production wells, or both. Production and reservoir data acquired with downhole sensors can improve the understanding of reservoir behavior and assist in the appropriate selection of infill-drilling locations

石油和天然气行业正在向认识到智能完井技术的最终潜力——提高油田开发效率和油井产能迅速发展。智能完井技术除了因能在高成本的海底油井和深海油井中免除井筒干预作业而颇具吸引力之外,现已证明智能完井技术还能够用更少的井数来改善油气生产(效果)、提高采收率。当应用于注水井和(或)生产井时,智能完井技术可以在开发非均质油藏或多层油藏时提高水驱效率和气驱效率。利用井下传感器采

and well designs. Intelligent-well technology can leverage investment and enable a single well to do the job of several wells, whether through controlled commingling of zones, monitoring and control of multilaterals, or even allowing the well to take on multiple simultaneous functions (i. e., injection, observation, and production).

集到的生产数据和油藏数据可以提高对油藏开采动态的了解,并且对加密井井位合理选择及油井设计提供帮助。智能完井技术可以降低成本并且使一口井起到多口井的作用,既可以控制多层油藏和多分支井监控,又可以使油井同时实现如注水、检测及生产等多种功能。

Exercises/练习题

1. How does the intelligent well completion differ from conventional well completion?/智能完井与常规完井的主要差别有哪些?

2. List elements of intelligent well completion. What are the roles played by each element?/列出智能完井系统的组成部件,每个部件的作用有哪些?

3. What are major players in the global intelligent well Market and their products?/全球智能井市场有哪些服务商?他们的产品是什么?

4. How was the dump flooding implemented in a West Kuwait well?/自流注水在科威特的那口井是如何实现的?

5. Describe the outlook for the future development of intelligent well completion. /简述智能井发展方向。

References/参考文献

[1] Abdullatif A, Omair A L. Economic evaluation of smart well technology[D]. Texas: Texas A&M University, 2007.

[2] Robinson M. Intelligent Well Completions[J]. Journal of Petroleum Technology, 2003, 55(8):57-59.

[3] Opeoluwa O. Nanocompletions versus Intelligent Well Completions, Investigating the Future of Nanotechnology in Well Completions – A Case Study[C]. SPE Nigeria Annual International Conference and Exhibition. Society of Petroleum Engineers, 2014.

[4] 王广宇. Near-Miss 定义及管理流程分析[J]. 中国安全生产科学技术, 2012, 08(6):197-200.

[5] Rawding J, Al-Matar B S, Konopczynski M R. Application of Intelligent Well Completion for Controlled Dumpflood in West Kuwait[C]. Intelligent Energy Conference & Exhibition. Society of Petroleum Engineers, 2008.

[6] Denney, Dennis. Intelligent-Well Completions in Agbami: Value Added and Execution Performance[J]. Journal of Petroleum Technology, 2010, 62(5):49-51.

[7] Glandt C A. Reservoir Management Employing Smart Wells: A Review[J]. Spe Drilling & Completion, 2005, 20(4):281-288.

[8] Gao C H, Rajeswaran R T, Nakagawa E Y. A literature review on smart well technology[C]. Production and Operations Symposium. Society of Petroleum Engineers, 2007.

[9] 杨莹娜. 智能井井下层间流体控制系统研究[D]. 西安:西安石油大学, 2013.

[10] 佘舟. 智能井完井管柱优化设计[D]. 西安:西安石油大学, 2017.

[11] 王洪英. 大庆深层水平井井眼轨迹设计与控制技术研究[D]. 大庆:东北石油大学, 2010.

[12] MikeRobson, 王新英, 赵炜, 等. 智能完井技术[J]. 石油石化节能, 2004, 20(2):29-31.

[13] Da Silva M F, Jacinto C C, Hernalsteens C, et al. Cableless Intelligent Well Completion Development Based on Reliability[C]. OTC Brasil. Offshore Technology Conference, 2017.

Unconventional Oil and Gas Production

6 非常规油气开采技术

"Unconventional" oil and gas is not chemically different from "conventional" oil and gas. The distinction stems from their position underground or from the unusual nature of their reservoirs. These conditions require the use of new, often complex, extraction methods. This chapter will start from the definition of unconventional oil and gas, introduce the geological characteristics and reservoir properties of unconventional oil and gas, and the technical means of developing unconventional oil and gas resources.

Audio 6.1

"非常规"油气与"常规"油气在化学上没有区别,区别在于它们在地下的位置或其储层特殊的性质。非常规油气要使用新的、通常是比较复杂的开采方法。本章从非常规油气的定义出发,介绍非常规油气的地质特征、储层性质,以及开发非常规油气资源的技术手段。

6.1 Overview

Conventional oil and gas is a category that includes crude oil and natural gas and its condensates. Unconventional oil consists of a wider variety of liquid sources including oil sands, extra heavy oil, Gas-to-Liquids and other liquids. In general conventional oil is easier and cheaper to produce than unconventional oil.

6.1 概述

常规油气资源是包括原油、天然气及其凝析油在内的一类油气资源。非常规油气资源包括油砂、超重油、天然气合成油及其他液态油在内的更广泛的液体油气来源。一般来说,常规油气资源比非常规油气资源更容易开发,开发成本也更低。

6.1.1 Definition

Unconventional oil and gas refers to the oil and gas resources which cannot be explored, developed and produced by conventional processes just in using the natural pressure of the wells and pumping or compression operations.

6.1.1 定义

非常规油气资源是指仅利用油井的自然压力、泵送或压缩作业这些常规工艺技术不能有效勘探开发的石油和天然气资源。

Until the 1990s the oil and gas prices were at a level leaving no space to develop technologies to explore and produce unconventional oil and gas resources. In the years 2000s, the oil and gas prices started to increase slightly and long term perspectives wee indicating that they would not come down soon. In the same time large consuming countries of fossil energy like USA, Canada, China started to realize to hold massive quantities of unconventional oil and gas. This conjunction justified research and development efforts for the exploration and production of unconventional oil and gas.

20 世纪 90 年代之前，石油和天然气价格较低，没有任何空间来发展勘探、开发非常规油气资源的技术。21 世纪初，石油和天然气价格开始小幅上涨，长期前景显示油价不会在短期内回落。与此同时，美国、加拿大、中国等化石能源消费大国开始意识到自己国内蕴藏着大量非常规油气资源。这样的机遇使得世界对非常规油气资源的勘探开发开始制定合理的研究和开发计划。

6.1.1.1 Unconventional Oil

Unconventional oil refers to hydrocarbons that are obtained through techniques other than traditional vertical well extraction. Unconventional oil may simply refer to tight oil where the sediments holding the oil must be hydraulically fractured to free the oil. The oil extracted in this case is no different than the oil from a reservoir that is tapped by a vertical well. Not only do these hydrocarbons require special extraction, but they will need additional processing and refining to extract traditional petroleum products from them. Sources of this type of unconventional oil include oil sands, oil shale, shale oil, tight oil, and heavy and extra-heavy oil.

6.1.1.1 非常规石油

非常规石油是指通过使用常规直井开采技术以外的技术获得的烃类资源。非常规石油可以简单地指存在于沉积物内的致密油，且必须采用水力压裂的方法才能采出。在这种情况下得到的石油与用直井开发的油藏中的石油没有什么不同。这些烃类资源不仅需要采取特殊的方法进行开采，而且需要额外的加工和精炼过程来提取传统的石油产品。这类非常规石油包括油砂、油页岩、页岩油、致密油、重质油和超重油。

Historically, the exploration and production of oil and natural gas focused on the sources that were easiest to access. Conventional oil wells are vertical shafts into pools of oil and gas that are under pressure, making them easy to bring to the surface. Unconventional oil sources of oil do not flow near the surface and sometimes do not flow at all in that they are in a solid or near solid state.

从历史上看，石油和天然气的勘探开发侧重于最容易获得的资源。常规油井垂直进入承压的油气藏，使油气资源易于到达地面。而一些非常规石油资源不会在地表附近流动，有时根本就不流动，因为其一般处于固体状态或接近于固体的状态。

Unconventional oil sources remained relatively untapped compared to conventional sources while conventional sources were plentiful. This is simply due to the technical requirements and costs associated with production making an unconventional oil reservoir less profitable than a comparably sized conventional one. Over time, however, most of the conventional sources were already tapped and producing, drawing down the conventional oil reserves around the world. So the desire to keep growing production turned to the unconventional sources. In the meantime, advances in technology, particularly steam assisted gravity drainage, horizontal drilling and hydraulic fracturing, made unconventional oil sources more accessible and reduced the cost of extraction while increasing extraction efficiency.

与常规石油资源相比，当常规资源比较丰富时，非常规石油资源处于相对尚未开发的状态。这主要是因为技术规范及与生产相关的成本使得开发非常规油藏比开发同等规模的常规油藏利润较低。但随着时间的推移，大多数常规石油资源已经被开采，全球范围内的常规石油储量降低。因此，石油产量持续增长的希望就寄托在了非常规石油资源上。与此同时，随着技术的进步，特别是蒸汽辅助重力泄油技术、水平井钻井技术和水力压裂技术的发展，非常规石油资源变得更容易获得，开采成本降低，开采效率提高。

As conventional sources of oil are exhausted, unconventional sources are making up a larger share of fossil fuel production. Moreover, unconventional oil production methods are being used on conventional wells to increase production or restart production on wells that were previously deemed depleted due to the extraction technology constraints of the time.

随着常规石油资源的枯竭，非常规石油资源在化石燃料的产量中所占比重不断提高。此外，正在使用非常规石油的生产方法来提高常规油井产量，或者由于受限于当时的开采技术而在之前被认为已经枯竭的油井也重新开始生产。

As mentioned, tight oil and shale oil is basically conventional oil that is hard to get out. Oil sands and heavy and extra-heavy oil, however, are highly viscous deposits of oil that have degraded overtime due to mixing with other materials. Shale oil is a mixture of rock and organic matter that is in the process of turning into an oil reservoir, but requires heating to finish the process. Depending on the depth of these heavier deposits, they can be mined or extracted using in situ processing to heat up the material to separate out the oil prior to extraction.

如前所述,致密油和页岩油是主要的非常规石油。油砂、重质油、超重油是高度黏稠的含油沉积物,这些含油沉积物与其他物质混合,随时间推移发生降解。页岩油是岩石和经过加热形成油藏过程中的有机物质的混合物。根据这些较重的含油沉积所处的深度,可以利用井下(原位)开采的方式挖掘或开采出这些含油沉积物,在开采前对其进行加热使其分解出所含的石油。

6.1.1.2 Unconventional Gas

6.1.1.2 非常规天然气

Unconventional gas is natural gas obtained from sources of production that are, in a given era and location, considered to be new and different. A new era has emerged in the petroleum industry in which substantial recovery of natural gas is expected from unconventional reservoirs based on innovative technology. In the broadest sense, unconventional gas is the natural gas that is more difficult to recover because the technology has not been fully developed, or is not economically feasible. A major category of unconventional resources stems from extremely tight reservoirs having permeability in microdarcies and nanodarcies. Typical recovery from unconventional resources of gas is quite low compared to that of conventional reservoirs. Sources at times considered to be unconventional include:

(1) coalbed methane;

(2) methane hydrate (gas hydrate);

(3) shale gas;

(4) tight gas.

非常规天然气是从特定的地质年代和地点下通过新的、不同的生产来源获得的天然气。基于技术创新,石油工业进入了一个新的时代,人们开始从非常规油气藏大量开采天然气。从广泛的意义上讲,非常规天然气是更难以采出的天然气,因为相关技术还没有完全成熟,或者开采的经济成本太大。非常规天然气资源大多储存在非常致密的储层,通常具有微米和纳米级的渗透率。与常规油藏相比,非常规天然气采收率非常低。非常规天然气主要包括:

(1)煤层气;

(2)甲烷气水合物(天然气水合物);

(3)页岩气;

(4)致密砂岩气。

Actually, the term "unconventional" has lost its original meaning. As of 2013, gas from shales, tight sands, and coalbeds accounts for 65% of U. S. natural gas production. By 2040 that share is expected to rise to 79%. The unconventional has become the conventional. Fig. 6. 1 displays the U. S. natural gas production from 1990 to 2035.

实际上,"非常规"这个词已经失去了原有的意义。截至2013年,来自页岩、致密砂岩和煤层的天然气占美国天然气产量的65%。预计到2040年,这一占比将上升至79%,非常规已成为常规。图6.1为1990—2035年美国天然气产量。

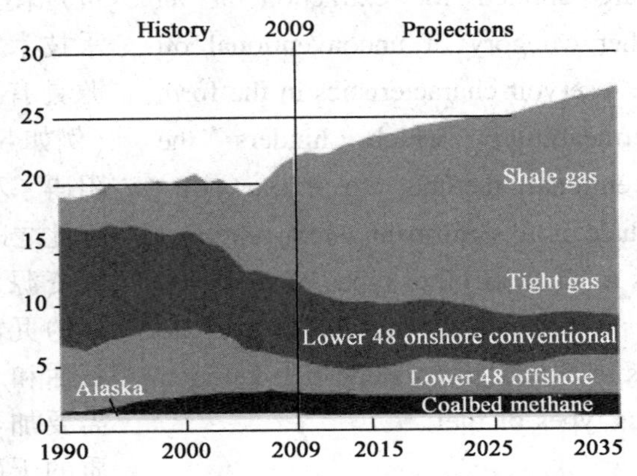

Fig. 6. 1 U. S. Natural Gas Production, 1990—2035 (Trillion Cubic Feet per Year)

(Source: https://www.2b1stconsulting.com/unconventional-oil-and-gas/)

6.1.2 Types and Locations of Unconventional Oil and Gas Resources

6.1.2.1 Unconventional Oil Resources

Unconventional oil reservoirs cannot be produced economically by traditional methods. The reservoirs require innovative technologies to develop, produce, and manage. Unconventional oil and gas have drawn worldwide attention due to the ever increasing demand for energy and dwindling resources of conventional oil. Based upon geologic and other evidences, many analysts believe that most of the conventional giant oil fields have already been discovered worldwide and conventional oil production is expected to reach a peak within the foreseeable future. On the other side of the spectrum, unconventional oil is more difficult to extract due to the relatively high cost per barrel of production and unique environmental issues. Above all, the various extraction technologies employed to produce unconventional oil are evolving. However, the development and production of unconventional reservoirs are becoming the center of attention

6.1.2 非常规油气资源的类型和区域

6.1.2.1 非常规石油资源

传统方法不能经济地开采非常规油藏,这些油藏需要新技术来开发、生产和管理。由于对能源需求的不断增长和传统石油资源的日益减少,非常规油气已得到全世界的关注。许多分析师认为,全球范围内大多数传统的大型油田已被发现,传统石油的生产预计将在可预见的未来达到顶峰。另外,由于生产成本相对较高及特殊环境问题,非常规油难以采出。但重要的是,生产非常规油的各种采收技术正在不断发展。然而,随着石油和天然气行业引入创新技术,并且只要油价能够支持商业化生

as innovative technologies are introduced in the oil and gas industry and as long as the price of oil supports commercial production. For example, Canada became a world leader in heavy oil production by unconventional methods in recent times.

产,非常规油藏的开发和生产会成为人们关注的焦点。例如,近年来加拿大通过非常规开采方法成为开采重油的世界领先者。

Unconventional oil resources can be viewed in two major categories. In the first category, oil having extremely high viscosity is hardly mobile unless certain thermal and nonthermal techniques are applied for extraction at an economic scale. The other category of unconventional oil resources has unfavorable reservoir characteristics in the form of ultralow rock permeability, which hinders the implementation of conventional methods to extract oil. Besides the above, oil shale is a significant unconventional resource. Oil shale refers to shale or other types of rock rich in kerogen, which is the precursor of oil and gas . Oil shale requires thermal processes to extract the organic-rich kerogen for conversion into various types of fuel.

非常规石油资源可分为两大类。一类是指原油具有极高黏度,几乎不可流动,除非在经济规模上应用某些热采和非热采技术。另一类非常规石油资源具有不利于开采的储层特征(例如超低的岩石渗透率),这阻碍了利用常规方法采出石油。除此之外,油页岩是重要的非常规资源。油页岩是指富含干酪根的页岩或其他类型岩石,它是石油和天然气的前体。油页岩需要加热过程来提取富含有机质的干酪根以转化为各种类型的燃料。

Unconventional oil resources are shown in Table 6.1, Unconventional oil resources, reserves and production are shown in Table 6.2.

非常规石油资源分类见表6.1,其储量和产量分布见表6.2。

Table 6.1 Unconventional Oil Resources

Unconventional oil resource	Attributes	Major extraction technologies
Heavy oil	Oil specific gravity ranges between 10° API and 20° API	Thermal, nonthermal, horizontal drilling
Extra heavy oil	Oil specific gravity is 10° API or heavier. Typically reservoirs are at a shallower depth with high permeability	Thermal, nonthermal
Oil sands, tar sands, bitumen	Oil specific gravity is 10° API or heavier. Typically reservoirs are at a shallower depth with high permeability	Thermal, nonthermal
Tight oil	Reservoir permeability in fractions of mD. Oil characteristics are similar to conventional oil	Horizontal drilling, multistage fracturing
Shale oil	Reservoir permeability in fractions of mD. Oil characteristics are similar to conventional oil	Horizontal drilling, multistage fracturing
Oil shale	Kerogen-enriched rock	Thermal. Oil extracted by retorting and distillation

Table. 6.2 Unconventional Oil Resources, Reserves and Production

Extra Heavy Oil	Resources bbl	Reserves bbl	Cumulative Production bbl
L. America	2,448	59	14
Venezuela	2,446	59	14
Asia	18	0.8	0.9
Others	19	0.3	1.4
World	2,485	61	16.5
Nat. Bitumen	Resources bbl	Reserves bbl	Cumulative Production bbl
N. America	2,451	174	5
Canada	2,397	174	5
Asia	427	42	0.0
Europe	349	29	0.0
Others	6	1.5	0.0
World	3,272	246.5	5
Oil Shale	Resources bbl	Reserves bbl	Cumulative Production bbl
N. America	2,100	NA	0.0
United States	2085	NA	0.0
Africa	159	NA	0.0
Europe	368	NA	2.5
Others	198	NA	2.5
World	2,826	NA	5

6.1.2.2 Unconventional Gas Resources

Unconventional gas resources include reservoirs from where the natural gas is difficult to recover as the technology has not matured or nontraditional methods are required to produce the reservoirs. Moreover, the cost of production is often higher than that of conventional gas reservoirs. Major deposits of unconventional gas that are produced currently include shale gas, tight gas, and CBM. Other types of unconventional gas, such as deep basin gas, have seen limited developments due to relatively high cost, dry hole occurrences, presence of sour gas, and various technological challenges. Methane hydrates, found in deep sea or in arctic regions, have not been extracted at a commercial scale. However, unconventional gas production is substantial in the

6.1.2.2 非常规天然气资源

非常规天然气资源包括难以采出的天然气,目前技术尚未成熟或需要非传统方法来进行生产,并且生产成本通常高于传统气藏。目前生产的非常规天然气主要包括页岩气、致密气和煤层气。其他类型的非常规天然气(例如深盆气),由于成本相对较高、干井的出现、酸性气体的存在及各种技术挑战,发展受到一定限制。在深海或北极地区发现的甲烷水合物尚未以商业规模开采。然而,截至2019

United States and constitutes 85% of total production as of 2019. The number is expected to increase in coming years. Unconventional gas resources in the world exceed 32,000 TCF (32×10^{12} ft^3), with the three major deposits located in North America, the former Soviet Union, and China. Global estimates of shale gas reserves are about 5700 TCF.

年,美国的非常规天然气产量巨大,占总产量的85%,预计未来几年这一数字将会持续增加。世界非常规天然气资源超过32000 TCF(32×10^{12} ft^3),其中三个主要矿床位于北美、前苏联地区和中国。同时,全球页岩气储量估计约为5700 TCF。

Currently, types of unconventional gas resources identified in various regions of the world are shown in Table 6.3.

目前,世界各地区探明的非常规天然气资源类型见表6.3。

Table 6.3 Types of Unconventional Gas Resources

Unconventional gas resource	Description	Reservoir development methods
Shale gas	Dry and wet natural gas trapped in ultratight shale reservoir. Typical shale permeability is in nanodarcies (10^{-9}D). Lithology is predominantly organic-rich shale low in clay content	Shale gas has been developed in significant quantities based on advances in horizontal drilling and multistage fracturing
CBM	Gas deposits in micropores and seams of coalbed. Formation permeability is usually low, between 1 mD and 25 mD. Production mainly occurs through coal cleats or seams	Reservoir development is based on vertical and horizontal well drilling combined with hydraulic fracturing
Tight gas	Gas trapped in very low permeability formations, chiefly sandstones and some carbonates. Reservoir permeability typically ranges in microdarcies (10^{-6}D). Lithology is predominantly marine shale rich in clay content	Reservoir development methods are similar to those of shale gas
Arctic and subsea hydrates	Molecules of methane trapped in the ice lattices. Abundant in arctic and subsea environments	Methane is released from ice under heat or reduced pressure. The resource is not yet commercially developed
Deep reservoir gas/ geopressured gas	Gas accumulations in various basins at significant depths, below 15,000 ft	Poses technological and economic challenges to drill and produce

The estimated major unconventional resources in various regions of the world are shown in Table 6.4.

表6.4列出了世界各地区估计的主要非常规天然气资源量。

Table 6.4 Unconventional Gas Resources in Trillion Cubic Feet

Region	Coalbed methane	Shale gas	Tight sand	Total resources
North America	3,017	3,840	1,371	8,228
Former Soviet Union	3,957	627	901	5,485
China and Central Asia	1,215	3,526	353	5,094
Pacific (OECD)	470	2,625	1254	4,349
Latin America	39	2,116	1,293	3,448
Middle East and North Africa	0	2,547	823	3,370
Sub-Saharan Africa	39	274	784	1,097
Western Europe	275	548	431	1,254
World	9,012	16,103	7,210	32,325

6.2 Tight Oil and Gas

6.2.1 Characteristics and Definition of Tight Oil and Gas

6.2.1.1 Tight Oil and Gas Definition

Tight oil (abbreviated LTO) is light crude oil contained in petroleum-bearing formations of low permeability, often shale or tight sandstone. But the term of tight oil is preferred to shale oil in order to avoid the confusion with oil shale. Tight oil or shale oil is different from oil shale. Oil shale is composed of solid hydrocarbon (kerogen), a decomposed organic matter still at solid state, in a sedimentary rock unit while tight oil is light crude.

Audio 6.2

Tight gas: Gas trapped in very low permeability formations, chiefly sandstones and some carbonates. Reservoir permeability typically ranges in microdarcies (10^{-6} D). Lithology is predominantly marine shale rich in clay content. Significant accumulations of natural gas are found worldwide in sandstone, carbonate, and shale formations with characteristically low to ultralow permeability. In fact, such accumulations exceed in volume in comparison to what is found in conventional gas reservoirs having higher permeability. Typically, rock permeability ranges from a fraction of a millidarcy in tight sandstones down to nanodarcies in shale. Extraction of gas is feasible

6.2 致密油气

6.2.1 致密油气定义及特征

6.2.1.1 致密油气定义

致密油(缩写为 LTO)是指低渗透含油气层(通常为页岩或致密砂岩地层)中所含的轻质原油。致密油原被称为“页岩油”,为了避免与“油页岩”混淆,逐渐改为“致密油”。致密油不同于油页岩,油页岩是由固态烃(干酪根,一种仍处于固体状态的已分解有机质)组成的沉积岩单元,而致密油则是一种轻质原油。

致密气是指被束缚在极低渗透率的地层中(主要是砂岩和一些碳酸盐岩)的气。储层渗透率通常以微达西(10^{-6}D)为单位,岩性主要为富含黏土的海相页岩。在世界范围内的砂岩、碳酸盐岩和页岩地层中发现了大量天然气,这些地层具有低到超低渗透率的特征。事实上,非常规气的储量超过了存在于较高渗透率地层中的常规气。通常,岩石包括渗透率大至致密砂岩

due to the low inherent viscosity of natural gas coupled with the high initial pressure of the reservoir. "Tight gas" is the term commonly used to refer to very low permeability reservoirs that are known to produce mainly dry natural gas.

的毫达西级别小到页岩的纳达西级别。由于天然气的低固有黏度及储层的高初始压力,开采此类天然气是可行的。"致密气"这个术语通常用来指存在于具有非常低渗透率储层的天然气,这些储层主要生产干气。

In the 1970s, gas reservoirs having a permeability of 0.1 mD or less were defined as tight gas reservoirs by the relevant authorities in the United States. However, the definition was politically aligned as it has been used to determine which operators would receive tax credits for producing gas from tight reservoirs.

在20世纪70年代,渗透率低至0.1mD或更低的气藏在美国被定义为致密气藏。该定义与政治政策保持一致,因为它是用于确定哪些运营商将获得从致密储层生产天然气的税收减免。

6.2.1.2 Characteristics of Tight Oil and Gas

6.2.1.2 致密油气特征

Tight oil, similar to tight gas, is a petroleum resource produced from ultra-low permeability shale, siltstone, sandstone, and carbonate, which are closely related to oil-source shales. This resource is defined as "light tight oil" by IEA and "tight oil" by Statoil, EIA, and some Chinese scholars.

与致密气相似,致密油是从超低渗透页岩、粉砂岩、砂岩和碳酸盐岩中开采出的石油资源,与油源页岩密切相关。这类资源被国际能源署定义为"轻质致密油",而挪威国家石油公司和一些中国学者将其称为"致密油"。

Tight oil is most prolific in North America, South America, North Africa, and Russia, but less prolific in Asia and Oceania. The hydrocarbon mainly accumulates in foreland basins, continental rift basins (Mesozoic strata), and craton basins (Paleozoic strata), and less in passive margin basins (Mesozoic strata) and back-arc basins (Cenozoic strata).

在北美、南美、北非和俄罗斯,致密油最为多产,但在亚洲和大洋洲的产量则一般。烃类主要聚集在前陆盆地、大陆裂谷盆地(中生代地层)和克拉通盆地(古生代地层),而被动边缘盆地(中生代地层)和弧后盆地(新生代地层)则较少。

The quality of source rocks is the most significant aspect for evaluating the unconventional resource abundance. In this study, the tight oil basins are selected by TOC higher than 1%, vitrinite reflectance R_o of 0.7% ~1.2%, and crude oil API higher than 38 °. Therefore, 84 basins (137 tight oil strata series totally) are selected from 468 basins globally for evaluation, and their tight oil potential is more than 240 billion barrels preliminarily estimated by volume method.

烃源岩的质量是评估非常规资源丰度的最重要方面。在一项研究中,致密油盆地的选择标准为总有机碳含量TOC高于1%,镜质体反射率 R_o 为0.7% ~1.2%,原油API度高于38 °API。因此,全球468个盆地中84个盆地(共计137个致密油层系列）被选中进行评估，并且通过体积法初步估算其致密油储量超过 0.24×10^{12}bbl。

Tight oil mainly accumulates in or near source rocks under the control of one or more sets of high-quality source rocks without trap boundaries. Therefore, according to the spatial relationships between tight oil reservoirs and high-quality source rocks, tight oil plays can be classified into eight types, above-source play, below-source play, beside-source play, in-source play, between-source play, in-source mud-dominated play, in-source mud-subordinated play, and interbedded-source play. For above-source, below-source, and beside-source plays which generally have conventional hydrocarbon features, high-quality source rocks and reservoirs are completely separated, and hydrocarbons migrate from source rocks to and accumulate in reservoirs; obvious segmentations with low gamma high resistivity of reservoirs and high gamma low resistivity of source rocks are found in well-logging curves. For insource plays, reservoir rocks are not developed but source rocks serve as reservoirs; the reservoir space mainly consists of organic pores with high gamma in the whole section. For between-source play, hydrocarbon can be supplied from both the upper and lower source rocks of reservoirs, and the monolayer of reservoir rocks is generally very thick (usually greater than 2m); in the well logging, reservoir rocks are often characterized by low gamma and high resistivity, which can be easily identified and can be developed as a separate reservoir.

致密油主要在一组或多组没有圈闭边界的优质烃源岩的控制下积聚在烃源岩中或其附近。因此，根据致密油藏与优质烃源岩之间的空间关系，致密油聚集区可分为八种类型，即烃源上部聚集、烃源下部聚集、烃源旁聚集、烃源中聚集、烃源间聚集、烃源中泥岩为主聚集、烃源中泥岩为次聚集和烃源夹层聚集。对于具有常规烃类特征的烃源上部聚集、下部聚集和旁聚集类型，高品质烃源岩和储层完全分离，烃类从烃源岩迁移并积聚在储层中；在测井曲线中发现了明显的储层低伽马高电阻率和源岩高伽马低电阻率的分段。对于烃源内聚集类型，储层岩石未被开发，烃源岩则为储层；储集空间主要由整个区域中具有高伽马值的有机孔组成。对于烃源间聚集类型，烃类存在于储层上部和下部烃源岩，而储层岩石的单层通常非常厚（一般大于2m）；在测井中，储层岩石通常具有低伽马和高电阻率的特征，可以很容易地被识别并且可以作为单独的储层开发。

The structural gentle-slope areas with favorable play types are the favorable zones for tight oil development because the structural slope areas are generally characterized by proximity to the mature source rocks, relatively better reservoir space, weak structural activities, and more "sweet spots".

具有良好油气聚集类型的结构性缓坡区是致密油开发的有利区域，因为结构坡面区域通常接近成熟烃源岩，具有相对较好的储层空间，构造活动弱同时存在更多“甜点”区域。

6.2.2 Production of Tight Oil and Gas

The oil which is produced or extracted from tight reservoirs is the same type of oil which can be produced from conventional reservoirs. It is the application of advanced technologies which make these developments unconventional. Different technologies are used for different plays but the most common methods used today are horizontal drilling and multi-stage hydraulic fracturing.

The purpose of drilling a horizontal well is to increase the contact between the reservoir and the wellbore. Tight oil reservoirs require some form of stimulation once the well has been drilled. The most common type of stimulation used by the oil and gas industry is referred to as hydraulic fracturing or fracking. In conventional oil reservoirs the reservoir permeability is sufficient that hydraulic fracturing may not be needed to achieve economic production rates. In unconventional oil, the reservoir permeability is typically very low and additional pathways must be created to enable the flow of hydrocarbons.

Advanced hydraulic fracturing technology means recovering tight oil from shale plays is becoming more and more economical and companies are rushing to acquire and exploit acreage in the formations where this oil is contained.

Natural gas with low permeability (below 0.1 mD) does not flow easily. Low permeability natural gas is called tight gas when it is contained in oil rock and shale gas when it is in shale rock. This resource cannot be developed profitably by vertical wells because of low flow rates. Production of tight and shale gas require hydraulic fracturing or horizontal wells. Horizontal wells techniques provide greater surface area in contact with the deposit compared to vertical wells, and enables more effective gas transfer and recovery of the gas in place. Today's technology is only suitable for onshore production and offers a maximum recovery rate of 20% of the volume in place. These production technologies have significant potential for improvements as there is a lack of basic research on tight and shale gas production. So far,

6.2.2 致密油气开采

从致密储层生产或提取的油气与常规储层生产的油气是同一类型,是先进技术的应用使这些油气成为非常规。不同的技术应用于不同的储层,但目前最常用的方法是水平井和多级水力压裂。

钻水平井的目的是增加储层与井筒之间的接触面积。一旦钻井完成,致密油藏就需要实施增产作业。石油和天然气工业使用的最常见的增产方法是水力压裂或压裂。在常规油藏中,储层渗透率足够大,因此可能不需要水力压裂来实现经济生产。而非常规油藏储层渗透率通常非常低,必须造额外的通道以使原油流动。

先进的水力压裂技术使从页岩地层中开采致密油的成本越来越低,石油天然气勘探开发公司期望从含有致密油的地层中获取更多储量。

低渗透率(低于0.1mD)储层中天然气不易流动。油岩中的低渗透率天然气称为致密气,在页岩中则称为页岩气。由于其流速低,垂直井无法经济开发低渗透率天然气。致密气和页岩气的开采需要实施水力压裂或钻水平井。与垂直井相比,水平井增加了井与储层的接触面积,并且能够实现更有效的气体流动和开采。现在的技术仅适用于陆上生产,并且能够达到的最大采收率为20%。由于缺乏对致密气和页岩气生产的基础

current production techniques have been developed based on empirical approaches.

研究，这些生产技术尚有巨大的改进空间。到目前为止，现有的生产技术都是基于过去开采经验开发的。

6.3 Coalbed Methane and Shale Gas

6.3 煤层气和页岩气

6.3.1 Geological Characteristics of Coalbed Methane and Shale Gas

6.3.1 煤层气和页岩气地质特征

6.3.1.1 Geological Characteristics of Coalbed Methane

6.3.1.1 煤层气地质特征

Coalbed methane (CBM) refers to the natural gas, chiefly methane, which is stored in the cleats and micropores of coal. It is produced from the coal bed deep down the earth surface. It is an unconventional source of gas. CBM is a byproduct of the coalification process, which consists of pure methane in majority of the fields. CBM deposits are found in over 60 countries. According to one estimate, CBM reserves in the top 20 countries are 1800 TCF. Large CBM development projects have been undertaken in China, India, and Australia, among others. Total reserves are quite substantial. The unconventional resource of CBM provides about 7% of the total natural gas consumption in the country.

Audio 6.3

煤层气(CBM)指天然气，主要成分是甲烷，储存在煤的内生裂隙和微孔中。它是一种产自地表深处煤层的天然气，属非常规天然气。煤层气是煤化作用过程的副产品，在大多数气田中由高纯度甲烷组成。煤层气藏遍布60多个国家。据调查，储量前20个国家的煤层气总储量已达1800 TCF。在中国、印度和澳大利亚等地相继开展了大型煤层气开发项目，总储备相当可观。煤层气这一非常规资源约占这些国家天然气总消费量的7%。

Coal bed methane gas is generally of two types:

(1) Biogenic (formed by bacterial action on coal);

(2) Thermogenic (formed by heat and pressure applied on coal).

一般来说，煤层气有两种类型：

(1)生物成因煤层气(通过细菌对煤的作用形成)；

(2)热(解)成因煤层气(通过施加在煤上的热和压力形成)。

Coal bed methane requires very little processing in the majority of instances and is pipeline quality gas, especially the biogenic gas. CBM is found mainly in bituminous and sub-bituminous coal which has better permeability but less gas content as compared to lignite coal which ranks less than bituminous and sub-bituminous coal. Permeability decreases and gas content increases with increasing coal rank.

在大多数情况下，煤层气几乎不需要处理，符合管道运输标准，尤其是生物成因煤层气。煤层气主要存在于烟煤和次烟煤中，其次是褐煤，与褐煤相比，烟煤和次烟煤具有更好的渗透性，但气体含量更低。煤层渗透率随着煤阶的增加而降低，含气量随着煤阶的增加而增加。

Coal bed methane exists almost exclusively in a condensed, liquid state. CBM reservoirs are different from sandstone or other conventional natural gas reservoirs, as the methane in CBM reservoirs is stored within the coal, as coal adsorbs the methane gas. Coal has high porosity therefore, it is able to absorb and retain large amount of gas. Unlike conventional natural gas, CBM contains very little amount of heavier hydrocarbons like propane or butane. Though, it may contain a few percent of carbon dioxide.

煤层气几乎全部以凝结的液态存在。不同于砂岩储层或其他常规天然气储层,煤层气储层中的煤层气储存在煤中而甲烷气吸附在煤上。煤具有较高的孔隙度,因此能够吸附和储存大量煤层气。不同于常规天然气,煤层气中只含有极少量丙烷或丁烷等重烃,但可能含有百分之几的二氧化碳。

The quality of gas depends on the depth of the gas reservoir. Deep CBM reservoirs, generally with depth greater than 3, 000 feet contain methane generated by cracking of kerogens and oil in coal to gas and is considered to be of superior quality compared to the gas found in shallow depths (< 3,000 feet). Fig. 6. 2 shows the cleats in coal that provide pathways to the production of gas.

煤层气的质量取决于煤层气储层的埋藏深度。深煤层气藏,深度通常超过3000ft,含有由干酪根及煤中的石油裂解生气产生的甲烷气,相比于产自浅层(深度小于3000ft)的煤层气,其质量更高。图 6. 2 为煤的内生裂缝,即割理,其存在为煤层气运移提供了通道。

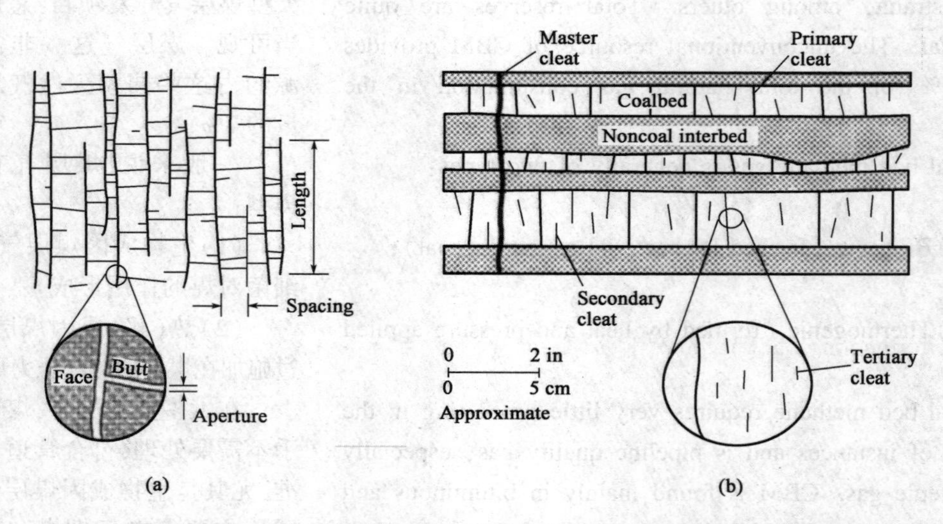

Fig. 6. 2 Cleats in Coal, which Provide a Natural Pathway to the Production of Gas

The storage of methane in the coalbed, the mechanism of fluid flow, rock characteristics, and reservoir development are unique to CBM reservoirs. The major differences between CBM reservoirs and conventional gas reservoirs are as follows:

煤层气储层中甲烷的储存、流体流动机制、岩石特征和储层开发均与常规储层不同。煤层气储层与常规天然气储层的主要区别如下:

(1) In contrast, source rock and reservoir rock are not the same in conventional gas reservoirs. Following migration from source rock, conventional gas is accumulated in reservoir rock by one or more trapping mechanisms; however, the process takes place over a long period of time and distance.

(1)煤层气储层的源岩和储层岩石与常规气藏不同。常规天然气从烃源岩迁移后,通过一个或多个圈闭机制积聚在储层岩石中,这一过程耗时久且距离远。

(2) In conventional gas reservoirs, natural gas is stored in a free state in pores and fractures; however, CBM is trapped mostly in an adsorbed state. Free gas in cleats and fractures of coal only amounts to a few percent of total volume of gas in place. According to a study, coalbeds can contain as much as 98% of the total volume of gas in an adsorbed state. Hence, in estimating the total gas in place in a CBM reservoir, the amount of gas adsorbed in micropores must be known.

(2)传统气藏中,天然气以自由态储存在孔隙和裂缝中,而煤层气主要为吸附态。煤的内生裂隙和裂缝中的游离气体仅占地下气体总量的小部分。研究表明,煤层含有多达98%的吸附态的气体。因此,在估算煤层气储层中的总气量时,必须知道微孔中吸附的气体量。

(3) In conventional reservoirs, the volume of gas in place can be calculated by applying the real gas law requiring the knowledge of hydrocarbon pore volume and prevailing reservoir pressure. In CBM reservoirs, the quantity of natural gas stored in coalbeds cannot be determined by using traditional methods of volumetric analysis. The adsorption capacity of coal is needed to estimate the gas in place. CBM reservoirs can store several times more gas in an adsorbed state than a conventional gas reservoir.

(3)在常规储层中,气体体积可以用真实气体定律、已知烃孔隙体积和主储层压力来计算。煤层气储层中,传统的体积分析法无法确定储存在煤层中的天然气量,需要根据煤的吸附能力来估算地下气体量。与常规气藏相比,煤层气储层可以在吸附状态下储存数倍于常规气藏的气体。

(4) Unlike conventional sandstone and limestone reservoirs where free gas is stored in relatively large pores, CBM is stored in the micropores of coal. Cleats or fractures in coal formations facilitate the flow of gas as the wells are drilled.

(4)煤层气储层与常规砂岩和石灰岩储层不同,常规储层中游离气体储存在较大的孔隙内,而煤层气的游离气储存在煤的微孔中。煤层中的裂隙或裂缝在钻井时可以促进气体的流动。

(5) The pore sizes in CBM reservoirs are smaller by orders of magnitude in comparison to conventional sandstone and carbonate reservoirs. Unconventional CBM reservoirs are usually characterized by relatively low porosity and permeability.

(5)与常规砂岩和碳酸盐岩储层相比,煤层气储层的孔径小几个数量级。因此非常规煤层气储层通常具有相对较低的孔隙度和渗透率。

(6) Transport of CBM in micropores of rock is characterized by diffusion governed by Fick's law; however, flow of CBM though fractures and seams obey Darcy's law. In contrast, the transport of fluids in conventional reservoirs is characterized by Darcy and non-Darcy flow.

(6)煤层气在岩石微孔中的运移一般通过扩散方式并且符合菲克定律;然而,通过裂缝和接缝的煤层气流又遵循达西定律。而常规储层中流体的运移以达西和非达西流动为特征。

(7) The unique characteristic of CBM production is the initial production of water contained in the cleats of coal formation. Water contained in cleats and fractures flows at a high rate during the initial production period.

(7)最初煤层裂隙中含有水,煤层气的生产特征是在初始生产期间煤的内生裂隙和裂缝中的水高速流动。

6.3.1.2 Geological Characteristics of Shale Gas

6.3.1.2 页岩气地质特征

Shale gas refers to unconventional natural gas that occurs in reservoir rock series dominated by organic-rich shale. It is a continuously generated biochemical origin gas, thermogenic gas or a mixture of the two, which can exist in free cracks in natural cracks. And in the pores, the adsorbed state exists on the surface of the kerogen and clay particles, and a very small amount is stored in the kerogen and the asphaltene in a dissolved state and the ratio of the free gas is generally 20% to 85%. Development of shale gas reservoirs has been spectacular since the dawn of the twenty-first century. Studies indicate that only a small fraction of hydrocarbons (10% ~ 20%) from source rock migrate to, and accumulate in, conventional reservoirs. However, the major portion of oil and gas generated in the process is trapped in source rocks, chiefly shale, where traditional means of production are not adequate. Hence, shale gas is referred to as an unconventional resource. Shale gas production has leaped several folds in just over a decade, accounting for a significant portion of total natural gas production in the United States (Fig. 6. 3). Significant increases in unconventional gas development are expected in the near future.

页岩气是指赋存于以富有机质页岩为主的储集岩系中的非常规天然气,是连续生成的生化成因气、热成因气或二者的混合,可以游离态存在于天然裂缝和孔隙中,以吸附态存在于干酪根、黏土颗粒表面,还有极少量以溶解状态储存于干酪根和沥青质中,游离气比例一般在20% ~ 85%。自21世纪初以来,页岩气藏一直被大规模开采。研究表明,源岩中只有小部分碳氢化合物(10% ~20%)会迁移并聚集到常规油藏中。该过程中生成的大部分石油和天然气会留在源岩(大部分为页岩)中,传统生产技术不足以开采。因此,页岩气被称为非常规资源。仅仅十多年的时间里,页岩气的产量已经跃升了几倍,占美国天然气总产量的很大一部分(图6.3),预计不久的将来非常规天然气开发量还将大幅增加。

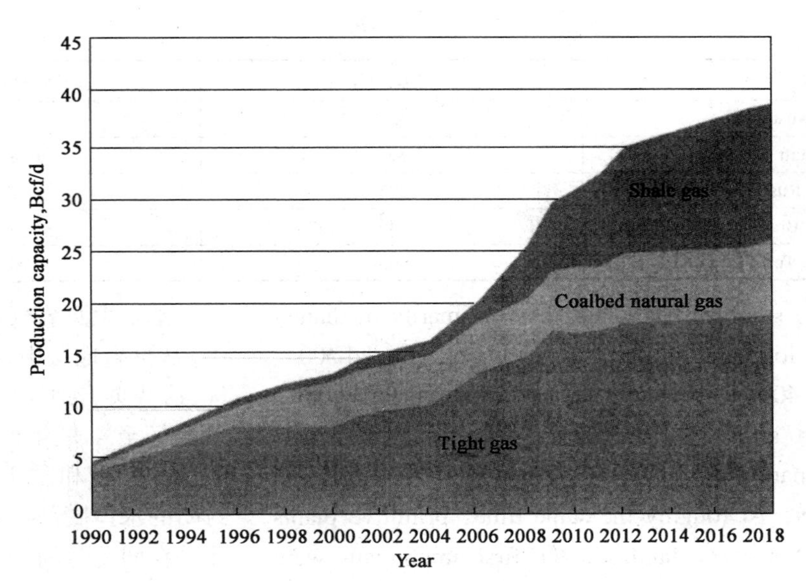

Fig. 6. 3　Shale Gas, CBM, and Tight Gas Production Forecast in the United States

Shale gas has become an increasingly important source of natural gas in the United States since the start of this century, and interest has spread to potential gas shales in the rest of the world. In 2000 shale gas provided only 1% of U. S. natural gas production; by 2010 it was over 20% and the U. S. government's Energy Information Administration predicts that by 2035, 46% of the United States' natural gas supply will come from shale gas.

自21世纪初以来,页岩气已成为美国日益重要的天然气来源,他们对页岩气的兴趣已经扩展到分布在世界上其他地区的潜在含气页岩。2000年美国页岩气产量仅占美国天然气总产量的1%;到2010年,页岩气产量在美国天然气总产量中所占的比例已经超过了20%,根据美国能源信息管理局的预测,到2035年,美国46%的天然气供应将来自页岩气。

Some analysts expect that shale gas will greatly expand worldwide energy supply. China is estimated to have the world's largest shale gas reserves(Table 6. 5).

一些分析人士预计页岩气将极大地促进全球能源供应。据估计,中国拥有世界上最大的页岩气储量(表6.5)。

Table 6. 5　Global Estimates of TRR of Shale Gas

Country	TRR, TCF	Percentage
China	1115	19. 3
Argentina	802	13. 9
Algeria	707	12. 3
United States	665	11. 5
Canada	573	9. 9
Mexico	545	9. 5

Continued

Country	TRR, TCF	Percentage
Australia	437	7.6
South Africa	390	6.8
Russia	285	4.9
Brazil	245	4.3
Total	5764	100

Basically, shale gas is natural gas (primarily methane) found in shale formations, some of which were formed 300 - million - to - 400 - million years ago during the Devonian period of Earth's history. The shales were deposited as fine silt and clay particles at the bottom of relatively enclosed bodies of water. At roughly the same time, primitive plants were forming forests on land and the first amphibians were making an appearance. Fig. 6.4 shows a shale gas geological trap.

基本上页岩气是(以甲烷为主)从页岩地层中开采出的天然气,这些页岩地层中的一些形成于3亿至4亿年前的泥盆纪时期。细粉砂和黏土颗粒在相对封闭的水体底部沉积形成页岩。几乎在同一时间,原始植物在陆地上形成森林,并且出现了最早的两栖动物。图6.4为一页岩气聚集的地质圈闭。

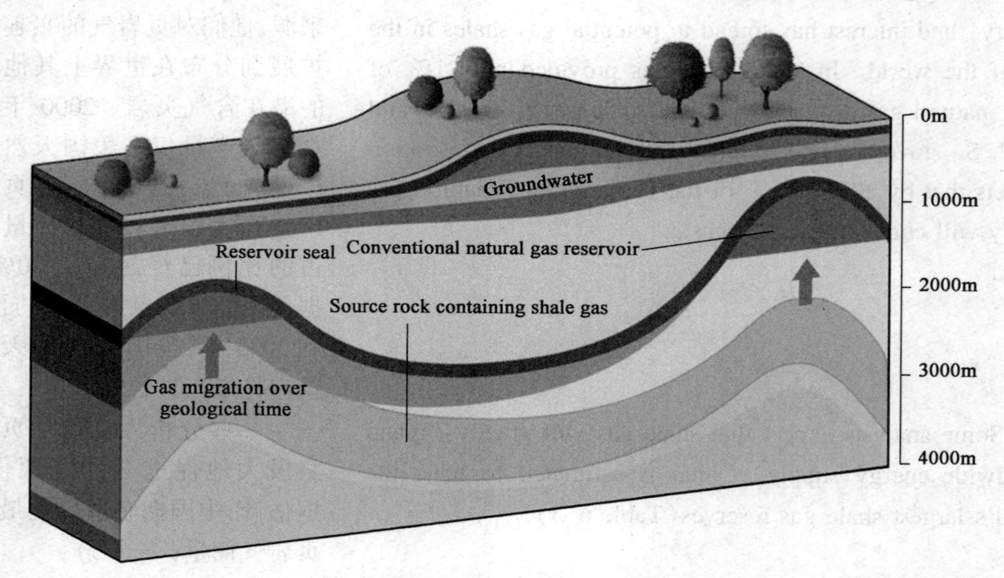

Fig. 6.4 A Shale Gas Geological Trap

(Source: https://www.2b1stconsulting.com/shale-gas/)

Some of the methane that formed from the organic matter buried with the sediments escaped into sandy rock layers adjacent to the shales, forming conventional accumulations of natural gas which are relatively easy to extract. But some of it remained locked in the tight, low permeability shale layers, becoming shale gas.

一部分与沉积物一起埋藏的由有机质形成的甲烷逃逸到与页岩毗邻的砂质岩层中,形成相对容易开采的常规天然气聚集(常规天然气藏);一部分甲烷则仍然储集在致密、低渗透的页岩层中,成为页岩气。

Shale gas is usually found in compact, low-permeability rocks deep beneath the surface. Reserves of this gas are very substantial and well distributed over the globe: they are estimated to represent between 120 and 150 years' worth of natural gas supply at the current rate of consumption. While geologists have known that shale gas existed deep beneath many areas of the North American continent, traditional vertical oil and gas drilling methods were able to access only a small fraction of the gas within these formations. But recently, operational efficiencies and proven technology have come together to make shale gas both accessible and economically competitive.

页岩气通常存在于地表深处致密、低渗透的岩石中。页岩气的储量相当可观且在全球范围内分布均匀:据估计,按目前的消耗水平,页岩气资源可满足120~150年的天然气供应需求。虽然地质学家都知道在北美大陆许多地区的深处有页岩气,但常规的垂直油气井只能开采出这些地层中的一小部分页岩气。目前,由于作业效率和成熟技术的相互结合,页岩气变得既容易开采,又具有经济竞争力。

Shale gas formations are "unconventional" reservoirs-i. e., reservoirs of low permeability. This contrasts with a "conventional" gas reservoir produced from sands and carbonates (such as limestone).

页岩气储层是非常规储层,即具有低渗透率的储层,这与由砂岩和碳酸盐岩(如石灰石)形成的常规储层形成了鲜明对比。

The bottom line is that in a conventional reservoir, the gas is in interconnected pore spaces, that allow easier flow to a well; but in an unconventional reservoir, like shale, the reservoir must be mechanically "stimulated" to create additional permeability and free the gas for collection.

最重要的一点是,在常规储层中,气体处于互相连通的孔隙空间内,可以更容易流入井筒中;但在如页岩地层等非常规储层中,必须采用一定的工艺措施对储层进行机械性改造以提高储层渗透率并使气体易于开采。

Geomechanical properties of shale reservoirs. The presence of natural fractures plays a significant role in effectively producing shale gas reservoirs. Young's modulus, Poisson's ratio, and fracture stress are some of the geomechanical parameters determined to evaluate the fractures in shale. Furthermore, data related to the length, height, orientation, and conductivity of hydraulically created fractures are needed.

页岩储层的地质力学特征。天然裂缝对有效生产页岩气藏起着关键作用。杨氏模量、泊松比和破裂应力是评估页岩裂缝的几个地质力学参数。此外,还需要与水力压裂后产生裂缝的长度、高度、方向和导流能力等相关的数据。

Sweet spots of shale. Conventional petroleum reservoirs have distinct geologic or hydrodynamic boundaries and are limited in extent. In contrast, shale formations with unconventional gas deposits are continuous in nature and may extend over a vast area. However, not all the areas of gas accumulation are capable of sustainable production. The economic success of shale gas production depends on the identification of sweet spots, where wells are drilled based on state-of-the-art technology. Sweet spots are of keen interest to the industry in the exploration and production of shale gas. Sweet spots are expected to have favorable geological and geochemical characteristics as in the following:

页岩的甜点区。常规石油储层有明确的地质或流体力学边界并且范围有限。页岩中非常规气体沉积本质上是连续的，并且可以延伸较大的区域，但并非所有天然气聚集区域都能够持续生产。页岩气开采的成功与否取决于对甜点区的识别，以及是否使用最先进的钻井技术。在页岩气的勘探和生产中，甜点区得到了业界的关注。甜点区一般被认为具有良好的地质和地球化学特征，具体如下：

(1) Relatively high TOC to yield significant quantities of gas, referred to as "black shale";

(1)相对较高的总有机碳含量来产生大量天然气，被称为"黑色页岩"；

(2) Desirable thermal maturity for rock to produce gas as indicated by vitrinite reflectance (gas window);

(2)通过镜质体反射率(气体窗口)反应出岩石具有理想热成熟度来产生气体；

(3) Better porosity and permeability for storage and mobility;

(3)更好的孔隙度和渗透率，方便气体储存和流动；

(4) Presence of natural fractures in the formation;

(4)地层中存在天然裂缝；

(5) Good hydraulic fracturing characteristics of rock as indicated by Young's modulus, Poisson's ratio, and fracture stress.

(5)杨氏模量、泊松比和破裂应力反映出岩石良好的水力压裂特性。

Navarette proposes the following guidelines to identify sweet spots:

Navarette提出用以下方法来识别甜点区：

(1) Reservoir thickness >200 ft;

(1)储层厚度大于200ft；

(2) TOC >1.0%;

(2)总有机碳含量大于1.0%；

(3) Porosity for storage of free gas >4%;

(3)储存游离气体的孔隙率大于4%；

(4) Permeability of shale >100 nD;

(4)页岩渗透率大于100nD；

(5) Brittleness index >25%.

(5)脆性指数大于25%。

In addition, sweet spots must have sufficient reservoir pressure and appropriate thermal maturity of rock.

此外,甜点区必须具有足够的储层压力和适当的岩石热成熟度。

6.3.2 Production Technology of Coalbed Methane and Shale Gas

6.3.2 煤层气和页岩气开采技术

6.3.2.1 Production Technology of Coalbed Methane

6.3.2.1 煤层气开采技术

To extract the gas, after drilling into the seam, it is necessary to pump large amounts of water out of the coal seam to lower the pressure. It is often also necessary to frac the seam to extract the gas. There is a similar catalogue of negative environmental and social effects as with Shale Gas. This includes methane migration, toxic water contamination, air pollution, increased carbon emissions and a general industrialization of the countryside. Impacts that are specific to CBM include depletion of the water table and potentially subsidence.

为开采煤层气,钻入煤层后必须从煤层中抽出大量的水以降低煤层压力。通常还需要对煤层进行水力压裂以开采煤层气。在煤层气的开采过程中,也有类似于开采页岩气的负面环境影响和社会影响,包括甲烷运移、有毒物质对水的污染、空气污染、碳排放增加及农村普遍工业化。煤层气开采带来的特有影响包括地下水位下降和潜在的地面沉降。

Coalbed methane production passes through three phases during the life-time of the reservoir, drain desorption, gas diffusion, and flow into the wellbore. This behavior differs significantly from the normal decline curve of conventional gas wells. The production profile of coalbed methane well is shown in Fig. 6.5.

在煤层气藏开采寿命内,煤层气的开采过程可以划分为三个阶段:排水降压解吸、气体扩散、流入井筒。其产量历史曲线与常规气井的正常衰减曲线有很大的差别。煤层气井的生产情况如图6.5所示。

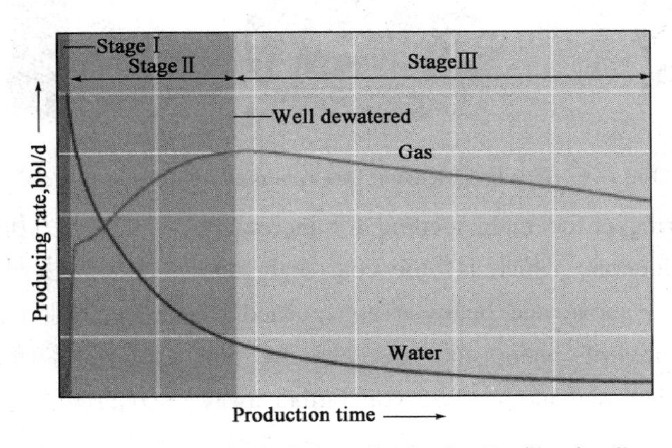

Fig 6.5 Typical Coalbed Methane Production Profiles for Gas and Water Rates: Three Phases of Producing Life

The first step is to drain and desorb. Since the coalbed methane is adsorbed on the surface of the coal, and the pores and cracks of the coal are filled with water, the coalbed methane cannot be separated from the surface of the coal. Therefore, it is necessary to first discharge the water filled in the coal, and the gas molecules are free of water, you can detach from the surface of the coal, which is called "desorption", after the coalbed methane enters into the pores of the coal.

第一步是排水降压解吸。由于煤层气吸附在煤的表面上,而煤的孔隙、裂缝中充满水,煤层气无法脱离煤的表面,因此就需要先将煤里充填的水排出,气体分子没有了水的"压制",就可以从煤的表面上脱离,这就叫作解吸,之后煤层气进入到煤的孔隙中。

The second step is diffusion. As more and more coalbed methane is detached from the coal surface, the concentration of coalbed methane in many small holes and seams is getting higher and higher, and it will spread from a high concentration to a low concentration. Finally, the cracks and pores are filled with coalbed methane.

第二步是气体扩散,由于从煤表面脱离下来的煤层气越来越多,许多小的孔、缝中煤层气的浓度越来越高,气体就会从浓度高的地方扩散到浓度低的地方,直到最后裂缝、孔隙中都充满了煤层气。

The third step is that coalbed methane flows from the cracks in the coal into the wellbore. Due to the diffusion of the previous step, the coalbed methane eventually merges into the large crack and finally flows from the large crack into the wellbore connected to it. Fig. 6. 6 is a schematic of coalbed methane desorption and diffusion process.

第三步是煤层气从煤的裂缝中流入井筒。由于上一步的扩散作用,煤层气最终汇入大的裂缝中,从大的裂缝流入与其相连的井筒中。图 6. 6 为煤层气解吸、扩散过程示意图。

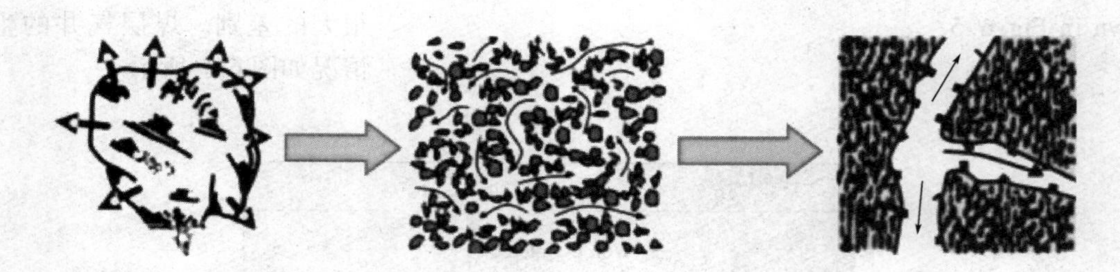

Fig. 6. 6 Coalbed Methane Desorption and Diffusion Process

Hydraulic fracturing is the main method for increasing coalbed methane production, which is, the coal seam and the mudstone sandstone above and below it are crushed by the force and pressure of water to cause cracks and communicate the coalbed methane gas reservoir farther away from the wellbore.

水力压裂是煤层气增产的主要方法,即通过水的压力将煤层及其上下的泥岩砂岩压开,使其产生裂缝,连通离井眼更远地方的煤层气储层。

6.3.2.2 Production Technology of Shale Gas

Shale gas mining technologies typically include seismic exploration techniques, drilling techniques, logging techniques, shale gas recording and field testing techniques, cementing techniques, completion technologies, and reservoir modification technologies. Hydraulic fracturing and acidification are included in shale gas reservoir modification techniques. Construction can be carried out by conventional or coiled tubing. In foreign countries, coiled tubing is often used for re-injection or secondary completion. It can be used for branch horizontal well fracturing to increase production, including nitrogen foam fracturing, gel fracturing, multi-stage fracturing, and clean water. Fracturing, simultaneous fracturing, hydraulic jet fracturing and repeated fracturing, multi-stage fracturing, fresh water fracturing, simultaneous fracturing, hydraulic jet fracturing and repeated fracturing are commonly used techniques for hydraulic fracturing of shale gas.

For shale gas, hydraulic fracturing of a reservoir is the preferred stimulation method.

This typically involves injecting pressurized fluids to stimulate or fracture shale formations and release the natural gas. Sand pumped in with the fluids (often water) helps to keep the fractures open. The type, composition and volume of fluids used depend largely on the geologic structure, formation pressure and the specific geologic formation and target for a well. If water is used as the pressurized fluid, as much as 20 percent can return to the surface via the well (known as flowback). This water can be treated and reused-in fact, reuse of flowback fluids for subsequent hydraulic fracture treatments can significantly reduce the volume of wastewater generated by hydraulic fracturing.

Fig. 6.7 represents the common equipment at a nature gas hydraulic fracturing drill pad.

6.3.2.2 页岩气开采技术

页岩气开采技术通常包括地震勘探技术、钻井技术、测井技术、页岩含气量录井和现场测试技术、固井技术、完井技术和储层改造技术。页岩气储层改造技术包括水力压裂和酸化,可以通过常规油管或连续油管进行施工。国外经常采用连续油管进行增产或二次完井。分支水平井压裂增产措施有多种,包括氮气泡沫压裂、凝胶压裂、多级压裂、清水压裂、同步压裂、水力喷射压裂重复压裂等,多级压裂、清水压裂、同步压裂、水力喷射压裂和重复压裂是目前页岩气水力压裂常用的技术。

对于页岩气开采来说,水力压裂是首选的增产措施。

页岩气储层水力压裂通常指注入高压流体来改造页岩地层来释放天然气。泵入携砂液有助于使裂缝保持张开,所用流体的类型、组成和用量在很大程度上取决于地质构造、地层压力、井的目标层地质特征。如果选用水作为压裂液基液,则多达20%的水可以通过井筒返出地面(称为返排),这些返排的水经过处理可以再次利用。将回流废水再次用于后续的水力压裂作业可以显著减少水力压裂产生的废水量。

图 6.7 为进行天然气水力压裂施工的井场上常见的压裂设备。

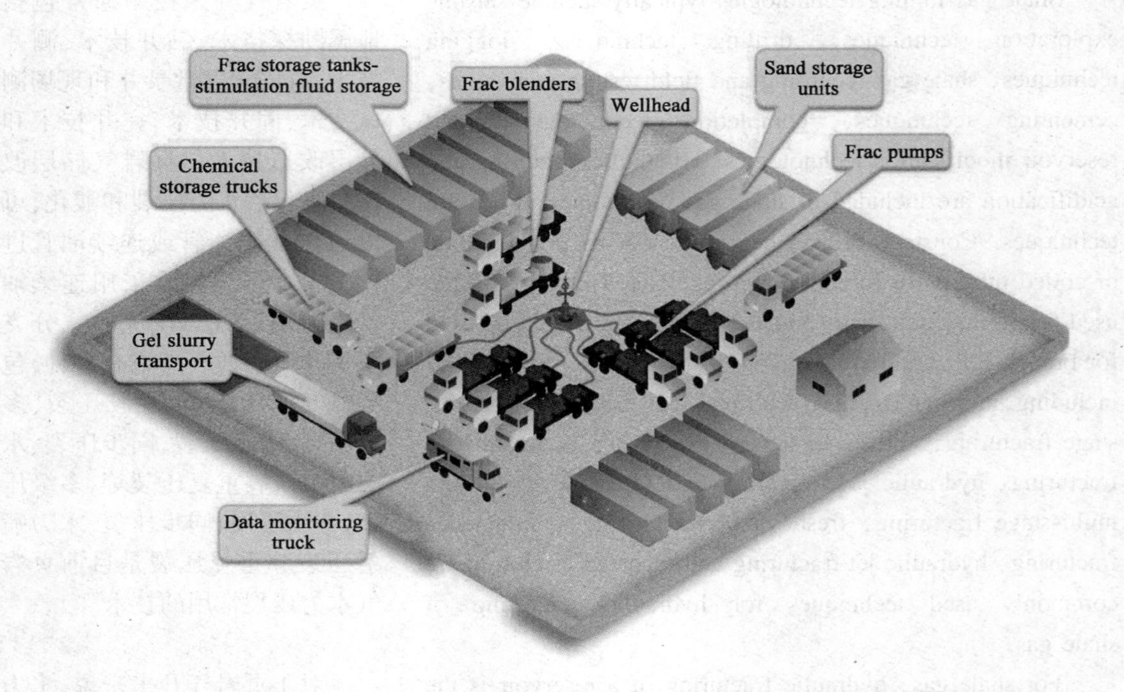

Fig. 6. 7 Representation of Common Equipment at a Natural Gas Hydraulic Fracturing Drill Pad
(Source: https://www.energy.gov/sites/prod/files/2013/04/f0/how_is_shale_gas_produced.pdf)

Another major technology often employed in producing natural gas from shale is horizontal drilling(Fig. 6. 8). The shallow section of shale wells are drilled vertically (much like a traditional conventional gas well). Just above the target depth-the place where the shale gas formation exists-the well deviates and becomes horizontal. At this location, horizontal wells can be oriented in a direction that maximizes the number of natural fractures intersected in the shale. These fractures can provide additional pathways for the gas that is locked away in the shale, once the hydraulic fracturing operation takes place.

水平井钻井技术(图 6.8)是另一个用于页岩气开发的主要技术。页岩气井位于浅层的部分类似于常规气井(直井)。在目标深度,即页岩气储层上方,井眼轨迹按预定方向偏斜并最终成为水平井。在这个位置,可以将水平井的延伸方向确定为能够最大程度地增加页岩中相交的天然裂缝数量的方向。一旦进行水力压裂作业,这些裂缝就可以为储集在页岩中的气体提供额外的运移通道。

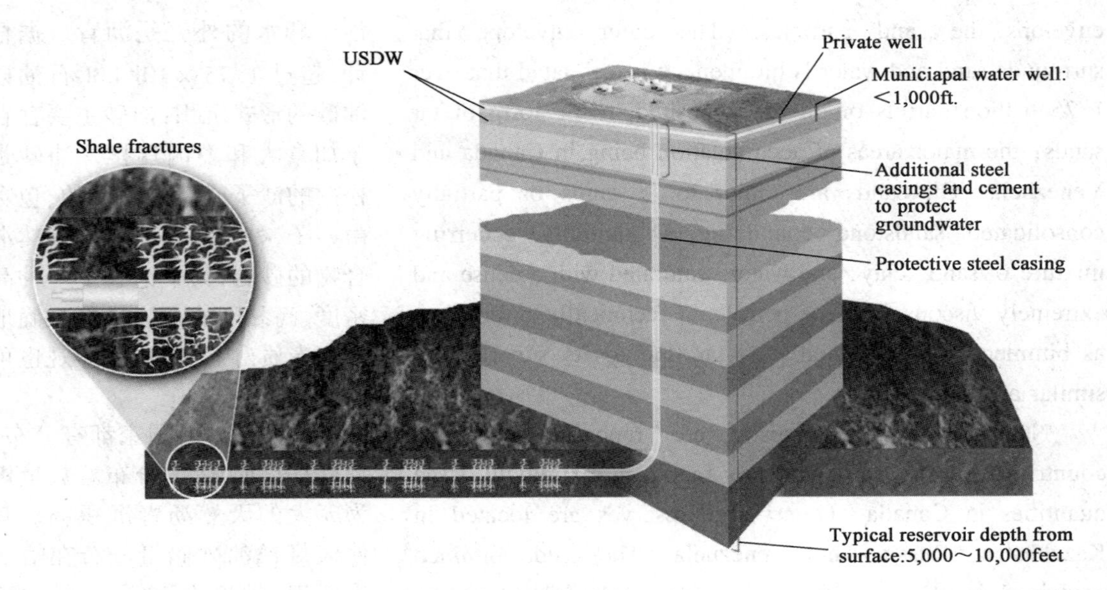

Fig. 6.8　Horizontal Drilling

(Source: https://www.energy.gov/sites/prod/files/2013/04/f0/how_is_shale_gas_produced.pdf)

Developing any energy resource-whether conventional or non-conventional like shale-carries with it the possibility and risk of environmental, public health, and safety issues. Some of the challenges related to shale gas production and hydraulic fracturing include:

无论是常规能源还是页岩气等非常规能源,对其进行开发都有引发环境问题、公共卫生问题及安全问题的风险。与页岩气开发和水力压裂相关的一些挑战包括:

(1) Increased consumption of fresh water (volume and sources);

(1)淡水消耗量增加;

(2) Induced seismicity (earthquakes) from shale flowback water disposal; Chemical disclosure of fracture fluid additives;

(2)页岩气开发过程中废水返排处理诱发地震活动,压裂液添加剂化学成分的泄漏;

(3) Potential ground and surface water contamination;

(3)潜在的地面和地表水污染;

(4) Air quality impacts;

(4)影响空气质量;

(5) Local impacts, such as the volume of truck traffic, noise, dust and land disturbance.

(5)地方性影响,如大的货车运输量、噪声、灰尘和占用耕地。

6.4　Oil Sands & Oil Shale

6.4　油砂和油页岩

6.4.1　Overview of Oil Sands and Oil Shale

6.4.1　油砂和油页岩概述

Oil sands, also known as tar sands or crude bitumen, or more technically bituminous sands, are a type of unconventional petroleum deposit. The constituents of tar sands are bitumen, clay, sand, and water. A thin film of water

Audio 6.4

油砂,也称焦油砂、天然沥青,或更准确地说是沥青砂,是一种非常规石油沉积物。油砂的成分是沥青、黏土、沙子和水。沙子外围包裹着一层水膜,围绕

envelops the sand particles. The outer envelope that surrounds sand and water is bitumen. It is estimated that over 1.75 trillion barrels of oil are deposited in the form of tar sands; the major areas of accumulation being in Canada and Venezuela. Oil sands are either loose sands or partially consolidated sandstone containing a naturally occurring mixture of sand, clay, and water, saturated with a dense and extremely viscous form of petroleum technically referred to as bitumen (or colloquially as tar due to its superficially similar appearance).

沙子和水的外壳是沥青。据估计,超过 1.75×10^{12} bbl 石油以油砂的形式沉积;油砂主要存在于加拿大和委内瑞拉。油砂是松散的沙子或部分固结的、包含由沙子、黏土和水组成的天然混合物的砂岩,其中饱和了一种高密度、高黏度的石油或更准确地说沥青质(因为相似的外观也可以称为柏油)。

Natural bitumen deposits are reported in many countries, but in particular are found in extremely large quantities in Canada. Other large reserves are located in Kazakhstan, Russia, and Venezuela. The crude bitumen contained in the Canadian oil sands is described by the National Energy Board of Canada as "a highly viscous mixture of hydrocarbons heavier than pentanes which, in its natural state, is not usually recoverable at a commercial rate through a well because it is too thick to flow". Crude bitumen is a thick, sticky form of crude oil, so heavy and viscous (thick) that it will not flow unless heated or diluted with lighter hydrocarbons such as light crude oil or natural-gas condensate. At room temperature, it is much like cold molasses. The World Energy Council (WEC) defines natural bitumen as "oil having a viscosity greater than 10,000 centipoise under reservoir conditions and an API gravity of less than 10 °API". The Orinoco Belt in Venezuela is sometimes described as oil sands, but these deposits are non-bituminous, falling instead into the category of heavy or extra-heavy oil due to their lower viscosity. Natural bitumen and extra-heavy oil differ in the degree by which they have been degraded from the original conventional oils by bacteria. According to the WEC, extra-heavy oil has "a gravity of less than 10 °API and a reservoir viscosity of no more than 10,000 centipoise".

油砂在许多国家都有分布,尤其是在加拿大,分布着数量极为庞大的天然沥青沉积物。其他大规模油砂储量分布在哈萨克斯坦、俄罗斯和委内瑞拉境内。加拿大国家能源局将加拿大油砂中含有的天然沥青描述为"一种比戊烷重的高黏度的烃类物质混合物,因为它实在是太黏稠以至于无法流动,通常无法在天然状态下通过井筒经济有效地将其开采到地面"。油砂是一种黏稠的原油,由于密度过大和较为黏稠,除非使用轻质原油或天然气凝析油等轻质烃类对其进行加热或稀释,否则油砂将不会流动。在室温条件下,油砂就像冰冷的糖浆一样(黏稠)。世界能源委员会(WEC)将油砂定义为"在原始油藏温度下脱气原油黏度超过 10000mPa · s、API 重度小于 10°API 的石油"。委内瑞拉的奥里诺科重油带有时被描述为油砂聚集区,但这其中的含油沉积物是非沥青质,由于黏度较低,应将它们归入重油类或超重油类。油砂和超重油的不同之处在于它们被细菌从原始状态下的常规石油中降解的程度不同。根据 WEC

对超重油的定义，超重油“API重度小于10 °API，在原始油藏温度下脱气原油黏度不超过10000 mPa·s”。

Due to the increasing global demand for oil, the extraction of bitumen from oil sands has become one of the growing alternatives to conventional crude oil production. Unlike the extraction of oil from a well, the extraction of bitumen is distinct and requires a more energy intensive process due to the viscous nature of the bitumen.

由于全球石油需求不断增加，从油砂中开采出的沥青已成为常规原油重要的替代能源之一。与通过油井开采石油不同，沥青的开采方法是独特的，并且由于沥青的黏性性质，开采沥青需要耗费更多能源。

There are two ways to extract bitumen from the oil sands: either mine the entire deposit and gravity separate the bitumen, or extract the bitumen in-place (or in-situ) using steam without disturbing the land. The technique used depends on the depth of the deposit.

从油砂中开采沥青的方法有两种：开采整个油砂并通过重力沉降分离得到沥青，或者在不对当地耕地造成影响的情况下通过注蒸汽原地开采沥青。具体使用何种技术取决于油砂的埋藏深度。

About 10% of the world's oil reserves are located in the Alberta oil sands. These deposits are estimated to hold almost 2 trillion barrels of oil, but less than 10% (about 166 billion barrels) are recoverable with current technology. Over 96% of Canada's total oil reserves are contained in the oil sands.

世界上约有10%的石油储量集中在加拿大阿尔伯塔油砂区。据估计，这些油砂沉积蕴藏着近2×10^{12} bbl石油，但以现有的技术只能开采出不到10%（约合0.166×10^{12} bbl石油）。加拿大96%以上的石油总储量蕴藏在油砂中。

Alberta's oil sands contain on average about 10% bitumen, 5% water and 85% solids. Most of the solids are coarse silica sand. Oil sands also contain fine solids and clays, typically in the range of 10% ~30% by weight.

阿尔伯塔油砂区油砂的平均沥青含量约为10%，平均含水量约为5%，平均固体含量约为85%，大多数固体是粗硅砂。油砂也含有细小固体及黏土，通常占重量的10%~30%。

6.4.1.1 Process Overview of Production from Oil Sands

6.4.1.1 油砂开采流程概述

Petroleum products are produced from the oil sands through 3 basic steps:

通过以下3个基本步骤可以从油砂中生产石油产品：

(1) Extraction of the bitumen from the oil sands, where the solids and water are removed.

(2) Upgrading of the heavy bitumen to a lighter, intermediate crude oil product.

(3) Refining of the crude oil into final products such as gasoline, lubricants and diluents.

Traditionally, a majority of the bitumen produced in Alberta was upgraded into synthetic crude oil before being sold to refineries on the open market. However, some bitumen is good enough to send directly to a high-conversion refinery that has the ability to process heavy/sour crude. Examples of diluted bitumen sold directly to refineries includes product from in-situ facilities around the Christina Lake and Cold Lake areas, diluted bitumen from Imperial Oil's Kearl Lake Mine and from Suncor's Fort Hills project.

Currently about 40% of the bitumen produced in Alberta is upgraded to synthetic crude oil. The remaining 60% is diluted with condensate and sold directly to market, without an intermediary upgrading step.

6.4.1.2 Bitumen Extraction

The method used to extract bitumen from the oil sands depends on the depth of the deposit. If the deposit is near the surface, the oil sands is mined and sent to a bitumen processing plant. For deposits that are deep below the surface, bitumen is extracted in-situ (or in place) (Fig. 6.9). Both facilities produce a relatively clean bitumen product which is mostly free of sand and water.

(1)从除去固体和水的油砂中开采沥青。

(2)重质沥青加工成(改质为)轻质的中间原油产品。

(3)将原油精炼成汽油、润滑剂及稀释剂等最终产品。

传统上,大部分产自阿尔伯塔省的沥青在市场上出售给炼油厂之前会被加工成(改质为)合成原油。但是,若部分沥青的质量足够好,则可以直接将其送到有能力加工重质高硫原油的高转化率炼油厂进行进一步加工。直接将稀释沥青出售给炼油厂的例子有:产自 Christina Lake 和 Cold Lake 地区附近的稀释沥青、产自 Imperial Oil 公司的 Kearl Lake Mine 的稀释沥青,以及 Suncor's Fort Hills 项目的稀释沥青。

目前,产自阿尔伯塔省的沥青中约有 40% 被加工成(改质为)合成原油。其余的 60% 在无需中间加工(改质)步骤的情况下使用冷凝物进行稀释并直接在市场出售。

6.4.1.2 沥青开采

使用何种方法从油砂中开采沥青取决于油砂沉积的埋藏深度。如果油砂沉积靠近地表,则用挖掘机开采油砂并将其送往沥青提炼厂进行下一步处理;对于深埋在地表以下的油砂沉积,则采用原位开采的方式对沥青进行开采(图 6.9)。这两种开采方法都可以开采出相对清洁的沥青产品,这些沥青产品绝大多数都不含沙子和水。

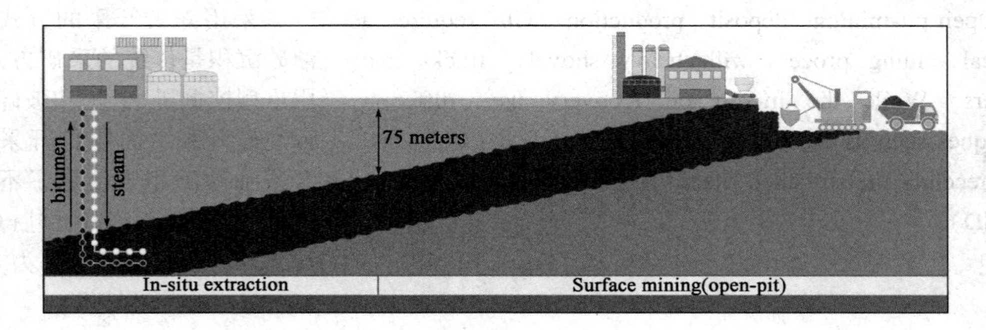

Fig. 6.9 Oilsands Mining vs In-Situ Bitumen Extraction

(Source: http://www.oilsandsmagazine.com/technical/oilsands-101)

6.4.1.3 Deep Deposits

Steam is injected into the oil sands deposit, reducing the viscosity of the bitumen. The mixture of bitumen and water is pumped to the surface where the water is recovered and recycled.

6.4.1.3 深部沉积油砂

为了降低沥青的黏度，通常要向深部油砂中注入蒸汽。沥青和水的混合物被泵送到地面后水在地面被回收并加以循环利用。

In-situ production is based on the injection of steam in the deposit through a horizontal drilling. The steam will cause the oil to melt within the sands and to drain down. The melted oil is then collected and pumping up the melted oil back to the surface.

原位开采技术基于水平井钻井并在油砂沉积中注蒸汽。蒸汽使沥青在油砂中熔化并流入生产井。之后，熔化的沥青被收集起来并泵送至地面。

The advantage of the in-situ process is that it leaves all the sand in the ground.

原位开采技术的优点是不需要挖掘地下的油砂沉积。

Hot water extraction was tested first in Canada in the 1920s, and then the first steam injection was developed in the 1960s.

20 世纪 20 年代，首先在加拿大对油砂进行了热水洗分离试验，之后在 20 世纪 60 年代第一次试验了注蒸汽开采油砂。

Since then all the companies are working to improve its efficiency and reduce its carbon and water footprint.

从那时起，所有的公司都在努力提高采收效率并且减少碳水排放。

6.4.1.4 Shallow Deposits

The deposit is mined and trucked to a main processing plant. Hot water is added to the oil sands, producing a pumpable slurry. Bitumen is recovered through a gravity separation process.

6.4.1.4 浅部油砂沉积

浅部油砂沉积用挖掘机开采出来后被送往主要的沥青提炼厂。向油砂中添加热水混合搅拌形成可泵送的混合浆料，最后通过重力沉降分离过程得到沥青。

Open-pit-mining deposit production will require a classical mining process with heavy shovels, trucks and crushers. While the in-situ oil recovery uses different techniques such as the Cyclic Steam Stimulation (CSS) and more recently the so called Steam Assisted Gravity Drainage (SAGD).

采用露天开采的方式开采油砂沉积是传统的采矿方式，运用重型挖掘机、卡车和破碎机进行油砂开采。而原位开采方式开采油砂沉积则使用了不同的技术，例如循环蒸汽吞吐(CSS)技术和蒸汽辅助重力泄油(SAGD)的前沿技术。

The mining method is considered to be very damaging to the environment, as it involves leveling hundreds of square miles of land, trees, and wildlife. Oil companies using this method are required to return the area to its original environmental condition after completing operations, adding further to costs.

由于涉及数百平方英里的土地，生长在内的树木和野生动物会受到影响，油砂露天开采方法被认为对环境非常有害。使用此种开采方法的石油公司必须在完成作业后将该地区的环境状况恢复到原来状态，这进一步增加了开采成本。

The in situ method is more costly than the surface mining method, but it is much less damaging to the environment, requiring only a few hundred meters of land and a nearby water source to operate. After drilling holes, a mining solution is pumped into the soil. At times explosions or hydraulic fracturing may be utilized to open pathways.

油砂原位开采方法相比露天开采方法成本更高，但对环境的破坏性要小得多，只需几百平方米的土地及附近的水源即可作业。钻井完成后，将采矿溶液泵入地下以获取沥青。有时也可以在地层内进行爆炸压裂或水力压裂来打开流动通道。

It is estimated by the Alberta government that 70 ~ 80 percent of oil in the oil sands is buried too deep for open pit mining; therefore, in situ methods will likely be the future of extracting oil from oil sands. The most common form of in situ is called Steam Assisted Gravity Drainage, or SAGD. Although in-situ production traditionally had lower recovery rates, recent advancements in technology have significantly improved bitumen recovery. In-situ extraction is expected to lead the growth in bitumen production over the coming decades.

据阿尔伯塔省政府估计，油砂资源中70% ~80%埋藏较深，不适合露天开采，因此，油砂原位开采方法很有可能成为从油砂中开采石油的未来趋势。最常见的原位开采方式为蒸汽辅助重力泄油。尽管原位开采技术通常具有较低的采收率，但随着该技术不断进步，沥青采收率已经显著提高。预计在未来几十年内，原位开采技术有望提高沥青产量。

6.4.1.5 Products Derived From the Oil Sands

Bitumen naturally contains a large fraction of complex long chain hydrocarbon molecules. The fraction depends on the geology of the reservoir and process used to extract the bitumen.

About 40% of bitumen produced from the oil sands requires an intermediate upgrading step for partial removal of the heavy hydrocarbon fractions and conversion into light synthetic crude oil (SCO). The SCO is then sold to refineries on the open market. Fig. 6.10 displays the process from produced oil sands to the final product.

6.4.1.5 油砂产品

沥青本身含有大量复杂的长链烃分子。最终得到的长链烃分子的比例取决于油藏地质条件和沥青开采工艺。

产自油砂的沥青中约有40%需要经过中间加工(改质)步骤,才能部分去除重质烃组分并且转化为轻质合成原油(SCO)。之后,轻质合成原油出售给炼油厂。图6.10展示了油砂经过加工成为最终产品的工艺流程。

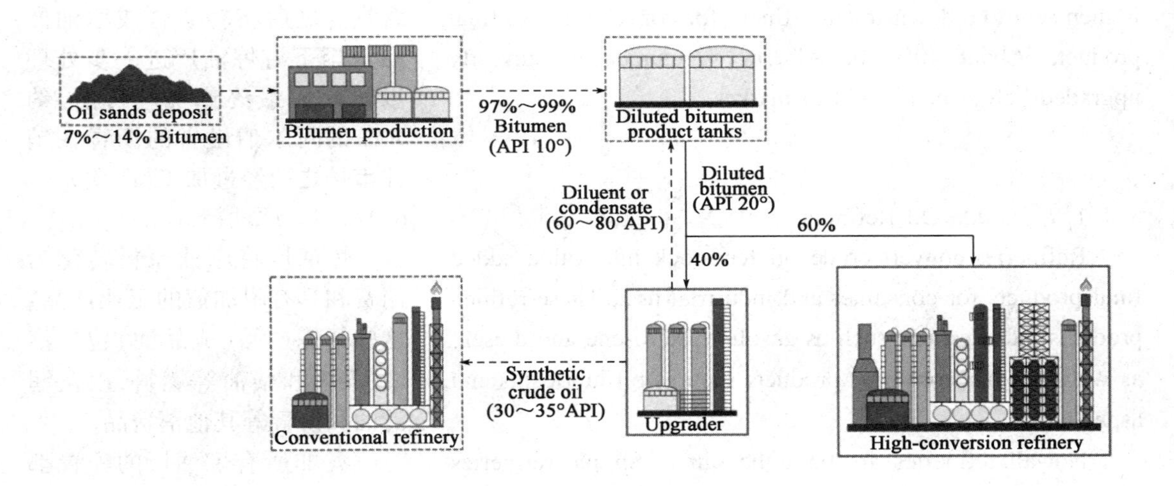

Fig. 6.10 From Oil Sands to the Final Product
(Source: http://www.oilsandsmagazine.com/technical/oilsands-101)

The remaining 60% of bitumen produced is blended with a lighter hydrocarbon (diluent) and sold directly to the open market. Many North American refineries are now designed to accept heavy oil streams, as long as the solids and water content is kept relatively low (less than 0.5%). Non-upgraded bitumen is also sour, containing relatively high sulphur content.

剩下60%的沥青与轻质烃类(稀释剂)混合并直接销往市场。许多北美地区的炼油厂现在可以接受重质原油,只要其固体含量和水含量保持在相对较低(小于0.5%)的水平。未经加工(改质)的沥青呈酸性,具有相对较高的硫含量。

6.4.1.6 Bitumen Upgrading

Bitumen produced from the oil sands is too heavy to be sent directly to a conventional refinery due to its high asphaltene and sulphur content. Depending on the extraction process used, bitumen can sometimes contains as much as 2% water and solids, which does not meet pipeline specifications for long distance transport. This product is therefore upgraded into light synthetic crude, which is then sold on the open market.

Upgrading is a process by which bitumen is transformed into an lighter and sweeter crude by fractionation and chemical treatment. This improves the quality of the oil, reducing its viscosity and sulphur content. The SCO product is then sent to a downstream refinery for conversion into final product. About 40% of Alberta's bitumen is currently upgraded before being sold to market.

6.4.1.6 沥青加工(改质)

由于具有高沥青质含量和高硫含量,产自油砂的沥青密度太大而不能直接送到常规的炼油厂;根据所使用的开采工艺,沥青有时可能含有高达2%的水和固体,这不符合长距离管道运输规范要求。因此,应将沥青加工成(改质为)轻质合成原油,然后出售。

加工(改质)就是通过分馏和化学处理将沥青转化为更轻、硫含量更低的原油的过程,这提高了油品质量,降低了黏度和硫含量。之后将轻质合成原油产品送至下游炼油厂进一步处理以转换成最终产品。目前,约40%的阿尔伯塔省的沥青在销往市场之前经过加工(改质)。

6.4.1.7 Crude Oil Refining

Refineries convert crude oil feedstock into value added final products for consumer and industrial use. These refined products include fuels such as gasoline, kerosene and diesel, as well as other consumer products such as oil lubricants and asphalt.

Not all refineries are built the same. Simple refineries can only process light crudes with a low sulphur content (sweet). Complex refineries have the ability to process heavier feedstock with a much higher sulphur content (sour).

Due to the rising production of heavy crude oil from Canada, Venezuela and Mexico, most Gulf Coast refineries have been modified to better handle heavier crude streams, which are generally less expensive than light/sweet crude. These refineries are known as high-conversion refineries and have the ability to accept sour crude oil streams with up to 10% heavy fractions, as long as the water and solids content is kept relatively low.

6.4.1.7 原油精炼

炼油厂将原油原料转化为消费和具有附加值的工业产品。这些精炼产品(成品油)包括汽油、煤油和柴油等燃料,以及润滑油和沥青等其他消费品。

并非所有炼油厂的配置都是一样的。配置简单的炼油厂只能处理硫含量较低的轻质原油(甜原油),而配置复杂的炼油厂则有能力处理硫含量较高的较重质原油原料(酸原油)。

由于加拿大、委内瑞拉和墨西哥的重质原油产量增加,大多数墨西哥湾沿岸的炼油厂都进行了改造,以更好地处理比轻质(甜)原油便宜的重质原油。这些炼油厂称为高转化率炼油厂,只要水含量和固体含量保持在相对较低的水平,它们就能够处理含有高达10%的重质组分的酸原油。

As more and more refineries around the world convert to heavy oil feedstock, there is less of a demand for stand-alone bitumen upgrading. The economics of upgrading therefore lies in the price differential between heavy diluted bitumen and light crude oil.

随着世界各地越来越多的炼油厂改用重质原油原料，单独对沥青改质的需求越来越少。因此，沥青改性的经济性在于重质稀释沥青和轻质原油之间的价格差异。

6.4.1.8 Transportation

6.4.1.8 运输

Crude oil from the Alberta oil sands (including diluted bitumen) is commonly transported to US and Canadian refineries by pipeline. About two-thirds of Alberta's exports to the US are destined for the Midwest area, centered around Chicago.

阿尔伯塔油砂区的原油（包括稀释沥青）通常通过管道输送到美国和加拿大境内的炼油厂。约三分之二的阿尔伯塔原油出口到美国并销往以芝加哥为中心的中西部地区。

Due to constraints in pipeline capacity and routing, crude oil transport by rail has become increasingly popular in recent years, climbing from near-zero in 2011 to almost 200,000 bbl/d by the end of 2014. Shipment by rail to the US was mostly destined for the US Gulf Coast. However, expansion of pipeline capacity to Quebec, Ontario and the US Midwest has reduced demand for rail transport. Nonetheless, crude-by-rail remains an important back-up mode of transport if new pipeline capacity does not come online in the next few years.

受限于管道容量和管道线路，近几年来，铁路运输原油的方式越来越受欢迎，从 2011 年接近于零的运输量增加到 2014 年底的近 200000bbl/d。通过铁路运往美国的原油大多运往美国墨西哥湾沿岸。通往魁北克省、安大略省和美国中西部的输油管道输油能力的扩展降低了对铁路运输的需求。尽管如此，如果未来几年新的输油管道未能投入使用，铁路运输原油仍将是一种重要的备用运输方式。

Oil shale is a rock that contains significant amounts of organic material in the form of kerogen. Up to 1/3 of the rock can be solid kerogen. Liquid and gaseous hydrocarbons can be extracted from oil shale, but the rock must be heated and/or treated with solvents. This is usually much less efficient than drilling rocks that will yield oil or gas directly into a well. Extracting the hydrocarbons from oil shale produces emissions and waste products that cause significant environmental concerns. This is one reason why the world's extensive oil shale deposits have not been aggressively utilized.

油页岩是一种富含干酪根形式有机质的岩石（固体干酪根含量高达 1/3）。通过使用溶剂加热和（或）处理油页岩，可以从油页岩中提取出液态烃和气态烃。这种开采方式的效率远远低于直接钻井将油气开采到井筒中。从油页岩中提取烃类资源会产生排放物及引起严重环境问题，这就是世界上丰富的油页岩资源没有得到积极利用的原因之一。

Oil shale usually meets the definition of "shale" in that it is "a laminated rock consisting of at least 67% clay minerals". However, it sometimes contains enough organic material and carbonate minerals that clay minerals account for less than 67% of the rock.

油页岩通常符合"页岩"的定义,因为它是"由至少67%的黏土矿物组成的层状岩体"。但有时油页岩含有大量的有机物质和碳酸盐矿物,导致黏土矿物在岩石中所占比例不到67%。

Deposits of oil shale occur around the world, including major deposits in the United States. A 2016 estimate of global deposits set the total world resources of oil shale equivalent of 6.05 trillion barrels (962 billion cubic meters) of oil in place.

油页岩沉积在世界范围内都有分布,主要分布于美国。根据2016年的全球油页岩资源储量估算,全球油页岩资源总量达到6.05×10^{12}bbl,(相当于$9.62 \times 10^{11} m^3$石油储量)。

Oil shale deposits exist in 37 countries globally, main countries with significant oil shale resources include the United States, Australia, Jordan, Morocco, Israel, Brazil, China, Russia and Estonia. But the largest and highest quality oil shale deposits are in sparsely populated areas of Colorado, Utah and Wyoming.

油页岩沉积分布在全球37个国家,拥有丰富油页岩资源的国家主要包括美国、澳大利亚、约旦、摩洛哥、以色列、巴西、中国、俄罗斯和爱沙尼亚。规模最大、质量最高的油页岩沉积集中在人口稀少的科罗拉多州、犹他州和怀俄明州。

The Eocene Green River Formation of Colorado, Utah, and Wyoming contain the largest oil shale deposits in the world. Oil shale, despite the name, does not actually contain oil, but rather a type of organic matter called kerogen, a precursor of oil that is converted to a type of crude oil when heated to about 450 – 500℃. The oil shale deposits are in three structural and sedimentary basins that have been recently assessed.

科罗拉多州、犹他州和怀俄明州的始新世绿河组拥有世界上最大的油页岩沉积。尽管叫油页岩,油页岩实际上并不含油,而含有一种称为干酪根的有机质。干酪根是一种当被加热至450~500℃时会转化为原油(页岩油)的生油母质。油页岩沉积分布于以下三个地质构造和沉积盆地中。

(1) Piceance basin in western colorado. Using a geologic-based assessment methodology, the U. S. Geological Survey estimated an in-place oil shale resource of 1.07 trillion barrels under Federal mineral rights, or 70 percent of the total oil shale in place, in the Piceance Basin, Colorado. More than 67 percent of the total oil shale in-place resource, or 1.027 trillion barrels, is under Federal surface management.

(1)位于科罗拉多州西部的Piceance盆地。美国地质调查局采用地质评价方法评估位于科罗拉多州的Piceance盆地,联邦矿业权下的油页岩资源量为1.07×10^{12}bbl,占当地油页岩资源总量的70%。超过67%的当地油页岩资源总量,即1.027×10^{12}bbl受联邦地表管理。

(2) Uinta basin in eastern Utah and western Colorado. Using a geology-based assessment methodology, the U. S. Geological Survey estimated a total of 1.32 trillion barrels of oil in place in 18 oil shale zones in the Eocene Green River Formation in the Uinta Basin, Utah and Colorado.

(2)位于犹他州东部、科罗拉多州西部的Uinta盆地。美国地质调查局采用地质评价方法估计位于犹他州和科罗拉多州的Uinta盆地的始新世绿河组有18个油页岩分布区,发现总共有相当于1.32×10^{12}bbl石油的油页岩资源量。

(3) Greater Green River basin in southwest Wyoming and northwest Colorado. The U. S. Geological Survey 2011 completed an assessment of in-place oil shale resources, regardless of grade, in the Eocene Green River Formation of the Greater Green River Basin in southwestern Wyoming, northwestern Colorado, and northeastern Utah. Green River Formation oil shale also is present in the Piceance Basin of western Colorado and in the Uinta Basin of eastern Utah and western Colorado. No attempt was made to estimate the amount of oil that is economically recoverable because there has not yet been an economic method developed to recover the oil from Green River Formation oil shale.

(3)位于怀俄明州西南部、科罗拉多州西北部的Greater Green River盆地。2011年,美国地质调查局完成了对怀俄明州西南部、科罗拉多州西北部、犹他州东北部的Greater Green River盆地的始新世绿河组的油页岩资源评价。绿河组油页岩也出现在科罗拉多州西部的Piceance盆地和犹他州东部、科罗拉多州西部的Uinta盆地。由于当时尚未提出对绿河组油页岩开采的经济有效的方法,因此并没有对经济可采储量进行评估。

Estimated total in-place resources are about 1.5 trillion barrels of oil for the Piceance Basin, about 1.3 trillion barrels of oil for the Uinta Basin and 1.4 trillion barrels of oil in the Greater Green River Basin. The Piceance Basin is the smallest of the three principal basins of the Green River Formation in terms of area covered and contains the highest concentration of high-grade oil shale (capable of generating at least 25 gallons of oil per ton of rock), with approximately 352 billion barrels of in-place oil resource.

据估计,Piceance盆地的油页岩资源总量约为1.5×10^{12}bbl,Uinta盆地的油页岩资源总量约为1.3×10^{12}bbl,Greater Green River盆地的油页岩资源总量约为1.4×10^{12}bbl。Piceance盆地是绿河组三个主要盆地中占地面积最小的,但含有最高浓度的高质量油页岩(1t含油岩石能产生至少25gal的石油),其资源量大约为3.52×10^{12}bbl。

Getting crude oil from rock represents perhaps the most difficult process of extraction. Oil shale must be mined using either underground or surface-mining methods. After excavation, the oil shale must undergo retorting. This is when the mined rock is exposed to the process of pyrolysis-applying extreme heat without the presence of oxygen to a substance, and producing a chemical change. Between 650 and 700 degrees Fahrenheit, the kerogen-the fossil fuel trapped within-begins to liquefy and separate from the rock. The oil-like substance that emerges can be further refined into a synthetic crude oil. When oil shale is mined and retorted above ground, the process is called surface retorting.

从油页岩中获得原油可能是最困难的油气资源开采过程。油页岩必须采用井下开采或露天开采的方法进行开采。开采出来以后，必须对油页岩进行干馏处理。干馏处理就是使开采出来的油页岩发生高温裂解，即在隔绝氧的情况下对其进行高温加热，使其产生化学变化。在650～700℉，储集在油页岩里的化石燃料干酪根开始液化并从岩石中分离出来(即裂解出页岩油)。出现的油状物质可通过进一步精炼制成合成原油。这种在地面对开采出来的油页岩进行干馏处理的油页岩开发方法称为地表干馏。

The problem is that this process adds two extra steps to the conventional extraction process in which liquid oil is simply pumped from the ground. In addition to mining, there's also retorting and refining of the kerogen into synthetic crude. Oil shale presents environmental challenges as well. It takes two barrels of water to produce one barrel of oil shale liquid. And without cutting-edge water treatment technology, the water discharge from oil shale refining will increase salinity in surrounding water, poisoning the local area.

问题是，这一开采过程比从地面简单地泵出液态油这种常规的石油开采过程增加了两个额外的步骤。除了采矿外，还需将干酪根干馏、精炼，从而制成合成原油。油页岩开采也带来了许多环境挑战。每生产 1bbl 合成原油，就需要消耗 2bbl 水。而且由于缺少先进的水处理技术，排放油页岩精炼过程产生的废水会使周围水体的含盐量增加，对当地造成污染。

There's also the matter of the rocks. Every barrel of oil produced from shale leaves behind about 1.2 to 1.5 tons of rock. What should be done with this remaining rock? There are certainly projects that require loose rock, but the demand may not meet the supply if oil shale production is ever conducted on a massive scale.

其次还有岩石的问题。每利用油页岩生产 1bbl 合成原油，就会留下大约 1.2～1.5t 的岩石。这些剩下的岩石该如何处理？虽然有一些工程项目需要这些岩石碎块，但如果大规模开发利用油页岩，对岩石碎块的供应可能无法与需求相匹配。

Royal Dutch Shell Oil Company has come up with an answer to some of the problems with oil shale refining. The company calls it In Situ Conversion Process (ICP). In ICP, the rock remains where it is; it's never excavated from the site. Instead, holes are drilled into an oil shale reserve, and heaters are lowered into the earth. Over the course of two or more years, the shale is slowly heated and the kerogen seeps out. It's collected on-site and pumped to the surface. This cuts out the mining aspect, and further reduces costs since there's no need to transport or dispose of spent rock.

荷兰皇家壳牌石油公司提出了解决油页岩精炼方面的问题的办法，称为原位开采技术(ICP)。在原位开采过程中，油页岩仍处在地下原来的位置，无需将其开采到地面。取而代之的是在油页岩储层中钻加热井，下放加热器对油页岩储层进行加热。在之后两年或更长的时间里，油页岩被缓慢加热，干酪根渗出。页岩油等烃类物质经过收集，通过生产井被泵送到地面。原位开采技术免除了采矿过程，并且不需要运输或处置废弃的岩石，进一步降低了开发成本。

Because of current obstacles, oil shale hasn't been commercially produced on a large scale. Simply put, it's currently more expensive and environmentally harmful than conventional drilling. But as the supply of crude oil diminishes and the price of petroleum rises, oil shale-especially under Shell's plan-is becoming increasingly attractive.

由于目前存在的种种障碍，尚未进行油页岩大规模商业化开发。简单地说，油页岩开发比常规钻井成本更高，对环境破坏也更严重。但是随着原油供应减少及石油价格上涨，特别是在壳牌公司提出新的油页岩开采方式后，油页岩开采将变得越来越有吸引力。

6.4.2 Production Technology of Oil Sands and Oil Shale

The viscosity of oil sands is extremely high and runs into thousands of centipoise or more. Hence, oil sands cannot be pumped to the surface utilizing conventional oil well technology. For in-situ mining, the most widely used methods are as follows:

(1) Cyclic steam enhancement method (CSS);

(2) Steam-assisted gravity drainage method (SAGD);

(3) Sand production and cold mining technology;

6.4.2 油砂和油页岩开采方式

油砂的黏度非常高，可达到数千厘泊或更高。因此，利用传统的开采技术不能将油砂泵送到地面。对于原位开采，广泛采用的方法主要有以下几种：

(1)循环蒸汽强化法(CSS)；

(2)蒸汽辅助重力泄油法(SAGD)；

(3)出砂冷采技术；

(4) Underground horizontal well injection gas solvent extraction technology (VAPEX);

(5) Underground in-situ catalytic upgrading technology;

(6) Hydrothermal cracking mining technology.

Syncrude Canada is the world's largest manufacturer of oil from oil sands. It is engaged in open pit mining of oil sands in Athabasca, and its open pit mining technology is leading the world.

Oil shale mining includes direct mining including open pit and downhole mining. Downhole mining technology means that oil shale buried underground is not mined and is heated directly underground, converted to shale oil or shale gas and transported to the surface. Open-pit mining is suitable for mining shallow deposits with low cost and high safety factor. Downhole mining has two modes: vertical shaft and horizontal tunnel mining, which is suitable for burying deep deposits. Direct mining is a relatively primitive method of mining. The limitations are relatively large, and the damage to the ecological environment is also very serious, mainly in three aspects:

First, ecological and water quality are seriously damaged. Whether it is open pit mining or underground mining, it is necessary to reduce the groundwater level below the horizon of the oil-bearing shale formation, and to extract 1 cubic meter of oil shale, generally need to extract 25 cubic meters of groundwater; mining water greatly increases the content of sulphate in the medium. In Brazil, oil shale mining has long damaged the ecological balance of the mine and its vicinity and the stability of water quality.

(4)地下水平井注气体溶剂萃取技术(VAPEX);

(5)井下原位催化改质开采技术;

(6)水热裂解开采技术。

加拿大 Syncrude 公司是全世界最大的从油砂中生产石油的公司,主要开采来自阿萨巴斯卡的油砂,其油砂的露天开采技术在世界上处于领先水平。

油页岩的开采方式有直接开采,包括露天和井下两种开采方式。井下开采技术是指埋藏在地下的油页岩没有开采,直接在地下进行加热转化为页岩油或页岩气,并将其输送到地面。露天开采适合于埋藏较浅的矿床,成本低,安全系数高。井下开采有竖井、水平坑道开采两种方式,适合于埋藏较深的矿床。直接开采是较原始的开采方式,局限性比较大,对生态环境的破坏也十分严重,主要表现在三个方面:

一是生态环境及水质破坏严重。无论是露天开采还是井下开采,都需要把地下水位降低到含油页岩层的层位以下。开采 $1m^3$ 油页岩,一般需要抽出 $25m^3$ 的地下水。采矿水极大地增加了地表水、地下水中硫酸盐的含量。在巴西,油页岩采矿长期破坏着矿山及其附近的生态平衡和水质的稳定。

Second, the ash residue pollution is serious. The oil shale obtained by direct mining is used to refine shale oil or directly burn, and a large amount of ash is generated. If it is not recycled, it will not only cause air pollution, but also occupy a large area of waste ash, in which metal elements and trace elements penetrate. The groundwater body harms people's production and life.

二是灰渣污染严重。通过直接开采得到的油页岩主要用于提炼页岩油或直接燃烧,因此产生大量灰渣,如果不回收利用不仅会造成空气污染,而且废弃灰渣占地面积大,其中金属元素和微量元素渗入地下水体,危害人类的生产生活。

Third, direct mining covers a large area and cannot be completely repaired once it is opened.

三是直接开采占地较多,一旦开垦就无法完全修复。

The Underground Conversion Process (ICP) is a patented technology developed by Shell to develop oil shale and other unconventional resources, which is particularly beneficial for the development of deep oil shale. The basic principle of ICP oil shale mining is to heat and crack the oil shale deposit in the underground, and to convert it into high-quality oil or gas, and then extract the oil and gas separately through the relevant channels; these high-quality oils will be extracted. After the gas is collected on the ground for processing, it can produce oil products such as naphtha and kerosene. The outstanding advantages of this technology are: improving the efficiency of resource development and utilization; reducing the damage to the ecological environment during the mining process, that is, less land occupation, no tailings waste, no air pollution, less groundwater pollution and minimizing harmful by-products. Although the technology is not yet fully commercialized, key technical issues such as processes and equipment have been resolved, and commercial demonstrations have been conducted in Colorado and Alberta, Canada.

地下原位转化工艺(ICP)是壳牌公司投入巨资研发出的开采油页岩及其他非常规资源的专利,对开发深部油页岩尤其有利。ICP 开采油页岩的基本原理是在地下对油页岩矿层进行加热和裂解,促使其转化为高品质的油或气,再通过相关通道将油、气分别提取出来;将这些高品质的油、气采集到地面进行加工后,可生产出石脑油、煤油等成品油。该技术的突出优点是:提高了资源开发利用效率,减少了开采过程中对生态环境的破坏,即占地少、无尾渣废料、无空气污染、少地下水污染及最大限度地减少有害副产品的产生。尽管该项技术现在还未完全商业化,但关键的工艺、设备等技术问题都已解决,并在美国科罗拉多州和加拿大阿尔伯塔省进行了商业示范。

Therefore, with the increase of environmental protection pressure, oil shale in-situ mining has become the development trend of large-scale commercial exploitation of oil shale in the future.

随着环保压力的增大,油页岩井下原位开采成为未来油页岩商业化大规模开采的发展趋势。

Exercises/练习题

1. Define unconventional oil. Distinguish it from conventional oil. Explain the role played by

reservoir characteristics and oil properties in defining unconventional oil. /阐述非常规油资源的定义,说明其与常规油资源的差异,解释油藏特征和原油特性在定义非常规油资源的作用。

2. Describe the major technologies in producing unconventional oil. /简述开采非常规油资源主要技术。

3. What are main differences between shale oil and oil shale? /油页岩和页岩油的主要差别有哪些?

4. How does multistage fracturing facilitate tight oil production? /如何用多段压裂开采致密油?

5. What are the major resources of unconventional gas? /有哪些主要的非常规气资源?

6. Describe the factors that may influence hydraulic fracturing design. /简述影响水力压裂设计的因素。

7. How does the mechanism of shale gas flow differ from conventional gas? /与常规气相比,页岩气流动机理有哪些不同?

8. What is CBM and how it is produced? /什么是煤层气? 如何开采煤层气?

References/参考文献

[1] Office USGA. Unconventional Oil and Gas Development: Key Environmental and Public Health Requirements [J]. Government Accountability Office Reports, 2012.

[2] Mercier T J, Johnson R C, Brownfield M E, et al. In-Place Oil Shale Resources Underlying Federal Lands in the Piceance Basin, Western Colorado[J]. Annals of Diagnostic Pathology, 2010, 16(3):219 - 223.

[3] Johnson R C, Mercier T J, Brownfield M E, et al. Assessment of In-place Oil Shale Resources in the Eocene Green River Formation, Uinta Basin, Utah and Colorado [J]. Archives of Virology, 2010, 148 (9): 1851 - 1862.

[4] Johnson R C, Mercier T J, Ryder R T, et al. Assessment of in-place oil shale resources of the Eocene Green River Formation, Greater Green River Basin, Wyoming, Colorado, and Utah [J]. Family & Consumer Sciences Research Journal, 2011, 44(2):159 - 171.

[5] Al-Nakhli A. R. Chemically-Induced Pressure Pulse: A New Fracturing Technology for Unconventional Reservoirs [C]. SPE Middle East Oil & Gas Show and Conference. Society of Petroleum Engineers, 2015.

[6] Johnston R. Seismic Technologies for Unconventional Reservoir Development[C]//Abu Dhabi International Petroleum Exhibition & Conference. Society of Petroleum Engineers, 2016.

[7] Wilson A. Focus on Unconventional Reservoirs Requires Advancements in Technology [J]. Journal of Petroleum Technology, 2013, 65(7): 103 - 107.

[8] Van Sickle S, Galloway J, McClellan C, et al. Economic and Operational Analysis of Systematically Deploying New Technologies in an Unconventional Play [C]//SPE/IATMI Asia Pacific Oil & Gas Conference and Exhibition. Society of Petroleum Engineers, 2015.

[9] Arthur J D, Langhus B, Alleman D. An overview of modern shale gas development in the United States[J]. All Consulting, 2008, 3: 14 - 17.

[10] Zhang X S, Wang H J, Ma F, et al. Classification and characteristics of tight oil plays[J]. Petroleum Science, 2016, 13(1): 18 - 33.

[11] Santos R G, Loh W, Bannwart A C, et al. An overview of heavy oil properties and its recovery and transportation methods[J]. Brazilian Journal of Chemical Engineering, 2014, 31(3): 571 - 590.

[12] Li Q. Characteristics of Tight Gas Reservoir in the Upper Triassic Sichuan Basin, Western China[C]. AAPG Annual Convention and Exhibition, 2015.

[13] Satter A, Iqbal G M. Reservoir Engineering: The Fundamentals, Simulation, and Management of Conventional and Unconventional Recoveries[M]. Gulf Professional Publishing, 2015.